JEWISH TOPOGRAPH

Heritage, Culture and Identity

Series Editor: Brian Graham,
School of Environmental Sciences, University of Ulster, UK

Other titles in this series

(Dis)Placing Empire
Renegotiating British Colonial Geographies
Edited by Lindsay J. Proudfoot and Michael M. Roche
ISBN 978 0 7546 4213 8

Preservation, Tourism and Nationalism
The Jewel of the German Past
Joshua Hagen
ISBN 978 0 7546 4324 1

Culture, Urbanism and Planning
Edited by Javier Monclus and Manuel Guardia
ISBN 978 0 7546 4623 5

Tradition, Culture and Development in Africa
Historical Lessons for Modern Development Planning
Ambe J. Njoh
ISBN 978 0 7546 4884 0

Heritage, Memory and the Politics of Identity
New Perspectives on the Cultural Landscape
Edited by Niamh Moore and Yvonne Whelan
ISBN 978 0 7546 4008 0

Geographies of Australian Heritages
Loving a Sunburnt Country?
Edited by Roy Jones and Brian J. Shaw
ISBN 978 0 7546 4858 1

Living Ruins, Value Conflicts
Argyro Loukaki
ISBN 978 0 7546 7228 X

Geography and Genealogy
Locating Personal Pasts
Edited by Dallen J. Timothy and Jeanne Kay Guelke
ISBN 978 0 7546 7012 4

Jewish Topographies

Visions of Space, Traditions of Place

Edited by

JULIA BRAUCH,

ANNA LIPPHARDT

ALEXANDRA NOCKE

Taylor & Francis Group

LONDON AND NEW YORK

First published 2008 by Ashgate Publishing

2 Park Square, Milton Park, Abingdon, Oxon OX14 4RN
711 Third Avenue, New York, NY 10017, USA

Routledge is an imprint of the Taylor & Francis Group, an informa business

First issued in paperback 2016

British Library Cataloguing in Publication Data
Jewish topographies : visions of space, traditions of
    place. - (Heritage, culture and identity)
    1. Jews - Social conditions 2. Jews - Identity 3. Jews -
    Civilization 4. Jews - Cross-cultural studies 5. Space and
    time - Religious aspects - Judaism 6. Sociology, Urba
    7. Jewish diaspora
    I. Brauch, Julia II. Lipphardt, Anne III. Nocke, Alexandra
    305.8'924

Library of Congress Cataloging-in-Publication Data
Jewish topographies : visions of space, traditions of place/[edited] by Julia Brauch, Anna Lipphardt and Alexandra Nocke.
    p. cm. -- (Heritage, culture and identity)
    Includes bibliographical references.
    ISBN 978-0-7546-7118-3
    1. Jews--Social conditions. 2. Jews--Identity. 3. Jews--Civilization. 4. Jews--Cross-cultural studies. 5. Space and time--Religious aspects--Judaism. 6. Sociology, Urban.
7. Jewish diaspora. I. Brauch, Julia. II. Lipphardt, Anna. III. Nocke, Alexandra.

    DS140.J49 2008
    909'.04924--dc22

                                                                    2007046582

ISBN 13: 978-0-7546-7118-3 (hbk)
ISBN 13: 978-1-138-25429-9 (pbk)

Contents

Part V Enacted Spaces

Epilogue

Foreword

In Hebrew the word Makom means place and has a twofold significance: on one hand Makom refers to the concrete physical place, on the other hand Makom is equivalent with God's name, and therefore refers to a metaphysical place. Between 2001 and 2007 a group of junior and senior scholars met regularly in the *Dachstube*, the attic of the Potsdam-based Moses Mendelssohn Center, in order to discuss the relevance of space and place in Jewish culture. This interdisciplinary research group at the University of Potsdam, "Makom: Place and Places in Judaism; The Meaning and Construction of Place References in European Jewish Culture from the Enlightenment to the Present", was financed by the German Research Foundation, *Deutsche Forschungsgemeinschaft (DFG)*. It offered junior scholars from diverse disciplines an academic framework for addressing the question of the meaning and construction of places, about 30 dissertations as well as several postdoctoral projects evolved, and the ground was laid for numerous articles, books, and even two exhibition projects. This creative output revolving around the core subject of *Jewish Topographies* shows the diversity as well as the significance of this research topic.

The three editors of this volume, Julia Brauch, Anna Lipphardt, and Alexandra Nocke, were all members of the Makom research group. They follow up on some of the central questions concerning Jewish place and space that had been raised in Potsdam and to discuss them in greater depth within this interdisciplinary collection of articles. On both the theoretical as well as the empirical level, this volume offers its readers a fascinating new perspective on Jewish places in all their diversity and multidimensionality.

The number of intensive academic discussions that have occurred in this area during recent years has demonstrated that the engagement with Jewish place and space is not just a passing trend, but poses essential questions for the curriculum of Jewish studies. I am pleased that the research group Makom could serve as a "greenhouse" for so many innovative ideas and approaches—among them this wide-ranging book, which provides this new field of study with a systematic framework from which it can move in many fruitful directions.

Julius H. Schoeps
Potsdam, September 2007

Acknowledgments

This anthology not only deals with places; its genesis owes tribute to the wide array of locations where the lively discussions of our editorial trio took place: in the Dachstube of the Moses Mendelssohn Center in Potsdam, at kitchen tables in Berlin-Kreuzberg, at Café Diwan, in a solitary cottage in the East German countryside, on long subway rides and extended walks in the park. These diverse locations, far removed from most of the places discussed in this book, provided the environment that allowed us to transform our ideas into this book project and made the time we spent together during the emergence of this volume a truly rewarding experience.

This collection has depended upon the support of many individuals. First of all, we would like to thank our contributors for their thought provoking articles, their attentiveness, and their patience. The opportunity to work with such a dynamic and inspiring circle of colleagues—from Israel, Hungary, Switzerland, Germany, Great Britain, Thailand, the US, and Canada—has been a great privilege.

In addition, we appreciate the many ways in which colleagues and friends have enriched the theoretical ideas that underlie this volume and the practical advice they gave throughout the editing process.

We are grateful to Joachim Schlör, whose initial concept resulted in the Makom program at the University of Potsdam, and who can rightfully be called the "father of Jewish space studies." The interdisciplinary research group, "Makom: Place and Places in Judaism", which ran for more than six years, represents an unequalled pioneering effort to institutionalize the engagement with Jewish place and space from an interdisciplinary platform.

We would further like to thank Julius H. Schoeps, the official spokesperson for Makom, who provided us with such a stimulating academic framework at the Moses Mendelssohn Center in Potsdam. Our anthology was nurtured by the intellectual exchange within this group of dedicated junior and senior scholars.

The German Research Foundation, *Deutsche Forschungsgemeinschaft (DFG)* and the DFG Graduate Research Group Makom, University of Potsdam, generously funded this book project.

We would also like to thank the *Centre Marc Bloch*, Berlin, for its moral and logistical support. Further we would like to thank Monika Hoinkis and her design class at the Universität der Künste (UdK), Berlin, who initiated a design competition for the cover of this book among her students. Many thanks to all her students for the creative drafts we received and especially to the winner, Clemens Jahn, for his convincing and beautiful cover design.

We are indebted to all those who waived individual copyright for this volume, especially the photographer Wilfried Dechau, who generously provided us with an image from the construction site of the Munich synagogue, as well as the management of Mini Israel for supplying us with map and images.

We also owe a debt of gratitude to Paula Ross of *Words by Design*, Berlin, who proofread the manuscript prior to its submission to our publishers. Her thoroughness, keen eye, insightful comments, and power of judgment far exceeded the task she was assigned.

And finally, a huge thank you to our editor Valerie Rose and Pamela Bertram and to everyone at Ashgate Publishing who encouraged and supported us throughout this project.

Julia Brauch, Anna Lipphardt, and Alexandra Nocke
Berlin, September 2007

When not paddling in the currents of time, I was gumming small green leaves to a paper tree pinned to the wall of my *cheder*, the Hebrew school. Every sixpence collected for the blue and white box of the Jewish National Fund merited another leaf. When the tree was throttled with foliage the whole box was sent off, and a sapling, we were promised, would be dug into the Galilean soil, the name of our class stapled to one of its green twigs. All over north London, paper trees burst into leaf to the sound of jingling sixpences, and the forests of Zion thickened in happy response. The trees were our proxy immigrants, the forests our implantation. And while we assumed a pinewood was more beautiful than a hill denuded by grazing flocks of goats and sheep, we were never exactly sure what all these trees were *for*. What we did know was that a rooted forest was the opposite landscape to a place of drifting sand, the exposed rock and red dirt blown by the winds. The diaspora was sand. So what should Israel be, if not a forest, fixed and tall? No one bothered to tell us which trees we had sponsored. But we thought cedar, Solomonic cedar: the fragrance of the timbered temple.

(Simon Schama, *Landscape and Memory*, 1995)

Exploring Jewish Space

An Approach

Anna Lipphardt, Julia Brauch, Alexandra Nocke

Time and history play central roles in our understanding of Jewish civilization. Place and space, by contrast, seem to be secondary categories at best. The interrelated motifs of the "people of the book" and the book as the "portable homeland"—together with the stereotype of "the wandering Jew," have conveyed the pervasive impression that the Jewish experience—except the Israeli one—is one of profound displacement, lacking not only a proper territory but also a substantial spatiality or attachment to place.

While according to Edward Soja, the tendency to privilege time over place is visible in most academic works produced in the historicist tradition,[1] this mode of perception seems to be particularly prevalent in the field of Jewish studies.

As for the Jewish spaces and places that *have* been studied in the past, the vast majority are related to religious tradition (e.g., the Temple, synagogues, eruvin, or cemeteries), to the Zionist project and the State of Israel, or to isolation, destruction, and remembrance (e.g., the ghettos in the Early Modern period, the death camps of the Nazis, or Holocaust memorials and museums).

The spatial turn that surfaced in the humanities and social sciences beginning in the 1980s, reached Jewish studies, as other turns before it, with a sound delay. Nevertheless, in the past decade there has been a notable increase in individual research projects, conferences, and research groups within Jewish studies that take space and place seriously: not only have these two categories become the focus of attention and serious subjects of research, but a critical spatial perspective has evolved and space has begun to be employed as an analytical category.

This volume ties into this trend and directs the reader's attention towards the production of Jewish space.

No space is a given—and Jewish space even less so when compared to the spaces of societies that have been more or less continuously settled within the boundaries of a stable territorial power. By looking at specific historical and contemporary case studies, the articles collected in this volume pursue the following questions: How have Jews experienced their spatial environments and how have they engaged with specific places? How do Jewish spaces emerge? Who is involved in the process of their emergence? How are Jewish spaces contested, performed, and used? And which features render a space Jewish at all?

This book explores the production of space rather than final spatial products, it focuses on "doing Jewish space" or "lived Jewish spaces," on Jewish spatial practices and experiences, and Jewish strategies of place-making. It intends to explore the location of Jewish presence rather than the construction and interpretation of Jewish spaces on the textual or metaphorical level, as these have already received wide scholarly attention. Henri Lefebvre, the founding father of critical spatial theory, pointed out the epistemic problem of a merely textual approach:

> Spaces contain messages, — but can they be reduced to messages? (...) Its mode of existence, its practical 'reality' (including its form) differs radically from the reality (or being-there) of something written, such as a book. Space is at once result and cause, product and producer. So, even if the reading of space (...) comes first from the standpoint of knowledge, it certainly comes last in the genesis of space itself. (...) [S]pace was *produced* before being *read*; nor was it produced in order to be read and grasped, but rather in order to be lived by people with bodies and lives in their own particular (...) context. In short, 'reading' follows production in all cases except those in which space is produced especially in order to be read.[2]

With this book we want to discover Jewish spaces and places that have received little attention so far, and to look at well-known Jewish spaces and places from new angles. By employing an interdisciplinary approach, with a strong foothold in cultural history and cultural anthropology, this collection also introduces new methodological and conceptual approaches to the study of the Jewish past and present.

Our aim is twofold: Along the lines of Edward Soja, we seek "to spatialize" the historical as well as the cultural narrative of the Jewish experience.[3] This spatialization does not imply a substitution of "space" for "time" nor a radical paradigmatic shift. Rather, it offers a broadening of perspective we consider to be of great epistemological value to the interdisciplinary field of Jewish studies and a holistic understanding of Jewish civilization. Furthermore, we want to adopt Yosef Hayim Yerushalmi's call to map not only the geography of Jewish expulsions and catastrophes, but also the geography of Jewish everyday life and the geography of Jewish hope.[4]

The research perspective chosen for this volume investigates Jewish topographies from within. Our particular interest is in shedding light on Jewish agency and *Eigensinn,* on the production of space rather than on the ways Jewish space has been perceived or controlled by the non-Jewish environment in the course of history, or the retrospective Judaization of alleged Jewish places. It thus explores the spatial strategies and experiences of a pronounced minority, reflecting the constraints encountered when trying to find and define its own space, as well as the creative solutions adopted in response.

Focusing on Jewish spaces from within does not imply looking at closed, separate entities or the "essence" of Jewish space. Jewish spaces are characterized by a deep-seated internal and external translocality[5] and transculturality,[6] entangled and interconnected with their respective environments as well as with other Jewish spaces throughout the world. In the era of globalization, as Wolfgang Welsch aptly put it,

> [t]he differences no longer come about through a juxtaposition of clearly delineated cultures (like in a mosaic) but result between cultural networks, which have some things in common while differing in others, showing overlaps and distinctions at the same time.[7]

This rather open-ended conception of transcultural and translocal Jewish space does not, however, imply that there are no borders or boundaries. Within the framework of this book these operate not so much as separation or demarcation lines, but as contact zones that allow for multifold transfers and entanglements, which, to be sure, are not devoid of tensions and conflicts, nor free of hierarchical power relations.

The Jewish Diaspora has existed over a longer time than any other diaspora and in a multitude of geographical and cultural settings. This allows for fruitful, intra-cultural (Jewish) comparisons as well as for variegated comparisons with the places and spaces of their respective co-existing societies.

Characterized by a high degree of mobility, by multilocal attachments and complex dynamics between life in the Diaspora and in Israel, the Jewish experience can be seen as a touchstone for the globalization process, providing ample insights into its premises, conditions, and perils as Clemens Kauffman has rightfully observed.[8]

Jewish spaces are 'Other' spaces, and from this very particular point of view Jews have produced a wealth of critical spatial thought. Other spaces challenge conventional conceptions of place and space, which, to a large extent—on the level of architecture, challenge spatial politics as well as theory—still rests on the paradigm of a homogeneous nation-state, rooted in a clear-cut territory. Exploring Jewish topographies thus also allows for new, subversive perspectives on the places and spaces of the majority society and for rethinking spatial assumptions that have been taken for granted.[9]

With this anthology we wish to broaden the dialogue for interdisciplinary exchange on Jewish place and space. As the academic disciplines and research fields brought together in this volume often employ different space-related definitions and concepts, the reader will encounter a diverse semantic field and "space" terminology. Because this book assembles authors as well as case studies from diverse cultural spaces and linguistic environments, it is neither possible nor does it seem reasonable to forge a uniform vocabulary or a new theoretical meta-vocabulary of space.[10]

While we have not imposed a consistent terminology on our authors, as editors we have oriented ourselves around the following conceptual distinctions: *Jewish places* are, in our understanding, sites that are geographically located, bound to a specific location, such as the Jewish quarter in Fez, Morocco, or the gravesite of Baba Sali in Netivot, Israel. *Jewish spaces* are understood as spatial environments in which Jewish things happen, where Jewish activities are performed, and which in turn are shaped and defined by those Jewish activities, such as a sukkah or a Bundist summer camp for children. Therefore, in our understanding *Jewish place* is defined by location, *Jewish space* by performance.[11] Both can be congruent or overlap, and the difference between them is not so much defined by *where* one can find them, but lies in their function, or, as Steve Harrison and Paul Dourish have put it, in the different roles they play: "space is the opportunity; place is the (understood) reality."[12] Rather than insisting on clear-cut distinctions between the two concepts, the focus of this book is attuned to their interdependences: Place and location are not eternally fixed or given in the positivist sense, but are always part of spatial dynamics, with changing significance in evolving contexts; on the other hand spatial performances are not devoid of real materiality and take place in specific local settings and geographical surroundings. Together they form what we understand as *Jewish topographies*.

In the following, we have made a first attempt to trace the historical development of Jewish space study and to illuminate the reasons for the prevailing focuses, biases, or omissions. In the second half of this introduction we present the conceptual framework of this book: that is the production of space, or Lived Jewish spaces, which falls in line with Lefebvre's concept of *espaces vécus*. It concludes with an overview of the structure of this anthology and of the individual articles.

Tracing the Historical Context

Concerning the research on place and space, Leopold Zunz, one of the founding fathers of Jewish studies (or rather of its forerunner, the Wissenschaft des Judentums) as an academic discipline, noted in his *Essay on the Geographical Literature of the Jews, from the Remotest Times to the Year 1841*, that:

> In the whole range of Jewish literature no branch of knowledge appears to be cultivated more scantily, or to be less known, than that of Geography. If any interest in the knowledge of the earth was excited now and then by books or by reflection, it was confined to that, which elevated the spirit to the creator, unalloyed by business or institutions of man.[13]

Already Zunz underscored the fact that Jews dealt intensively with places of religious importance—the Holy Land, above all—as well as with natural science, but not with the "political-statical element of geography," as he called it. The reason for this deficit, in his view, was the lack of professional opportunities for Jews "to devote their energies to geography."[14]

Since the beginnings of the Wissenschaft des Judentums some two hundred years ago, much has changed concerning the professional prospects for Jewish geographers, urban planners, and other "space" specialists; the integration of Jewish studies into mainstream academic institutions has seen significant progress as well. Jewish space, however, remains under-researched. It has neither been included into general space studies, such as geography, architecture, urban and landscape planning, nor systematically integrated into Jewish studies, as a subdiscipline or research field, as a research object or perspective, or a key concept.

There are three main reasons for this state of affairs: First of all, the multi-locale attachments, the long and variegated history of migration, and the complex interrelationship between biblical, real, and messianic topography that characterize Jewish space make it difficult to investigate within conventional spatial paradigms or by employing conventional means. As Neil Jacobs has pointed out,

> Jewish geography may have been harder to discern [a]s compared to the geography of peoples that are/were geographically concentrated—and who often possess(ed) the trappings of a geopolitical entity, i.e., a kingdom, state, etc. (…) The impression that Jews—unlike "normal peoples" were not geographically rooted has long existed at both the popular and the scholarly level. At the popular level, such impressions are seen in views of the "wandering Jew," rootless, without true underlying loyalty to the non-Jewish geopolitical entity in which s/he lives.[15]

From the architectural point of view, Bruno Zevi in his programmatic text, *Spazio et non-spazio ebraico,* has addressed the significance of process vis-à-vis the final product, which further complicates the investigation of Jewish space: "We privilege the accretion over the status-quo, the formation instead of the form as a concluded entity. Our history is antistatical and antispatial."[16]

The second reason for the spatial "blind spot" stems from the divisions and demarcations of academic disciplines. Even today there is no substantial dialogue between Jewish studies and the core disciplines of critical space studies such as geography, architecture, or urban planning. The "space" disciplines are usually located at technical universities or in the departments of geoscience or engineering, and only in rare cases are they connected with the humanities, and even less with Jewish studies, through academic teaching or primary research. One example was our own research group "Makom," which included two doctoral students (out of 30) from architecture but did not cooperate with any university department or research institute in the fields of architecture, urban studies or cultural geography.

In the United States, Robert D. Mitchell has even observed an increasing institutional disengagement between geography and Jewish studies in recent decades. While the 1980s saw as many as eight geography departments offering courses with a Jewish focus, by 1997 none of these courses were still being offered.[17]

Without exact figures at hand, we would argue that in Germany, and increasingly in Central and Eastern European countries, a different trend is occurring. More and more local history initiatives and architecture departments have joined forces in recent years to restore synagogues, cemeteries, and mikves destroyed during the NS era, or to explore their architectural history, as for example, the virtual reconstruction project of synagogues by architectural students at the Technical University of Darmstadt.[18] It should be emphasized however, that in the context of these projects, students and scholars usually deal with architectural relicts of destroyed Jewish communities and rarely study in depth the spatial conditions, practices, and challenges of contemporary Jewish life, or even engage in the planning and construction of new Jewish spaces.

The third reason for the marginalization of Jewish space in the academic realm can be found in the predominant epistemology of Jewish studies. The "forgetfulness" concerning place and space in Jewish studies is a consequence or result of the space-indifferent or even "anti-spatial" focus of its major sub-disciplines vis-à-vis their central research objectives—religion, history, and language. In general—and the academic perspective follows suit—Jewish religion is seen as *transcending* or even *substituting* specific places or spaces,[19] as the foundation uniting the Jewish people *despite* the differences that derive from the multitude of geographical localities in which Jewish communities and individual Jews have settled.

As already mentioned above, Jewish historiography, for its part, has been shaped epistemologically by the primacy of time, which went along with downplaying the role of place and space. The powerful topoi of "the People of the Book" or "Our Homeland, the Text" emanating from literary studies have also contributed to a view that de-emphasizes the empirical notions of space and place while focusing instead on strategies, which are, as literary scholar Amir Eshel put it, primarily concerned "with poetically inhabiting makom, with turning spaces of dwelling into places of meaningful existence, and not necessarily with physical housing or with places of worship as such."[20]

These non-spatial or *trans*-spatial characteristics have fundamentally shaped Jewish civilization throughout history, as well as our perception of it. They are certainly of central importance and distinguish it from other, territorially more bounded civilizations. Yet when we want to investigate the Jewish experience in a comprehensive, holistic way, it is not enough to focus on the transcendental dimension only.

Where Have Jewish Topographies Been Researched and Explored So Far?

Even though there has been no comprehensive, systematic research tradition dealing with Jewish space so far, there are numerous fruitful spatial approaches to be found in Jewish studies that go beyond the textual or philosophical notions.

Jewish Studies

The Wissenschaft des Judentums, Judaic studies, and Judaistik (i.e., scholarly traditions or programs with a strong focus on religion, the Hebrew language, and on the ancient and medieval period) have produced and integrated a wealth of knowledge of archeology, the study of biblical geography and localities, and the analysis of religious architecture and structures such as the Temple, the synagogue, and the eruv.

In the younger research tradition of Jewish studies (with a broader interdisciplinary approach and a stronger focus on the modern era) research fields such as local historiography, urbanity, the study of "the ghetto" (though more as an idea than in empirical terms), and art history, with its interest in Jewish architecture, have contributed to our understanding of Jewish place and space. In both research traditions, Eretz Israel—the idea, project, and state—has played a crucial role for the development of spatial thinking and analysis, which will be discussed in greater detail below.

A much stronger connection to place and space than that found in Judaic or Jewish studies of Western provenance can be found in another Jewish research tradition—in Eastern European Jewish cultural and historical research that was strongly influenced by Diaspora-Nationalism as it was promoted by the Yiddish cultural movement or the Jewish labor party *Bund.* To prove the geographical "rootedness" and a distinct Jewish geography or cultural space (*Kulturraum*) was of vital political interest to the Diaspora nationalists, as Neil Jacobs has detailed.[21] In the context of other national movements, academic research served as an important means by which to manifest and ground the "imagined community" in a way that was not only plausible to the ethnic group but also to the community of nations. The research and teaching of this scholarly tradition was genuinely interdisciplinary, and furthermore, was closely intertwined with lay initiatives in conducting "historiography and ethnography from below," as the ethnographic *zamler kraysn* YIVO maintained throughout the Jewish world, or the *Landkentenish* movement in Poland, which for the first time will be presented in detail by Samuel Kassow in this volume. The numerous local community studies that emerged from this research tradition did not limit themselves to the institutional history of a given community or a history of its prominent figures, but also dealt extensively—from a decisively Jewish perspective—with the history of the town or village in which it was located, its Jewish and non-Jewish localities, the surrounding landscape, Jewish and non-Jewish place legends, settlement structures, and urban sociology. Even during the Holocaust the protagonists of this research tradition, despite being interned in ghettos and concentration camps, carried on their work, this time documenting the isolation and eventually the ultimate destruction of their communities. Only to a very limited extent could this research tradition and its assets be transferred to the United States, where YIVO relocated in 1940 in New York.

The Israeli Spatial Turn

Without the Holocaust we would today probably credit the emergence of a new Jewish sensitivity to and awareness of Jewish place and space to the Diaspora nationalists. With the Holocaust, East European Diaspora-nationalism was put to an end and it was Zionism, more precisely, political Zionism, that came to dominate the perspectives on Jewish space.

As part of the nation building process and in search of consolidation of Israel's territorial identity, academic core disciplines dealing with space, like architecture, and urban and regional planning became well-established. From the beginning, they dealt with "Jewish space," which was experienced in the most natural way as in the Zionist self-image, everything done in Israel was (and is) Jewish. In this sense, one may say that Israeli academia is the only academic community in the world that indeed has an institutionalized Jewish space research tradition.

The "territorialization of identity,"[22] which Jonathan Boyarin refers to, is strongly connected to this Zionist return to the land of Israel (Eretz Israel) and it has far-reaching consequences. It affects the individual, as well as the public discussion on Jewish existence in Israel. Zali Gurevitch and Gideon Aran describe this phenomenon as follows:

> The Land, as an object of reflection for Israelis and as a means to converse with their past— with the Book—is not mere territory but rather the 'we' of a people with a common fate and perhaps even a common mission. The Israeli preoccupation with the Land is thus the conversation that lies at the heart of Israeli identity. In the Land—or, as it will be termed the Place (with a capital 'P') Israelis engage in the search for the source of their collective identity.[23]

With the foundation of the state of Israel, the old conflict—or the dialectics— between the purpose of the Jewish people as the "people of the land" (*am ha-aretz*) on the one hand, and as the "people of the book" (*am ha-sefer*) on the other, gained new significance.[24]

Along with the awakening of political Zionism, the question arose as to whether the projected sovereignty of Eretz Israel was of any theological significance. The diversity and intellectual elaborateness with which the relationship between Jewish sovereignty and religious sanctity is interpreted is indeed impressive. It is this relationship that lies at the heart of the Israeli-Jewish preoccupation with space, although it is no longer the only important factor that shapes it.[25] Political conjuncture in particular contributed to the fact that the question of the new, national Israeli space, the question of its physical dimensions, and especially the question of its borders and their raison d'être became the core of the Zionist self-conception.

In the struggle for a Jewish national home (as Balfour put it), and later for a state, territory and borders were not self-evident, and with the ongoing Israeli-Palestinian conflict, this is still true. With the conquest of biblical territory in the Six-Day War, a national-religious dimension was added to the discourse on land and territory. Opposite trends, like the disenchantment after the Yom Kippur War in 1973, the founding of Shalom Achshav in 1978, and eventually the Oslo peace process, contributed to an even more diversified debate among the Israeli public.

A broad social science literature covers this territorial discourse with its diverse ideological aspects in depth.[26] Baruch Kimmerling's *Zionism and Territory* (1983), which is today considered a classic, was one of the pioneering publications in this field. With the advent of postcolonial discourse, an entire body of literature evolved around the concern with territory, power relations, and discrimination in the context of the Arab-Israeli conflict. Prominent Israeli social scientists like Gershon Shafir, Ilan Pappe, and Ronen Shamir embrace the assumptions of a postcolonialist perspective on Zionism and use it as a point of departure for their research.[27]

Another thriving branch in the Israeli academic landscape is the field of political geography dealing with the politically and ideologically motivated appropriation of the land. In conjunction with this trend, the function of local myths like Massada or Tel Hai have been analyzed and deconstructed within the context of Zionist ideology.[28]

An important development of the last decade, which is related to the general "spatial turn" is the publication of several titles that use the terms "space" and "place" as explicit research perspectives. Those volumes often fall within the purview of cultural studies, but archeology, and even legal studies have also contributed to this trend. These studies do not represent a specifically Israeli development, but they do benefit from the spatial sensibility described above and focus on questions in which place and identity play a major role. Three examples: In a study by legal geographer

Issachar Rosen-Zvi, *Taking Space Seriously*, the production of political space through legal treatment is analyzed by looking at three groups (Arab-Jews, Bedouins, and Ultra-Orthodox) and the social and cultural consequences of their three different spatial states.[29] One of our contributors, Haim Yacobi, is the author of another important book investigating spatial phenomenona by employing interdisciplinary approaches. In his study *Constructing a Sense of Place,* Yacobi asks two intriguing questions: "What is the role of architects and planners as mediators between national ideology and the politicization of space?" as well as "What is the contribution of the very act of shaping the landscape to the construction of a sense of place?"[30]

Less theoretically inspired, but with a very strong and physically oriented space perspective, is the study by Meron Benvenisti, *Sacred Landscape*, which recounts the "buried history" of Arab places in Israel. In a reflective statement, Benvenisti recalls: "As long as I remember myself, I have moved within two strata of consciousness, wandering in a landscape that, instead of having three spatial dimensions, had six: a three-dimensional Jewish space underlain by an equally three-dimensional Arab space."[31]

Against this backdrop, we argue that the preoccupation of Israeli academia with space and place, which preceded the international spatial turn, played a crucial and pioneering role. Seen in a wider historical context, it is plausible to say that without the existence of the state of Israel the academic exploration of Jewish space and place in general would look quite different: the Israeli focus on territorial matters, the imagination and reconstruction of ancient Jewish places, and the challenge by Israeli and non-Israeli critics of the religious or ideological "givenness" of the Jewish relationship to the territories, within and beyond the Green Line, has also created a new sensibility for other Jewish places.

The Spatial Turn and the Emergence of New Discursive Arenas

In the wake of the general spatial turn, initiated by scholars of cultural geography, architectural theory, urban studies, migration studies, and anthropology, Jewish studies has witnessed a veritable boom during the past decade of projects that investigate questions of place and space. This new impulse, has, however, originated with individual scholars and so far has not led to the establishment of substantial networks or stable institutions. To date, one of the most dynamic spheres for academic exchange on Jewish space and place are the academic conferences that have become more and more frequent over the past years. This period has been witness to a growing conceptual awareness at these forums vis-à-vis "place and space"—a significant shift from what had previously seemed to serve as not much more than a fashionable label. The following list of conferences reflects the ever-growing spatial boom since the mid-1990s.[32]

Conferences on Jewish Space

Reclaiming Memory: Urban Regeneration in the Historic Jewish Quarters in Central Europe
International Cultural Centre, Krakow (Poland)
June 2007

Jewish Spaces: Die Kategorie Raum im Kontext kultureller Identitäten
Centre of Jewish Studies (CJS) of the Karl-Franzens University, Graz (Austria)
April 2007

No Direction Home: Re-imagining Jewish Geography
Lehigh University, Bethlehem (Pennsylvania, USA)
March 2007

Jewish Journeys
Isaac and Jesse Kaplan Centre for Jewish Studies and Research/Parkes Institute for Jewish/non-Jewish Relations, Cape Town (South-Africa)
January 2007

Jewish Space in Central and Eastern Europe: Day-to-day History
Center for Studies of the Culture and History of East European Jews, Vilnius, in cooperation with Lithuanian Institute of History, Vilnius (Lithuania)
May 2006

The Interplay between Real and Imagined Places in Jewish Civilization
DFG-Research Program "Makom," University of Potsdam (Germany)
June 2005

Place and Displacement in Jewish History and Memory: Zakor v'Makor
Kaplan Centre for Jewish Studies, University of Cape Town (South Africa)
January 2005

The Jew's Space: Architecture, Urbanism, and the Jewish Subject
Pennsylvania State University (USA)
March 2004

"Makom" Conference for Doctoral Students
DFG-Research Program "Makom," University of Potsdam (Germany)
November 2003

Jewish Conceptions and Practices of Space
Stanford University (California, USA)
May 2003

Place and Space in Modern Jewish Experience, 1989–2002
DFG-Research Program "Makom," University of Potsdam
(Germany)
June 2002

Sacred Spaces: Preserve or Abandon?
Organized to celebrate the Opening of the District of Columbia Jewish
Community Center Washington, DC (USA)
February 1996

Geography and Jewish Studies
University of Maryland at College Park (USA)
March 1995

**The Future of Jewish Monuments: An International Conference on
the Preservation of Historic Sites and Structures**
Sponsored by the Jewish Heritage Council, World Monuments Fund,
and Hebrew Union College, Jewish Institute of Religion, New York
(USA)
November 1990

**The Role of Geography in Jewish Civilization: Perceptions of Space,
Place, Time, and Location in Jewish Life and Thought**
Ohio State University (USA)
October 1990

This growing trend is also observable in the increase of space-related panels and individual presentations at the annual conferences of the Association for Jewish Studies (AJS) and the Association of Israel Studies (AIS), as well as in Rambi, the Index of Articles on Jewish Studies of the Jewish National and University Library.[33]

The new research field was further consolidated by special issues of Jewish studies journals and thematic anthologies that focused on broader conceptual questions, starting with *Land and Community*, edited by Harold Brodsky in 1998,[34] and the special issue on "Studies in Jewish Geography," edited by Neil Jacobs for *Shofar* in the same year.[35] They were followed by a special issue on "Jewish Conceptions and Practices of Space," edited by Charlotte E. Fonrobert and Vered Shemtov for *Jewish Social Studies* in 2005;[36]

a special issue of *American Jewish History* on "A Sense of Place" in 2007;[37] a special issue of Israel Studies "Territory and Space in Israeli Society and Politics" in 2008, as well as two German-language volumes that emerged out of the Makom research group in Potsdam.[38]

The first interdisciplinary research group to deal with Jewish space that has been created in the wake of the spatial turn was the Makom program in Potsdam, supported by the DFG. With its colloquia, thematic workshops, and conferences it brought increased visibility to the field in Germany and provided a home and a systematic framework for many innovative research projects. However, because Makom was never intended to be permanent—it operated over a period of six years, coming to the end of its tenure in 2007—it is still too early to assess whether and what lasting impact it may have had.[39]

Locating This Project

In the following, we will locate our project in the post-1989 European space. It is all but self-evident that this book on Jewish Topographies emerged out of Potsdam, *the* bastion of Prussian conservatism and a frontier town of the East German regime, a city whose Jewish community had been almost completely destroyed in the Holocaust and was finally erased from the register of associations in 1949 by the communists. Thus we wish to shed light on the various geo-political changes that have provoked a new awareness for space in general and for Jewish space in particular during the past two decades, and which also set in motion our own mental maps.

In the Anglo-American sphere, the origins of the spatial turn can be traced back to the effects of decolonization and globalization, which called into question the hegemony of national space. In Israel and Europe, the spatial perceptions of the world and these respective regions have been shook up by the changes that took place at the end of the 1980s.

As for Israel, it was the Middle East Peace Process, set off by the ongoing First Intifada, a weakening Soviet Union and the First Gulf War, which for the first time since the foundation of the state in 1948 opened up a perspective for regional integration. Even if much of what had been discussed during the bi- and multilateral negotiations in Madrid and Oslo never translated into political action, (and was in fact set back after the assassination of Itzhak Rabin in 1995 and by the outbreak of the Second Intifada in the fall of 2000) it made many Israelis aware of their own national space and made them look at their regional surroundings in a new light.

In the case of Europe it was the end of the Cold War in 1989 that catalyzed a major spatial reconfiguration. Until then, the awareness of geopolitical space had been characterized by experiences of separation and confrontation between two antagonistic blocks. This also profoundly shaped our individual horizons. In West Germany, where all three of us grew up, our generation was looking westwards and our geographical horizon, shaped by student exchange programs and individual travel adventures like the obligatory Inter-Rail tour, was tuned towards Western Europe or North America. The East was left blank—just like on the weather chart on West German television, which visualized the territory of the GDR as an empty space, a no-man's-land with a little red dot in the middle, indicating the location of West Berlin. With the fall of the Berlin Wall, we watched the spatial manifestations in which we had grown up, rapidly falling apart during our last years of high school, and a new world opened up before our eyes. The continent was stirred up and began to shift eastwards, as Karl Schlögel aptly put it.[40] This change in perception was paralleled by an ever-growing integration, mending the rift between Eastern and Western European countries on the political, economic, and cultural plane as well as in everyday life.

These geopolitical changes prepared the ground for the immense resonance the spatial turn has excited in German academia in the past couple of years. While the memorial turn is receding somewhat, "space" has become the new academic darling. In mainstream academia the trend first arrived, when the *Historikertag 2004*, the biennial Congress of German Historians chose *Raum* as its central theme. And ever since then, the "spatial turn" has sped up, with an abundance of conferences, publications, and large-scale research projects, as anyone listed on *H-Soz-u-Kult,* the most important humanities mailing list for Germany, will confirm: rarely a week passes in which not at least two or three space-related items are posted.[41] Apart from the many academic works that insert the term "space" in their titles in order to appear au courant, there have been many innovative and comprehensive works contributing to general spatial theory, but due to language barriers they have to date received little academic attention outside of Europe.[42]

Reemergence of the Jewish Dimension in the European Public Realm

Last but not least, the events of 1989 have led to a new awareness of Jewish space and Jewish places in Central and Eastern Europe, which have by now become *the* test case for a Europe that is historically conscious, diverse, and inclusive. Whereas the Cold War promulgated strong homogenizing dynamics and strengthened the (self-) perception of monolitic, clearly demarcated European nation-states on both sides of the Iron Curtain, the changes initiated by its fall in 1989 led to a change of perspective which allowed many to rediscover European diversity, historical interconnectedness, and transnational dynamics. This reevaluation process took place not only on the international level, but soon echoed within the national contexts of the individual European states (though not always at the same speed, nor with the same amount of enthusiasm). European societies became more and more aware of their internal diversity, historical depths, and the ways in which the transnational moment was playing out within the context of their individual nation-states. As a part of this process, Jewish space resurfaced in public discourse as well as in European cityscapes. More specifically, a new awareness began to emerge concerning historical Jewish spaces, those destroyed by the Nazis, and later on dispossessed, concealed and silenced by the regimes of postwar Europe—in the West (West Germany, Austria) as well as in the East (East Germany, Eastern European countries, former member states of the SU)—although these destructive and obscuring actions manifested themselves in very different forms and to very different degrees.

Today the public debates on the (re-)localization of Jewish space are perceived as the litmus test for a critical historical consciousness and inclusive identity of European societies: be it the controversies concerning memorial sites for the Holocaust (e.g., the Holocaust memorial in Berlin), the visualization of the historical Jewish dimensions within European cityscapes (e.g., the restoration project "historical Jewish Ghetto fragments" in the Old Town of Vilnius), the great public enthusiasm with which new Jewish spaces such as the newly constructed synagogues in Dresden or Munich have been met, or the popularity of pseudo- or "virtual" Jewish spaces, as Ruth Ellen Gruber has put it, in "Jewish Prague."[43]

While it remains to be seen how much of these debates on the open stage will be transformed into sustainable political substance and a deepened awareness, 1989 has also resulted in an increasing diversification of the Jewish communities in Europe and brought about fundamental inter-communal debates about the configuration of internal and public Jewish spaces, as our authors Brigitta Ezter Gantner and Matyás Kovacs demonstrate in the case of Budapest.[44]

Summing Up

Overall, the following tendencies should be noted:

- To date, there are no comprehensive, sustainable Jewish space studies—by sustainable we mean a solid institutional framework as well as integration into the Jewish studies curriculum.
- The academic discourse on Jewish space still tends to be strongly shaped by the different Weltanschauungen of its protagonists, and runs along dichotomic lines: Israel vs. Diaspora; religious vs. secular; leftist vs. rightist; Zionist vs. post-Zionist.

- After the Holocaust, an unbiased approach (or approaching an unbiased perspective at all) to European Jewish topography is not easily achieved. As a result of the Holocaust, the predominant focus, especially in Europe proper, is on spaces of death and remembrance, memorials and museums, voids and relicts, rather than on living Jews and their spaces and spatial strategies, past and present.
- Following the destruction of the Jewish communities in Europe, the importance Israel holds on a political level since the foundation of the state in 1948, and the prominence of Zionist points of view, conceals the importance that other places and spaces held and still hold for Jews throughout the world.

Lived Spaces, or: Approaching the Topography of Jewish Presence

In light of the previous focuses and omissions, the concept of *lived Jewish spaces* provides a particularly valuable contribution to the study of Jewish place and space. Focusing on *how* Jewish spaces are produced, we are not so much interested in an ideal objective, the "essence" of Jewish space or the built end-product but rather in the creation of Jewish space and in Jewish spatial experience. This approach ties into Henri Lefebvre's concept of the *espace vécu* or "lived space," which he conceives as a space of the inhabitants and users:

> This is the dominated—and hence passively experienced—space which the imagination seeks to change and appropriate. It overlays physical space, making symbolic use of its objects. (…) [It] is alive: it speaks. It has an affective kernel or centre: Ego, bed, bedroom, dwelling, house; or: square, church [or, in the Jewish case: the synagogue; the eds.], graveyard. It embraces the loci of passion, of action and of lived situations, and thus immediately implies time. Consequently it may be qualified in various ways: it may be directional, situational or relational, because it is essentially qualitative, fluid and dynamic.[45]

It is to these *espaces vécus* that Lefebvre devotes specific attention because for too long they have been neglected in general space studies as well. How this research perspective can be applied to Jewish space becomes evident from the articles collected in this book.

The temporal focus of the different contributions covers the nineteenth and twentieth centuries; however some authors also refer to places and spaces of earlier periods and go back in time as far as to the biblical and Talmudic period.

Our aim here has been to explore as wide a geographical spectrum as possible. While the geographic representation is not inconsiderable—Israel, North Africa and the Middle East, North America, Central and Eastern Europe, as well as Germany—it is far from being exhaustive or comprehensive. In several contributions the focus lays on the connection and correlation of two or more places e.g., Morocco and Netivot, an Israeli development town in the Negev desert, where many Jewish-Moroccan families have settled; the foodways that link the former Soviet Union and Jewish neighborhoods in Brooklyn and "Queenistan"; and the ties between Israel and backpacker enclaves in the Far East where young Israelis come together.

The selection of articles presented here is as much influenced by our own academic viewpoints and mental maps of Jewish spaces as it is by the topography of Jewish studies as an academic field. Policy decisions on what research receives funding and other support and what does not, the degree of access to the Jewish spaces themselves, as well as to archives and other relevant resources all play a role in determining the depth and breadth of the knowledge that explicates these sites. Thus, the world map of Jewish topographies in fact comprises two interconnected layers: the map of the actual, existing Jewish spaces, in the past and the present, as well as the map of those spaces that have captured the attention of the space-centered disciplines of academia.

Moving through Jewish Topographies

In its composition this volume aims to develop concepts of place through which the "perception" and "seeing" of Jewish places and spaces can be learned anew. The book starts out with concrete places in the sections "Construction Sites" and "Jewish Quarters," followed by the more conceptual "Cityscapes and Landscapes." Attention is next directed to the dynamic constructions and deconstructions in the sections "Exploring and Mapping Jewish Space" and "Enacted Spaces." The organization of the book thus follows the shifting meaning of Jewish topographies, from the microcosms of Jewish daily life to place-based explorations, virtual worlds, and enacted spaces.

Instead of reading Jewish space along the lines of the established dichotomies—Israel/Diaspora; religious/secular; internal/external—the focus is on the transitions and dialectic tensions in between and the resulting multidimensionality that lies at the heart of Jewish topographies.

The categories chosen to structure the individual sections are open and universal, and deliberately so. They constitute a framework that allows—in a first rapprochement—well-established conceptions of Jewish places and spaces to be transcended, such as the *ghetto* or the *shtetl*, both of which tend to reproduce clichéd and stereotypical perceptions of Jewish space instead of conveying the complex life and world of those who lived in predominantly Jewish quarters or small towns.[46]

Construction Sites

This section concentrates on the role of actual buildings, their construction process, and their role in contemporary urban life. Miriam Lipis examines the function performed by the symbolic house, the *sukkah*, in questions of belonging. Manuel Herz confronts the invisible constructed space, the *eruv*, with recent examples of "architecture with Jewish context" in Germany, and Haim Yacobi looks at the Israeli development town of Netivot as a diasporic place produced by the interplay of the sovereign state and its immigrants. All three of these contributions touch on questions of belonging while at the same time analyzing the role architecture and urban planning play as vehicles for identity formation.

Jewish Quarters

What makes a well-defined territory, predominantly inhabited by Jews, Jewish? The authors in this section radically challenge and enlarge the traditional understanding of Jewish quarters and offer new insights into the multifaceted processes of physical survival, everyday life routines, and identity construction. Moreover, they make visible a fascinating diversity within one and the same community. Kenneth Helphand explores the complex history of the gardens established by Jews imprisoned in ghettos during the NS period. Susan Miller investigates the spatial boundaries between Jews and Muslims in the historical *mellah* of Fez, whereas Etan Diamond's Jewish topographies center around the religious micro-spaces of Thornhill, a suburb of Toronto. Eszter Brigitta Gantner and Mátyás Kovács' article outlines the evolving Jewish subcultures within the urban landscape of Budapest and the complementary virtual space of the Internet.

Cityscapes and Landscapes

How do Jews relate to their environment? The city is usually seen as the prototypical Jewish place, and Joachim Schlör probes the urban qualities of modern Jewish life. Yet, other environments have also left their imprint on Jewish life. Haya Bar-Itzchak describes how Jews (re)created Poland by telling local legends and thus Judaized their places of settlement. While it is a commonplace that Zionists were eager to return to the countryside to revive the idea of physical Jewish labor, Yael Zerubavel looks at the role the desert played as a symbolic counter-landscape in pre-state Israeli culture. Gilbert Herbert directs his attention to the sea and the question whether there is any special relationship between Jews and the art of seafaring.

Exploring and Mapping Jewish Space

Up to this point we have seen how specific sites, quarters, and environments have been reflected and appropriated. But Jewish space can also be produced by the modes of moving through, exploring, and mapping. Samuel Kassow describes how, with the *Landkentenish* movement in Poland, an understanding of travel and local historiography as a national mission emerged. Looking at Katmandu or Goa, a first glance reveals nothing Jewish. But Erik Cohen conjures these locations—chosen by Israeli backpackers on their typical trips across the Far East—and how they have emerged as Israeli enclaves, becoming simultaneously counter-places and part of the Jewish-Israeli experience. Shelley Hornstein shows how diasporic space can be explored using cyberspace, revealing that the Diaspora as well as the Internet is characterized by itineraries.

Enacted Spaces

The last section shifts the focus to the performative side of space production, i.e., to all kinds of actions, activities, and stagings that are related to specific place/s or space/s. We encounter Eve Jochnowitz's evocation of culinary landscapes and place-making practices of Russophonic Jews in New York. With Galeet Dardashti we learn about the musicians of the Buena Vista Baghdad Club and the "mediascapes" that are created around their lives in Iraq and Israel. We follow Michael Feige in his deconstruction of the "authentic" Israeliness in Mini-Israel, an amusement park near Tel Aviv. These contributions not only illustrate the myriad ways in which Jewish space is interpreted and represented in a Jewish context, but how it is actually generated and lived.

***

And the future of Jewish topographies? In the epilogue, Julian Voloj concludes the journey undertaken here by entering into the virtual realm via the Internet platform Second Life, and introduces a fast-evolving, new Jewish space of opportunities, in which yet unimagined Jewish places are being created, enacted and performed.

Notes

1   Edward Soja, "History: Geography: Modernity," in id., *Postmodern Geographies: The Reassertion of Space in Critical Social Theory* (London: Verso Press, 1989), 10–42.
2   Henri Lefebvre, *The Production of Space* (Oxford: Blackwell, 1994), 131, 142–43.
3   Soja, "History: Geography: Modernity," 10.

4　See Yosef Hayim Yerushalmi," Exile and Expulsion in Jewish History," in *Crisis and Creativity in the Sephardic World, 139–1648*, ed. Benjamin R. Gampel (New York: Columbia University Press 1997), 3–21; here specifically: 11–14; id., "Ein Feld in Anatot: Zu einer Geschichte der jüdischen Hoffnung," in *Ein Feld in Anatot: Versuche über jüdische Geschichte* (Berlin: Wagenbach, 1993), 81–95; here specifically 90; also published in French: Yosef Hayim Yerushalmi, "Un champ à Anathoth: Vers une histoire de l'espoir juif," *Esprit* 104/105 (1985): 24–38.

5　The concept of translocality has been developed mainly by anthropologists concerned with the entanglements of regions in Africa, the Middle East and Asia, where in many areas national borders have been established only in the course of the 20th century. The concept refers simultaneously to exchanges, transfers and flows as well as to the production of locality and notions of delineation and structure, and takes into account that there are variegated borders within or beyond the nation state whose crossing, depending on the situation, might be more significant than the crossing of national borders, see Ulrike Freitag, "Translokalität als ein Zugang zur Geschichte globaler Verflechtungen," *geschichte.transnational*, 10 June 2005, http://geschichte-transnational.clio-online.net/forum/id=879&type=diskussionen (last accessed August 28, 2007).

6　Wolfgang Welsch, "Transculturality: The Puzzling Forms of Cultures Today," in *Spaces of Culture: City, Nation, World*, eds. Mike Featherstone and Scott Lash (London: Sage 1999), 194–215; here 199–200.

7　Ibid., 203.

8　Cf. Clemens Kauffmann, "Die politische Logik des Ortes," in *Raumdeutungen: Ein interdisziplinärer Blick auf das Phänomen Raum*, eds. Sabine Feiner, Karl G. Kick, Stefan Krauß (Münster: Lit, 2001), 281–317; here 282f.: "The Jewish experience is a touch stone for the conscious-raisers of globalization. For 5761 years, Jews lived as a people of the world in a transnational social space and do not quite fit into the new logics of placelessness. As the Diaspora of the Jews has taught, the decoupling of a people and territory and thus the annihilation of geographical and social proximity is a key experience not only of modern societies. Transnationality has for all times been the form of existence of Jews. By means of its historical example one can show the premises, conditions and endangerments [of globalisation]" [283], trans. by the eds.

9　On differing spaces see Michel Foucault, "Of Other Spaces," *Diacritics* 16 (1986): 22–7; Edward Soja, Barbara Hooper, "The Spaces that Difference Makes: Some Notes on the Geographical Margins of the New Cultural Politics," in *Place and the Politics of Identity*, eds. Michael Keith and Steven Pile (London: Routledge, 1993), 183–205; Akhil Gupta, "Space, Identity and the Politics of Difference," *Cultural Anthropology* 7, no. 1 (February 1992): 6–23.

10　An example in case are the very different meanings of linguistically closely related terms such as 'territoire' or 'terroir' in French, 'Territorium' in German, and 'territory' in English, which in the Israeli case e.g., has been turned into a proper toponym, signifying occupied the West Bank and Gaza.

11　In this approach we rely on the helpful conceptual distinction of Michel de Certeau who defines a place as "an instantaneous configuration of positions. It implies an indication of stability. (…) [S]pace is composed of intersections of mobile elements. It is in a sense actuated by the ensemble of movements deployed within it. Space occurs as the effect produced by the operations that orient it, situate it, temporalize it (…) In contradistinction to the place, it has thus none of the univocity or stability of a 'proper.' In short, *space is a practiced place.*" Michel de Certeau, "Spatial Stories," in id., *The Practice of Everyday Life* (Berkeley: University of California Press, 1988), 115–30; here 117 (emphasis in the original).

12   Steve Harrison, Paul Dourish, "Re-Placing Space: The Roles of Space and Place in Collaborative Systems," in *Proc. ACM Conf. Computer-Supported Cooperative Work CSCW'96* (Boston, MA), 67–76; here 67, 69.

13   Leopold Zunz, "Essay on the Geographical Literature of the Jews," in *The Itinerary of Rabbi Benjamin of Tudela*, ed. Abraham Asher, 2 vols. (London: Asher, 1841), vol. II, 230–317; here 230–31. [first published as id., "Geographische Literatur der Juden von der ältesten Zeit bis zum Jahre 1841," in *Sammlung von Schriften und Reden des Dr. Zunz*, ed. by Curatorium der Zunzstiftung, vol. I. (Berlin: 1875), 146–216.

14   Zunz, "Essay on the Geographical Literature of the Jews," 304.

15   Neil G. Jacobs, "Introduction: A Field of Jewish Geography," in *Shofar* 17, no. 1 (Fall 1998), special issue "Studies in Jewish Geography," 1–18; here 2–3, 4.

16   "Noi privilegiamo la crescita sull'essere, la formazione rispetto alla forma come entità conclusa. La nostra storia è antistatica e antispatiale." Bruno Zevi, "Spazio et non-spazio ebraico" [Jewish space and non-space], presentation (Confronto del compositore jazz Uri Caine con la musica di Mahler), Comunità Ebraica di Venezia, 29 novembre 1998, http://www.fondazionebrunozevi.it/19892000/frame/892000.htm (last accessed August 21, 2007).

17   Robert D. Mitchell, "Introduction: Genesis; Introducing Geography in Jewish Studies," in *Land and Community: Geography in Jewish Studies*, ed. Harold Brodsky (Bethesda, MD: University Press of Maryland, 1997), 1–5; here 2.

18   See http://www.cad.architektur.tu-darmstadt.de/synagogen/inter/menu.html (last accessed August 21, 2007). Another example is the Bet Tfila: Research Unit for Jewish Architecture in Europe, run by the Center for Jewish Art at the Hebrew University in Jerusalem together with the Department of History of Architecture of the Technical University in Braunschweig (Brunswick), which puts a strong emphasis on the involvement of students as part of its documentation and restoration activities, http://www.bet-tfila.org/en/start.htm (last accessed August 21, 2007).

19   See e.g. Abraham Joshua Heschel, *The Sabbath: Its Meaning for Modern Man* (1951; repr., New York: Farrar, Straus and Giroux, 1995) where he lays out in detail a theology that conceives of Jewish religious practice as an "architecture of time."

20   Amir Eshel, "Cosmopolitanism and Searching for the Sacred Space in Jewish Literature," in *Jewish Social Studies* 9, no. 3 (2003): 121–38; here 124.

21   For a discussion on the "ideological origins of modern Jewish Geography" and especially of Max Weinreich's role as the pioneering cultural geographer of Ashkenaz, see Jacobs, "Introduction," 7–18.

22   Jonathan Boyarin, "A Response from New York," in *Grasping Land: Space and Place in Contemporary Israeli Discourse and Experience*, SUNY series in Anthropology and Judaic Studies, eds. Eyal Ben-Ari and Yoram Bilu (New York: State University of New York Press, 1997), 217–29; here 218.

23   Zali Gurevitch and Gideon Aran, "The Land of Israel: Myth and Phenomenon," in "Reshaping the Past: Jewish History and the Historians," *Studies in Contemporary Jewry* (1994), 195–210; here 195.

24   Although Zionism strove to "reunify book, people, and place," Israel has still to live with the tension as Gurevitch notes: "With Zionism the old paradox of the place came to life, newly embodied in the Zionist act—calling to rebuke the book that is 'Jewish' for the land, yet through this very return to the land to enact the voice of the book." Zali Gurevitch, "The Double Site of Israel," in *Grasping Land*, 203–216; here 206.

25   For an original approach how to reconcile Jewish bonds to the whole of Erez Israel with political independence in times of the waning power of the nation state see Yosseph Shilhav, "Jewish Territoriality between Land and State," *National Identities* 9, no. 1 (2007): 1–25.

26 The spatial theme in recent Israeli academic writing (mainly by geographers and social scientists) is reviewed by David Newman, "Controlling Territory: Spatial Dimensions of Social and Political Change in Israel," in *Traditions and Transitions in Israel Studies*, Books on Israel, Vol. VI, ed. Laura Z. Eisenberg (Albany, NY: State University of New York Press, 2003), 67–87. The studies he discusses are: the aforementioned *Grasping Land*; Meron Benvenisti, *Sacred Landscape: The Buried History of the Holy Land since 1948* (Berkeley: University of California Press, 2000), Amiram Gonen, *Between City and Suburb: Urban Residential Patterns and Processes in Israel* (Aldershot: Avebury Press, 1995); Juval Portugali, *Implicate Relations: Society and Space in the Israeli-Palestinian Conflict* (Dordrecht: Kluwer Academic Publishers, 1993); Itzhak Schnell, *Perceptions of Israeli Arabs: Territoriality and Identity* (Aldershot: Avebury Press, 1994); Oren Yiftachel and Avinoam Meir, eds., *Ethnic Frontiers and Peripheries: Landscapes of Development and Inequality in Israel* (Boulder, CO: Westview Press, 1997).

27 Gershon Shafir, *Land, Labor and the Origins of the Israeli-Palestinian conflict 1882–1914* (Cambridge, UK: Cambridge University Press, 1989), Ronen Shamir, *The Colonies of Law: Colonialism, Zionism and Law in Early Mandate Palestine*, Cambridge Studies in Law and Society (Cambridge, UK: Cambridge University Press, 2000), Ilan Pappe, *A History of Modern Palestine: One Land, Two Peoples* (Cambridge: Cambridge University Press, 2003).

28 A landmark study in this context is Yael Zerubavel, *Recovered Roots, Collective Memory and the Making of Israeli National Tradition* (Chicago: University of Chicago Press, 1995). See also Myron J. Aronoff, "Myths, Symbols, and Rituals of the Emerging State," in Laurence J. Silberstein, ed., *New Perspectives on Israeli History: The Early Years of the State* (New York: New York University Press, 1991), 175–92.

29 Issachar Rosen-Zvi, *Taking Space Seriously: Law, Space, and Society in Contemporary Israel* (Aldershot: Ashgate, 2004).

30 Haim Yacobi (ed.), *Constructing a Sense of Place: Architecture and the Zionist Discourse* (Aldershot: Ashgate 2004), 4. See also Alona Nitzan-Shiftan, "The Israeli 'Place' in East Jerusalem: How Israeli Architects Appropriated the Palestinian Aesthetic after the '67 War," *Jerusalem Quarterly* 27 (Summer 2006), 15–27. Nitzan-Shiftan shows how the Israelization of the indigenous Palestinian architecture was used to find an authentic "architecture 'of the place'". By reading "it as biblical architecture, as an uncontaminated primitive origin of architecture, or simply as typically Mediterranean" Israeli architects succeeded to appropriate this style as if it were originally Israeli. Ibid., 22.

31 Meron Benvenisti, *Sacred Landscape: The Buried History of the Holy Land since 1948* (Berkeley: University of California Press, 2000), 1.

32 Starting point for this list was the very useful website by Bruce Janz "Research on Place and Space," http://pegasus.cc.ucf.edu/~janzb/place/. We are also indebted to all the colleagues who responded to our query on H-JUDAIC (June 26, 2007), asking for "any other conference, symposium or workshop which dealt with Jewish space and place." We list here only the events that have a clear spatial focus.

33 Comp. to "Past AJS Conferences," http://www.ajsnet.org/past.htm; past AIS conferences, http://www.aisisraelstudies.org/pastconferences.html; "Rambi," http://jnul.huji.ac.il/rambi/.

34 *Land and Community: Geography in Jewish Studies*, ed. Harold Brodsky (Bethesda, MD: University Press of Maryland 1997).

35 "Studies in Jewish Geography" (Special Issue), ed. Neil G. Jacobs, *Shofar* 17, no. 1 (Fall 1998).

36 "Jewish Conceptions and Practices of Space" (Special Issue), eds. Charlotte Elisheva Fonrobert and Vered Shemtov, *Jewish Social Studies* 11, no. 3 (2005).

37   "A Sense of Place" (Special Issue), eds. Deborah Dash Moore and Dale Rosengarten, *American Jewish History* 93, no. 2 (2007).

38   *Der Ort des Judentums in der Gegenwart, 1989–2002* [The Location of Jewishness in the Present, 1989–2002], series Sifria, eds. Michal Kümper, Anna Lipphardt et al. (Berlin: Be.Bra, 2004); *Makom: Orte und Räume im Judentum* [Makom: Places and spaces of Judaism], series Haskala, vol. 35 (Hildesheim: Georg Olms, 2007).

39   The Deutsche Forschungsgemeinschaft/DFG (German Research Foundation) is the central, self-governing research funding organisation in Germany and maintains a specific program sponsoring innovative interdisciplinary research topics for a period of up to six years.

40   See Karl Schlögel, *Die Mitte liegt ostwärts: Europa im Übergang* (München: Hanser, 2001).

41   The full-title of the mailing list reads "Humanities: Sozial- und Kulturgeschichte"; the moderated list is part of H-net, for more see, http://hsozkult.geschichte.hu-berlin.de/.

42   Representatively, we refer to the seminal studies of Karl Schlögel, *Im Raume lesen wir die Zeit: Über Zivilisationsgeschichte und Geopolitik* (München: Hanser, 2003); Martina Löw, *Raumsoziologie* (Frankfurt/M.: Suhrkamp, 2001) and the collection *Raumtheorie: Grundlagentexte aus Philosophie und Kulturwissenschaften*, eds. Jörg Dünne and Stephan Günzel (Frankfurt/M.: Suhrkamp, 2006), that assembles canonical and forgotten texts by German- and French-language authors to confront "the proclamation of the spatial turn in the anglo-saxonian shpere with its largely European roots" (13, trans. by the eds.).

43   Ruth Ellen Gruber, *Virtually Jewish: Reinventing Jewish Culture in Europe* (Berkeley: University of California Press, 2002).

44   See also, Diana Pinto, "Jewish Spaces versus Jewish Places? On Jewish and Non-Jewish Interaction Today," in *Der Ort des Judentums in der Gegenwart, 1989–2002*, 15–25; id., "A New Jewish Identity for Post-1989-Europe," in *JPR Policy Paper* 1 (1996).

45   Lefebvre, *The Production of Space*, 39, 42.

46   See Anna Lipphardt, "'Dos amolike yidishe geto.' Blick in das jüdische Viertel in Vilne" [The former Jewish ghetto. View onto the Jewish Quarter of Vilna], in *Simon Dubnow Institute Yearbook IV* (Göttingen: Vandenhoeck & Ruprecht, 2005), 481–505, especially 503–505. .

Part I

Construction Sites

1 A Hybrid Place of Belonging

Constructing and Siting the Sukkah

Miriam Lipis

Children in Beer Sheva posing with a sukkah that they constructed, 2005

The commandment to create a space as a ritual object, and most importantly to dwell in it, can already be found in the text of the Torah, the foundational text of Judaism:

> During [these] seven days you must live in thatched huts. Everyone included in Israel must live in such thatched huts. This is so that future generations will know that I had the Israelites live in huts when I brought them out of Egypt. I am God your Lord. (Leviticus 23:42–3)

This moves a spatial experience—an immersive environment—into center stage. The above passage from Leviticus is read in relationship to the practice of constructing a *sukkah*[1] during the one-week long Jewish holiday of *Sukkot*. A *sukkah* is a thatched booth with at least three walls, and must be built anew every year. Most are simple booths, but they can also be integrated into rooms whose roofs can be opened, as one must be able to see the stars through a sukkah's roof. The constructing of sukkahs is an ancient practice still performed today.[2]

As spatial practices influence society and vice versa, the way people inhabit the world, a room, or their bodies thus reveals something about the way they think. This article aims to explore the sukkah as a place-making tool that creates a symbiotic place of belonging.[3] Traditionally, a place of belonging is associated with a single geographical location; but this is not the case here. The symbiotic places of belonging evoked by the sukkah are composed of at least four real and imagined places: the local, the Land of Israel,[4] the Bible as the portable homeland, and God's presence. After a theoretically aimed introductory section, these four places of belonging will be discussed, followed by a description of how these real and imagined, physical and mythical places are constructed in modern urban contexts.

This analysis is mapped out at the intersection of several disciplines: architecture, Jewish studies, and cultural studies. The sukkah has been overseen by both architectural practice and theory although it is based on the iconology, function, and symbolic meaning of the house, which is at the center of architecture. However, architectural theory and history are opening up, particularly through the influence of gender studies, as Jane Rendell writes: "First, new objects of study—the actual material which historians choose to look at; and second, the intellectual criteria by which historians interpret those objects of study."[5] This move has made it possible to analyze a temporary ritual structure as architecture. To do justice to the unusual character of sukkahs, I applied criteria not only from full-scale permanent buildings but also developed new criteria for their interpretation from the objects themselves. This theoretical work has been deeply influenced by my empirical fieldwork, photographing sukkahs in Germany, England, France, Belgium, Holland, the United States, and Israel during the last several years.

The question of place, travel, and belonging was always present during those fieldtrips.[6]

Approaching the Sukkah as Architecture

Usually, the sukkah is only analyzed as a ritual object and looked at from within the religious context.[7] The sukkah is considered differently here—as architecture: forms, sites, materiality, and construction methods are all taken into consideration. The construction rules of a sukkah place it clearly in the realm of a building. For example, the book, *A Guide to the Laws of Succos*,[8] has a section discussing the construction of the walls, where an emphasis is placed on the appearance of enclosure. For a sukkah, three walls are required, two of which must be seven *tefachim* long and the third one must be only one *tefach* long (1 *tefach* is either 9.28 cm/3.66 in or 7.84 cm/3.01 in). However, this third wall must still appear as if it is a wall—it must give the entire structure a feeling of enclosure. This led the rabbis to investigate visual perceptions. How can a wall that has only one *tefach* of material appear to be a much wider wall?

> You should place this wall [the one that is only one *tefach* wide] within three *tefachim* of one of the other walls [that is seven *tefachim* wide] (…) As long as the space between them is less than three *tefachim* (…) it is considered as if it were filled up. As a result we now have a third wall with a length of four *tefachim*, this being the majority of the minimum length of a wall of a kosher *succah* (…). But in order to fulfill the requirement for seven *tefachim* an "entrance frame" should be made. This involves placing an erect board opposite the small wall, at the far end of the area of the *succah*, and then placing a crossboard extending from the top of the small wall to the top of the erect board.[9]

The book *A Guide to the Laws of Succos* contains further explanations about how to construct the walls, totaling another sixty-eight pages, elaborating on how much or little material a sukkah wall must have in order to make a sukkah valid, and how far apart walls can be and still create an enclosure. These are examples of an ancient rabbi's thoughts about how to create an enclosure with the minimum amount of materiality. This ancient and at the same time contemporary architecture is interpreted in this article in the context of discourses about the role spaces have in the production, maintenance, and transformation of identity formations, social interactions, and concepts of belonging, which are of great concern for many individuals and communities, whether they are in a diasporic condition or not.

The sukkah, like many other Jewish religious objects, is often only perceived as a museum exhibition piece or an image in a catalogue, where it is deprived of its original site and context and loses many of its architectural characteristics. It is relegated into a symbol for dead history and not lived practice.[10] However, it is in fact lived architecture, for longstanding demographic reasons—which are ultimately cultural reasons—most likely built in an urban context. One aim of this article is to increase the visibility of sukkahs, not least in order to enhance our perception and use of urban spaces. Analyzing the sukkah as a symbiotic place sets an example for looking at the multitude of meanings a single space may evoke—no matter how small or slight.

New Concepts of Diaspora in the Sukkah

The sukkah is situated in the tension between diaspora and belonging, oscillating between opening up to the local culture and environment and creating an enclosed and protective space for an individual, family or community. The Jewish Diaspora is often seen as the prototypical diaspora, although today diasporas are seen as a global phenomenon. In turn, thoughts and concepts developed in the context of different diaspora experiences have influenced reflections on the Jewish Diaspora. The notion of diaspora implies that there is a single, physical place of belonging at which the diasporic person cannot be. An antagonism is set up where the diaspora location—the local—is never the place of belonging. However, diaspora no longer has only the negative connotation of not being at this single place of belonging; diaspora is also described as being multivalent, hybrid, and full of opportunities.[11] A diasporic condition influences a person's identity and Stuart Hall remarks that "diaspora identities are those which are constantly producing and reproducing themselves anew, through transformation and difference."[12] If we assume that the local as a diaspora is a site of possibilities and can be a place of belonging, completely different concepts of home can arise. Bell Hooks argues that new interpretations of home are influenced by experiences elsewhere, which leads to a composite notion of home.

> Home is no longer just one place. It is locations. Home is that place which enables and promotes varied and everchanging perspectives, a place where one discovers new ways of seeing reality, frontiers of difference. One confronts and accepts dispersal and fragmentation as part of constructions of a new world order that reveals more fully where we are, who we can become: an order that does not demand forgetting.[13]

In turn, these new concepts challenge the notion of the counter space—the place of belonging—as being a singular, physical place.

A Painted Sukkah that Contains the Essence

A painted sukkah from southern Germany or Austria from around 1800, exhibited in the Jewish Museum in Paris in 2003,[14] initially led me to this project. This sukkah still embodies the essence of my project here, which explores how the sukkah is used as a tool for constructing places of belonging. Contemporary discourses about the function of architecture as a place of belonging independent of any specific geography, frame the analysis. I want to argue here that the sukkah, which is a prototypical yet symbolic house and temporary in character, is part of a survival strategy of the Diaspora. A sukkah constructs and expresses a hybrid concept of places of belonging, which overcomes the dichotomy of having or not having a place of belonging, by superimposing several real and imagined places. Thus, constructing and expressing both: a sense of belonging and a longing for another place.

The sukkah in the Jewish Museum in Paris exemplifies this. It is a small wooden structure measuring 2.7 x 2.7 meters (or 2.96 x 2.96 yards), with a door and two windows. From the outside it is brown, which gives it a rather simple exterior appearance. In contrast, the four interior walls are elaborately painted. The wall facing the door depicts an image of Jerusalem. On the adjoining wall to the right there is a painting of a stylized southern German or Austrian village set in a landscape that is linked visually to the one of Jerusalem. The third wall features a shield with the Ten Commandments. The fourth wall, where the door is located, combines German or Austrian floral still life paintings and columns, with two of the columns being twisted. The panels are visually joined by a tasseled curtain, which originates at the painting of the shield with the Ten Commandments and runs along the lower border of the otherwise separate paintings, thus forming one entity.

When I saw this sukkah for the first time, the interior paintings astonished me for two reasons. First, the elaborateness of the paintings seems to negate the transitory nature of the sukkah. The holiday *Sukkot* commemorates the Exodus from Egypt and the wanderings of the Jewish people in the desert. In addition, the sukkah is an explicit and symbolic reconstruction of the huts in the desert, which leads to their portable character. During Temple times, the holiday *Sukkot* was also one of the *Shalosh R'galim*, the three pilgrimage festivals for thanksgiving, and the sukkah bore reference to the huts constructed during the harvest season.[15] The command to construct a new sukkah every year, and to dwell in it for only one week, contributes to its transitory character.[16] The construction method used in the sukkah in the Jewish Museum in Paris reflects its transitory and portable character. The walls are made up of thirty-seven numbered wooden boards, so that the sukkah can be taken apart every year and reassembled the following year. Second, the imagery itself was striking, or rather the choice of the depicted places and objects: Jerusalem, a local village, the Ten Commandments, floral still lifes, and twisted columns. Each stands for a place of belonging: Jerusalem for the biblical homeland; the local German or Austrian village for the current place of residency; the Ten Commandments for the Bible as the portable homeland; the still lifes for the German or Austrian culture;

and the twisted columns, which are a typical representation in Jewish art of the Temple in Jerusalem.[17] Isn't it strange to have this accumulation of places of belonging in a symbol for the Exodus, in a transitory and portable structure? Even stranger is the grouping of seemingly contradictory places of belonging; Jerusalem and the local, the art and architecture of the Temple and of Germany or Austria, the physical and the immaterial.

The Four Places of Belonging

These four places of belonging are present in every sukkah, even though they are rarely represented this straightforwardly. The local becomes part of a sukkah the moment it shifts from a textual object to a physical one. In the texts dealing with the sukkah, the site is generic and only the site requirements are described. A physical sukkah is, however, always constructed at a given time at a specific site—the local. The extent to which the local becomes integrated into a sukkah depends on the individual designs. In the case of a sukkah inside an existing room, where only the roof is removed or opened up, the local is a substantial part of its architecture. On a visual level, the local can be merely a part of the structure, for example, when the walls of a sukkah are made of a fairly transparent material, or one can look out the entryway or a window.

The local merges with the architectural elements of the sukkah in front of the inner eye when the eye oscillates between the perception of the interior space delimited by the walls and the exterior space seen through the walls or openings. Even when all the surfaces of a sukkah—walls, ceiling, and floor—are constructed particularly for the purpose of a sukkah, one would still experience the local immediately before entering a sukkah. This direct succession also merges the sukkah with the local. The local can also be expressed in the specific architectural styles of a region. For example, carpets are often used for the walls in traditional Bukharan sukkahs.[18] The local may also be integrated by means of decorating traditions such as the custom of decorating the sukkah with curtains from the Torah ark, as was common in Turkey. The use of local fruits instead of the seven fruits of the Land of Israel is another way of bringing the local into the space.

Traditionally, it has been customary to decorate a sukkah with the seven fruits of the biblical Land of Israel: wheat, barley, grapevines, figs, pomegranates, olives, and dates. The use of fruits for the decorations establishes a simultaneous connection to Jerusalem and the Land of Israel. As already mentioned, *Sukkot* was one of the harvest festivals, and during the Temple period it was celebrated as a thanksgiving pilgrimage festival to the Temple in Jerusalem. This symbolism is strengthened by the notion that the architecture of the archetypal sukkah booth resembles the romantic concept of the harvest booths in the fields, even though many sukkahs differ from that image. Another string of associations connects the sukkah to the tabernacle in the desert, which is then linked to the tabernacle in the Temple in Jerusalem, which leads ultimately to a third messianic temple in Jerusalem.[19] This string of thoughts is sometimes backed by posters used to decorate a sukkah, which depict drawings or 3D animations of the Tabernacle or the Temple in Jerusalem.[20] In addition, the sukkah is a symbol for the Exodus and exile, as described in the Leviticus passage quoted at the beginning of this paper. As such, it is the counter site *par excellence* to Jerusalem and the Land of Israel, reconstructing the dwellings during the forty years of wandering in the desert. Either, if one believes that the Israelites lived in actual booths during the wanderings, the architecture is modeled on these booths, or, if one believes that the word 'sukkot' refers to the "clouds of glory," the *sekhakh* invokes the desert dwellings by creating shade and alluding in this manner to the clouds.[21] Sukkahs must be constructed in a temporary manner, and their transitory and unstable architecture is understood as a symbol for exile. The mythical Land of Israel is in a dialectical and often intimate relationship with the exile. Each place requires the other.

The Bible is an integral part of any physical sukkah, since the texts, not drawings, provide the blueprints for the space. The activity of constructing a sukkah transcribes the words into space. At the same time that it represents the Bible, the sukkah also contradicts it. One space is physical, experiential, and visual while the other is textual, intellectual, and auditory. The sukkah is a transcription of the text into lived experience, as are most Jewish rituals. It is, however, special in the sense that it creates an immersive environment. On a spiritual level, it offers an opportunity to enter the transcribed text, the portable homeland, with the entire body. The *eruv* and the *mikveh*[22] are also immersive environments, but only the commandment to construct a sukkah can be traced back to the foundational text of the Torah. In addition, one can interpret the order of the holidays as a way to strengthen the connection between the sukkah and the text. The holiday *Simchat Torah* takes place immediately after the holiday *Sukkot*. At *Simchat Torah*, meaning "Torah festival," the last portion of the Torah scroll is read aloud, directly followed by the first portion. The precarious huts are thus redeemed by the text.

When considering the Bible as God's sanctuary, the sukkah as a transcription of the text becomes indirectly a dwelling place for God's presence. A seed for this was already planted by the editors of the Bible in texts like I Kings 8:17, where we find God's name substituted for his physical body, "to build a house for the name of God." With the name—the word—as God's presence, the redactors made a self-reflexive reference to the Bible itself. Another immaterial part of the architecture, the shade on the floor, is also part of the reference system to God's presence. A sukkah's roof is made of *sekhakh*, cut plant material, which must be constructed in such a way that the floor of the sukkah is not completely shaded, meaning that the shade needs to be recognizable as shade. This is achieved through the contrast between the sun- or moonlit areas on the floor and those shaded areas. The shade is supposed to remind the dweller of the *clouds of glory* described in the Exodus narrative, which in turn are a symbol for God's presence.[23] This can be seen in the context of one of the names of God mentioned in the biblical texts—*Ha-Makom*—which can be translated as place, locality, or residence. This constructs a God who is at the same time transcendent and immanent, the pivoting point being spatiality. The *Tabernacle* and the Temple, which are also referenced by a sukkah, are called God's dwelling place, setting up an association of space and architecture with God's presence. The architecture can be seen as a physical manifestation while God's body always remains absent or elusive, like a cloud.[24]

The construction specifications—from the Talmud to the internet—and the meanings attached to the sukkah give it a distinctly spatial orientation, but they evoke a space that has no specific geographical orientation in the here and now. The beauty of the design is that it can be constructed on any site that provides access to the sky. In theory, in the text, it is a generic design. In practice, physically, it is a very specific design adjusted to the circumstances of a given site. A sukkah has many symbolic meanings, encoded in construction methods, material traditions, and decorating customs. Along with the architecture, one confers these meanings onto a site. These material elements and the performative aspects of the *Sukkot* rituals allegorize at least the above-mentioned four real and imagined places simultaneously. By constructing a sukkah one projects these places onto the site. In turn, the architecture of the sukkah is itself a projection surface for these places, sometimes literally when the four places are painted on its walls, and at other times more metaphorically.

Being Rooted in the Symbol for the Exodus

Some of the places referenced by a sukkah are associated with being rooted, others with being uprooted. The material specifications for the roof, the *sekhakh*, express the same contradictory character. The *sekhakh*, must be made of plant material, something that has roots in the ground, but the plants must be cut from the ground. The two meanings of the holiday play out the same way. They commemorate the wanderings in the desert, a nomadic experience, and celebrate the fall harvest, an experience of a sedentary community. During the holiday of *Sukkot*, the place of residency is supposed to be considered impermanent and the sukkah is supposed to be considered one's permanent dwelling. This is expressed in the *Shulhan Arukh* a sixteenth century compilation that is still a contemporary authority in Jewish law:

A person should eat, drink, sleep, walk, and live in a *succah* the entire seven days, by day and night just as he lives in his own home throughout the rest of the year. During the seven days of *Succos* a person should consider his house a temporary place and his *succah* as his permanent locale.[25]

The opposite might actually be the case. Even if one does not feel at home in a particular place, one's dwelling might feel rather permanent and stable compared to an insubstantial sukkah. On the other hand, the sukkah as an example of Jewish architecture connects the person dwelling in it to a long history of ancestral sukkahs. It might create a feeling of being at home within a religious culture. In some instances, a person or family might change their permanent place of residency but take their sukkah along. In such a case, the sukkah could present a space of greater continuity than the new apartment or house.

The majority of Jews living in cities, and more and more of those living in rural areas, reside in spaces they neither designed nor constructed. In most places, architects design the buildings and professional builders construct them. Today, it is rarely the case that the designer and the resident are the same person. One might renovate and change certain aspects of the space, but few start from scratch. Looking at it from this perspective, the sukkah extends back to a tradition of building one's own dwelling. During my photographic journey in Israel documenting sukkahs, I came across several sukkahs built by local children. In Ashdod, children who saw me photographing sukkahs came running to me and asked me to photograph their sukkah, too. They had built a sukkah on a vacant lot with little more than the plentifully available garbage from the site. They took great pride in their sukkah and when I arrived they were sweeping the ground. Then they all posed for me with their house. A similar scene took place in Beer Sheva. Some children that lived in the same apartment complex built a sukkah together in an adjacent area, which comprised a playground and an old bomb shelter. Again, the sukkah was made of scraps, which did not diminish the pride of the children in their work (see opening page of this article).

For some of them, this was their only sukkah; others had also built a sukkah with their families. In neither case were the neighborhoods very religious. Only some residents built sukkahs and probably very few literally dwelled in them for the entire week. The children did not build these dwellings for religious reasons but because it fulfilled other emotional, cultural, and sociological needs.

The Urban Context

As a building, a sukkah amalgamates profane and religious spaces into a single entity. The construction manuals focus on the religious symbolism and meaning when discussing the architecture—the size, materiality, construction method, site—of sukkahs. In the construction manuals they always remain site-less, pure religious structures, textual structures. However, one can explore the real architecture of sukkahs in order to find responses to contemporary concerns about their most common site—the urban environment. How do people deal with urban wastelands and leftover spaces? How do people bring nature back into high-density neighborhoods? How do people design transformable spaces? And how do people create spaces that tell a multitude of stories simultaneously?

A sukkah creates a temporal overlay with a different connotation. After the dismantling of the sukkah, one might still affiliate the site with the symbolism of its architecture and the experience that took place inside. This is of particular interest when it comes to urban wastelands, sites that are leftover and are underutilized. During my photographic journeys, I surprisingly often came across sukkahs located on urban wastelands. These sites were in the midst of residential or mixed-use areas in an otherwise built-up urban context: areas in parking lots where no cars parked, alleyways that led nowhere, between buildings, or yards that were neither part of a building's front yard nor of the sidewalk. These places were usually neglected and littered with waste. Such sites and sights made me aware of the ubiquitous presence of seemingly discarded urban areas and the possibility of reinscribing them into people's mental maps through the use of the sites for sukkahs. Only once the spaces were used did they become memorable sites. By occupying a site, the sukkahs revealed open spaces in the urban fabric. Sukkahs located on urban wasteland exposed the tension between people not having enough private outdoor space to construct a sukkah and the simultaneous presence of abundant space, which is not used beneficially during the remainder of the year. Although usually constructed on these less desirable sites out of a need, they present an active engagement with them. The urban wastelands are transformed into homes, from which memories emanate. The activity of dwelling on these sites shows their potential to function as a catalyst for urban transformation.[26]

In urban wastelands, nature often creeps back into the city by sprouting weeds and trees. While this kind of nature is rarely appreciated, there is often a longing for more nature in cities. When looking and describing urban spaces, a dichotomy is often constructed in which nature and buildings are seen as oppositional, and buildings replace nature. The architecture and symbolism of the sukkah integrate both sides of the dichotomy nature/building and thus neutralize its oppositional and hierarchical character. The sukkah is a building but by means of the roof nature is present. The inclusion of nature is not an afterthought, as are leftover spaces at a building site being transformed into a garden, but an integral part of it. The roof is theologically the most important element and visually the most distinctive feature of the structure. The *sekhakh*, made of cut plant material, together with the tradition of decorating the roof with fruits transforms the ceiling into a fertile and abundant garden. Thus the sukkah brings the fields and fertility symbols back into the city. The history of the holiday *Sukkot* as a thanksgiving holiday creates another link to agriculture and fertility. The sukkah as a building type is closely related to harvest huts. In the region of ancient Israel it was common to build huts in the fields so that the farmers did not have to return home daily during the fall harvest.[27] Detached from its original site and function, this building type is inserted into the opposite context — the urban — where it retains its old meaning and acquires new ones.

Engaging the Site

Being able to see the stars through the roof is a very important feature of a sukkah, hence one looks for a space that is open to the elements, or has the potential to be opened up. Sukkahs are often located on all kinds of outdoor spaces: balconies, roofs, sidewalks, in gardens, or on small leftover pieces of land. Finding a place at all can be a challenge within a dense urban context. During a fieldtrip to Paris, I saw, for example, a sukkah built in a niche on a busy sidewalk in the center of the city.

A sukkah built on a busy sidewalk in the Marais district, Paris, 2003

*An apartment building in Williamsburg, New York, fitted retroactively
with balconies for sukkahs, 2004*

On another occasion in Jerusalem, I encountered a sukkah that blocked a
public pathway almost completely. A person could squeeze by, but passing
with a stroller or bicycle was impossible. The owners posted a note on the
corner of their sukkah, saying that they had no other place where they could
construct a sukkah. They apologized for any inconvenience this might cause
and described an alternative route for those unable to pass. Since dwelling
in a sukkah is often interpreted as eating in it, the ideal site of a sukkah is
also in close proximity to one's regular dwelling place, preferably near the
kitchen. The need to have a space, which is open to the sky, and the wish
to have it close to one's kitchen has led some people to integrate sukkahs
into existing rooms. In England, for example, semidetached houses or row
houses are common in the residential neighborhoods of the suburbs. The
houses often have winter gardens or single story extensions. Those kinds of
spaces can easily be fitted with a roof that opens up, so that one can construct
a sukkah inside the house. From the desire for convenience, to issues of
security and privacy, and the lack of any other suitable space, the reasons
for integrating a sukkah into an existing space vary widely. Even sukkahs
constructed on balconies or sidewalks permanently change the meaning of
these spaces. A balcony is no longer only a site for breakfasts in the sun
but also the site where wanderings in the desert take place. As several real

and mythical places are referenced in each sukkah, their site of construction becomes layered with these places, too.

The architecture of the sukkah is of course influenced by the architecture of the site, for example the size of the site and the building elements surrounding it. However, the sukkah can also influence the architecture of residential buildings and Jewish community centers and the like. In Israel, the designs for apartment buildings that have been constructed for religious tenants take the need for an outdoor space into consideration and integrate the balconies into the overall designs. This limits the density, as a high-rise building could not offer an outdoor space with an unrestricted access to the sky for every tenant. In the Williamsburg neighborhood in New York, I came across generic multistory brick buildings where huge balconies had been added to the facade. Instead of being stacked one on top of the other, as is common with balconies, they were offset.

Despite the above-mentioned examples of how sukkahs can influence the architecture of permanent buildings, most people who are unaware of the phenomenon of such structures easily overlook them. Internally, however, i.e., within the community that brings them forth, sukkahs have an assertive and reassuring character, creating their own dwellings and offering a hybrid place of belonging. A sukkah's architecture has a fluid and open character that puts it in a constant exchange with the environment. Sukkahs exemplify how buildings can engage with a site without laying an exclusive and exclusionary claim to a place. They allow and even encourage simultaneous associations on two levels. First, the presence of a sukkah still permits the association of its site with other agendas. Second, a sukkah references and creates a relationship to several places at the same time. This does not weaken a sukkah's tie to a specific local site, but places it in the context of other places. In this way the sukkah has an oscillating character. It makes the beholder's heart and mind oscillate between the different real and imagined places, creating a multifaceted place of belonging to the inner eye. Where the two axes of diaspora and of place of belonging intersect, the kind of hybrid space that emerges depends on each individual sukkah and on the person perceiving it.

Notes

1   Note on spelling: The Hebrew plural of the word 'sukkah' is 'sukkot'. However, in order to differentiate the plural of the sukkah booths from the name of the holiday 'sukkot', the English plural—'sukkahs'—is used. Furthermore, the capitalized word God is considered a name and refers specifically to the Jewish god. Similarly, the capitalized word Diaspora refers to the Jewish Diaspora.

2   Texts from the Torah discuss how to construct a sukkah. See, for example, *Jeruschalmi: der Palästinesische Talmud; Sukkah die Festhütte*, trans. Charles Horowitz (Tübingen: J.C. Mohr, 1983). Jacob Neusner, *The Talmud of Babylonia: An Academic Commentary*, VI Sukkah (Atlanta:

Scholars Press, 1994); Jacob Neusner, *The Talmud of Babylonia: An American Translation; VI. Tractate Sukkah* (Chico California: Scholars Press Brown University, 1984). Jacob Neusner, *The Talmud of the Land of Israel: A Preliminary Translation and Explanation: Sukkah*, Vol. 17 (Chicago: The University of Chicago Press, 1988); Hans Bornhäuser and Günter Mayer, trans. and eds., *Die Tosefta: Übersetzung und Erklärung; Seder Moed*, Vol. 2 (Stuttgart: W. Kohlhammer, 1993); Shulchan Aruch, Ch. 44 ("The Sukkah") http://www.torah.org/advanced/shulchan-aruch/classes/orachchayim/chapter44.html; "Sukkot," Judaism 101, http://www.jewfaq.org/holiday5.htm (both accessed, August 9, 2007).

3   For a more thorough discussion of the topic of this article see Miriam Lipis, "Symbolic Houses and Hybrid Places of Belonging in Judaism" (PhD diss., Humboldt University of Berlin, 2007).

4   For the conflation, conflicts, and changes between the mythical and the geographical Land of Israel see: Jean-Christophe Attias and Esther Benbassa, *Israel, the Impossible Land* (Stanford, California: Stanford University Press, 2003).

5   Jane Rendell, "Section Introductions," in *Gender, Space, Architecture: An Interdisciplinary Introduction*, eds. Jane Rendell, Barbara Penner, and Iain Borden (London: Routledge, 2000), 232.

6   I would like to thank Jakim Notea for the photographs of sukkahs in Williamsburg, New York.

7   For a description of the symbolism of the sukkah see: Josiah Derby, "The Wilderness Experience," *The Jewish Bible Quarterly* 103 (1998); Daniel Epstein, "Suspended in Mid-air: the Sukkah, Unlike One's Permanent Residence, is Indeed a Fleeting Thing," *Ha'aretz*, September 29, 2004; Abraham Isaac Kook, "The Sukkah: Fortress of Defence," in *The Sukkot/ Simhat Torah Anthology*, ed. Philip Goodmann (Philadelphia: The Jewish Publication Society, 1973); Herbert Levine, "The Symbolic Sukkah in Psalms," *Prooftext: A Journal of Jewish Literary History* 7, no. 3 (1987): 259–67; Jeffrey Rubenstein, "The Symbolism of the Sukka," *Judaism* 43, no. 4 (1994): 371–87; Jeffrey Rubenstein, "The Symbolism of the Sukka (Part 2)," *Judaism* 45 (1996): 387–98.

8   Moshe Morgan, *A Guide to the Laws of Succos* (New York: C.I.S. Publishers, 1994).

9   Ibid., 187.

10 An interesting case was the exhibition *The Chicago Booth Festival: Architects Build Shelters for Sukkot* at the Spertus Museum in Chicago in 1995. Architects were asked to design sukkahs specifically for the exhibition. Those sukkahs were literally at home in the museum context.

11 In addition to Stuart Hall and Bell Hooks see for example: Daniel Boyarin and Jonathan Boyarin, *Powers of Diaspora: Two Essays on the Relevance of Jewish Culture* (Minneapolis: University of Minnesota Press, 2002). Nicholas Mirzoeff, ed., *Diaspora and Visual Culture: Representing Africans and Jews* (New York: Routledge Press, 2000).

12 Stuart Hall, "Cultural Identity and Diaspora," in *Diaspora and Visual Culture: Representing Africans and Jews*, ed. Nicholas Mirzoeff (London: Routledge, 2000), 31. See also: Stuart Hall, *Representation: Cultural Representations and Signifying Practices* (London: Sage Publications, 1997).

13 Bell Hooks, *Yearning: Race, Gender, and Cultural Politics* (Boston: South End Press, 1990), 148.

14 See: Naomi Feuchtwanger-Sarig, "Fischach and Jerusalem: The Story of a Painted Sukkah" in *Jewish Art* 19/20 (1993). *Christie's Amsterdam Catalogue* "Important Judaica" (1988).

15 Similarities can still be found today in the harvest huts of the Bedouins in the Sinai. See: Irit Zaharoni, ed., *Derekh Eretz: Man and Nature* (Tel Aviv: Misrad ha-bitahon, "Ba-mahaneh," 1983).

16 Sukkahs are portable in character but not all sukkahs are actually portable, though all must be reconstructed yearly. This ranges from completely reconstructing a sukkah every year, to just opening the ceiling of a room and covering it with *sekhakh*.

17 Bianca Kühnel, "The Use and Abuse of Jerusalem," in *Jewish Art: The Real and Ideal Jerusalem in Jewish, Christian, and Islamic Art* 23/24 (1997): XIX–XXXVIII; Bracha Yaniv, "The Origins of the 'Two-column Motif' in European Parokhot," in *Jewish Art* 15 (1989): 26–43.

18 The exhibition *Sukkahs from around the World*, which was held at the Israel Museum in Jerusalem in 2003, gave an overview of different construction and decorating styles.

19 Rachel Wischnitzer, *From Dura to Rembrandt: Studies in the History of Art* (Milwaukee: Aldrich; Wien: IRSA; Jerusalem: Center for Jewish Art, 1990), 55.

20 I found the same kind of posters in religious neighborhoods in Paris, London, Antwerp, and in Israel.

21 In the Talmud, there is a discussion about whether the word *sukkot* refers to huts or to the clouds of glory, as Josiah Derby writes: "The Talmud (Sukkah 11b) records a difference in opinion between R. Eliezer and R. Akiba. The former believed that the text should not be taken literally, and that the word 'sukkot' means the Clouds of Glory that shield the people. R. Akiba rejected this metaphysical interpretation, and insists that the Israelites actually built sukkot for themselves. Rashi accepts R. Eliezer's view, and Ibn Ezra, the realist, goes along with R. Akiba. The Ramban offers his own solution in the form of a compromise: They were, indeed, sheltered by the Clouds of Glory, but the clouds were in the shape of sukkot." Derby, "The Wilderness Experience," 196.

22 The *eruv* is a Shabbat demarcation that facilitates the carrying between different public and private spheres, which would otherwise be prohibited on Shabbat. A *mikveh* is a ritual bath that is used for ritual purification of objects and persons. For a detailed discussion of the *eruv*, see Manuel Herz's article in this volume.

23 See for example: Exod. 13:20–22: "And they took their journey from Succoth, and encamped in Etham, in the edge of the wilderness. And God went before them by day in a pillar of cloud, to lead them the way; and by night in a pillar of fire, to give them light; that they might go by day and by night: the pillar of cloud by day, and the pillar of fire by night, departed not from before the people."

24 While God's body is only vaguely described in the biblical texts, there are several books from late antiquity in which the body of God is described in great detail. Naomi Janowitz writes about one of the texts, the "Shi'ur Komah": "This very self-conscious equation of knowledge of 'the measurement' with securing a place in the world to come functions as an interpretive frame for the entire text. (…) In *Shi'ur Komah*, the names of the limbs embody the performative force of each particular bit of the deity. God is conceived of as a collection of powerful forces (limbs), each force being captured in the name of the limb. God's power is concretized and delineated and defined in each act of naming. Every single name embodies its particular limb, this time at the level of language. Thus the reciting of the list of names is a series for mini transformational moments." Naomi Janowitz, "God's Body: Theological and Ritual Roles of Shi'ur Komah," in *People of the Body: Jews and Judaism from an Embodied Perspective*, ed. Howard Eilberg-Schwartz (New York: State University of New York Press, 1992), 190–191.

25 Shulchan Aruch, section 1. Quoted in: Moshe Morgan, *A Guide to the Laws of Succos* (New York: C.I.S. Publishers, 1994), 441.

26 The Dutch project *Paradise Parasite* introduced very consciously temporary structures into a neighbourhood in Utrecht as a catalyst for urban transformations. Liesbeth Melis, *Parasite Paradise: A Manifesto for Temporary Architecture and Flexible Urbanism* (Rotterdam: NAi Publishers/SKOR, 2003).

27 Derby, "The Wilderness Experience."

2 "Eruv" Urbanism

Towards an Alternative "Jewish Architecture" in Germany

Manuel Herz

With the beginning of Diaspora, Judaism developed the *eruv*[1] as one of the oldest concepts of communal life and spatial organization. It is the model of the eruv that I will use to examine and evaluate "Jewish architecture," or the physical presence of Jewish communities, in Germany. The essay opens with an introduction to the concept of the eruv, placing it in the matrix of historical development and religious practice within Judaism, and traces the specific spatial understanding formulated in Jewish thought over time, especially after the beginning of Diaspora. The second part of the text uses the concept of the eruv as a means of analyzing and evaluating the specific condition of the Jewish presence in Germany, a country that could be described as embodying an archetype of contemporary Diaspora, using the recently completed Jewish community center in Munich as a case study. In the last part of the essay I consciously step outside of the descriptive and analytical realm that usually characterizes academic writing, and become more propositional. As a practicing architect, I would like to propose an alternative to the ways in which the visual and other physical manifestations of German Jewry occur in the urban fabric. Taken as a programmatic conclusion, it is a call for an urban strategy based on the eruv, which in its agenda and ideology is different from today's practice.

The "Eruv" as an Urban Model

Even though Judaism is usually—historically—perceived as being a very nonarchitectural religion, the eruv is part of a well-articulated lineage of spatial concepts that are thoroughly embedded in Jewish thinking. I want to argue that from ancient times, Jewish thought has always had a fundamental concern with space, and that a very specific understanding of space, and consequently of architecture, has developed.

It is precisely at a time of extreme "placelessness," that the question of what constitutes a place arises for the first time in Jewish history. During the exodus from Egypt, the period of the ultimate displacement, the search for a definition of a place arises for the first time: "Abide you every man in his place, let no man go out of his place on the seventh day." (Exod. 16:29) Because the term "place" is not further qualified, a set of rules has to be formulated that defines what does and does not constitute a place. The eruv is developed out of this command and is a method that can be used to qualify and define such a place. The *Masechet Eruvin* is the second tractate in the second order of the Talmud, the tractate just following the tractate *Masechet Shabbath*, with which it is very closely linked. Poignantly spoken, the *Masechet Eruvin*, encompassing hundreds of pages, is a commentary on this single word in the Bible: space.

Two main models of space are implicit in the Talmud and earlier in the Bible: the City and the Desert. They represent the two poles of existence that define the geographical matrix of Jewish history: kingdom and placelessness. Mediating between the two, the eruv creates a system that gives them a shared urban form and condition. The desert, the ultimate state of placelessness, becomes a unifying condition that leads to the pursuit of a place and stability. An aggregate of tribes, fleeing from slavery in Egypt and on their way to the "promised land," is transformed into a nation by a set of laws and commandments, the Torah. Placelessness was seen as a temporary and preparatory state in the process of transforming the nomadic tribes into a nation of settlers. The laws of the Torah were formed in the desert but aspired to a different condition: settlement and stability. Within the Jewish legal system, laws were always tied to places; in the desert they had to relate to places that were as yet unattainable.

The nomad's space—"his place"—is not tied to a specific location. Thus, the notion of place had to be divorced from a fixed geographical definition, and became a portable entity. The eruv is a means of creating such an abstract notion of space, space that can be deployed wherever it is needed—portable, dynamic, and private. After the settlement, the Temple that was built in Jerusalem became the focal point of the Jewish nation, around which its entire religious life revolved. The Temple epitomizes the significance of architecture as a means of unifying a nation. It became the Jewish symbol of settlement and urban life. The eruv became a means of reestablishing the Temple, in a conceptual form, in cities around the world. Jewish law created for its people an invisible and abstract set of spaces inhabited by nothing but memory. Indeed, the Talmud itself is a construction, a record of arguments between rabbis and scribes from different places and different times, speaking to each other across deserts and across generations. Remolded and reinhabited by ever-new generations, it forms a radical assembly engaged in an imaginary debate.

The *Masechet Eruvin* defines and distinguishes between different domains, and states the laws pertaining to each. The domains are defined in terms of signifiers relating to the two conditions of "city" and "desert." These definitions disregard the use and functions of the spaces as well as their legal ownerships. They rely completely on their representational and morphological aspects—the shapes, sizes, and elements that constitute their boundaries. A private domain is defined as "an enclosed area which is enclosed by partitions no less than ten-cubits (approximately 4.5m) high or bounded by a trench ten cubits deep and four wide."[2] In the Temple, a door measuring ten cubits separated the mundane from the holy, the public area from the private. The *Devir* was the utmost private space, where only the high priest was allowed to enter, once a year. A public domain is defined as, "an area like a public thoroughfare that is frequented daily by 600,000 people." In this case the original state of the command given by God becomes the defining condition of a public domain. The 600,000 people (which in fact is only the number of men within the group) in flight through the desert—boundless space as the direct opposite of defined place—is the model by which a public domain is defined. The city—referring to the displacement of the desert—is transformed by the eruv on the Sabbath into a representation of the Temple and thus from the public into the private domain. If the eruv area is understood as the Temple of Jerusalem, the outer area is the desert, and movement into the eruv is an act of wandering that culminates in the appropriation of a place.

The eruv uses a chain of signifiers to turn the city into a private space. As the ultimate private space is the Holy of Holies, it becomes necessary to "build" the Temple over the city. Because of the "technical" difficulties of doing so, the Temple was reduced to its roof, as a sign representing the Temple. The method used to signify a roof over the city is to make a wall around it. Thus the eruv proceeds from the absurd act of making a roof over the city by building a wall. Every walled space has openings in it. In the representation of the roof, doorways are therefore equivalent to walls: a series of doorways represents a continuous solid wall. The city is circumscribed and delimited by a "wall" made in "the shape of the door," with its measurements taken from those of the Gate of the *Devir*—ten cubits high and at least three wide. In this way each door signifies the Gate, and entering the eruv becomes a holy act. The shapes of the doors are made according to the techniques used to build the Temple, namely, two posts and a cross beam. The posts and the beams can be made of any material of any thickness, as long as they are capable of withstanding an ordinary wind: a light cord stretched over thin poles is adequate. What becomes evident is that the construction of the boundary approximates an infinite chain of symbols that function independently of material support: from a private place to the Temple, from the Temple to a roof, from a roof to a wall, from a wall to gates, from gates to the shape of a door, from a door to a post and beam, from a post and beam to a cord. Only by possessing the key for deciphering this chain of symbols, can we read the cords that are stretched from the Temple over the street junctions in contemporary urban environments. Yet this chain of references is not purely linear; the symbols connect to other entities within and outside the chain. Each also refers to the ideal city, and therefore a complex structure of references and a multiplicity of meanings are established.[3]

The definitions of spaces in the Talmud are bound to their morphology. The shapes and sizes of elements of the city determine the status of the spaces they enclose, without reference to the functions they perform. The recreation of an abstract entity in the city depends on the availability of props. If space is subject to law, the definition of a space can be changed temporarily in order to change the laws that govern it. Such a change is implemented through the introduction of signifying elements and the designation of existing physical elements as signifiers. This creates an apparently absurd situation: the city as a private space, with urban features representing walls and doors, and a cord serving as the entrance to the Temple. Paradoxically, the very precision of these laws, and their rigorous application are the source of their flexibility. The eruv symbolically changes the nature of urban space. As the definition of space transforms and mutates, so too do the laws bound to it. The eruv therefore demonstrates the direct relation between law and space: it is the point in space and time where the law is transgressed by an urban intervention and the city is revalued by the law.

Jewish law forbids a whole range of work on the Sabbath, formal employment as well as travel, the spending of money, and the carrying of objects outside the home. Movement in public spaces on the Sabbath is severely restricted by laws stated in both the Bible and the Talmud. Movement and the carrying of objects are not, however, restricted within the private domain. Law is bound to space, and when the definition of a space changes, so too, does its program. If an urban space is designated as private, movement and carrying become permissible within that space. By redefining the space, the eruv redefines the behavior that is permissible within it.

The eruv shifts the current notion and meaning of the private and the public in the urban landscape. Public space is not the space of exchange and activity but a restrictive space of limitation. Private space, expanded to include the public domain, becomes the space of liberation and interaction. This idea is represented in a text fragment by E. Jabès in his book, *The Book of Questions*: "(…) You think it is the bird which is free. Wrong: it is the flower."[4] Only through knowledge of one's defined space can one act freely. Only knowing one's limits and the realm of one's movement, can one truly be free. The bird, for whom only the sky is the limit, cannot be free as it has no location, no definition, and therefore lacks the tools to act. In this metaphor, the bird is likened to the Jewish nation, in flight through the desert before they were given the command concerning their place. Place and the private domain, in the case of the eruv, does not, however, entail ownership. The space is symbolically private in terms of Jewish law, but in terms of civil law it remains public.

Applied to the contemporary urban fabric, the eruv introduces a different understanding of space and territory, a pluralism where objects and spaces can be subjected to more than one reading. The eruv extends its boundaries and geography to other structures, streets, and objects, threatening to infiltrate the space of private property with the public signs of an alien practice. In fact, one group's use of objects and space as signs does not preclude their use by any other group. Yet it would be wrong to suggest that the eruv constitutes a form of signifactory imperialism, for, paradoxically, it is only imperialism that insists that an object can mean only one thing, and that a boundary must be observed by everyone. In the polyglot, multicultural city, readings of space and place do not have to be linked to a territory and urban organization; the act of communal interpretation brings to the urban fabric an increase of meaning rather than a reduction.[5] At the heart of the problem is not the question of imposing upon urban space an obscure religious practice, but rather the willingness of city authorities to sanction the city as the site of multiple readings. The eruv is the ritual reconstitution of a lost place, framed as law and embodied as a book. Within the Diaspora, therefore, it was never a particular urban site, but a means of siting. It links, as a proposal, the daily life and public policy of contemporary urban space with concerns that are particular to Jewish life and identity. The eruv offers a sign system that revaluates and indeed sanctifies objects as diverse as a façade of an office building, street furniture, or other mundane elements of the area, making them intelligible in terms of the lost Temple of Jerusalem.

The eruv proposes interventions in the city that are small-scale, strategic, and for the most part nonmaterial. It intervenes by means of decisions about readings of the city, rather than reconstructing it so that it may be reread. Thus it provides a model for pluralist uses of the city that do not exclude other readings of the same object. It opposes the idea of the fundamental equivalence of one function and one object, or of one meaning and one object. As such, it represents a contribution to contemporary urbanism. Leaving behind its religious origins in Talmudic interpretation, the eruv has lessons to teach the Western city in terms of the economy of significations, boundaries, and the distinction between inside and outside on the one hand, and the scarcity of buildings and land on the other.

Emmanuel Levinas writes: "These old texts teach precisely a universalism that is purged of any particularism tied to the land."[6] The eruv creates a modern urban form and condition out of the opposition that was established in the Talmud between the Temple and the desert, and temporarily defines the territories relating to them. Such a reading is made possible by the way the Talmud defines the city and defines architecture: it assumes that the city does not exist in its physical embodiment alone, and that its material elements are always pointing towards something else. Thus the eruv bridges two cities—one that is perceived and tangible, the other aesthetically ideal. The urban dweller appropriates the city she or he lives in. They decipher but must also write each new interpretative framework. A second metaphorical or "mobile" city is overlaid upon the existing one by the practice of moving through the city. Thus, the main architectural implication and "ideology" of the Talmud is the understanding that "place," and the construction of one's place is independent of a specific (geographical) location.[7] Using the methods employed on the one hand in the construction of the Talmud and also in the construction of the eruv itself, it provides the means to create a place in the diasporic condition. It registers in the visible world the outcome of an encounter that would otherwise remain intangible.

This system thus described has been formulated within one of the most traditional and Orthodox communities, firmly embedded in a religious framework. Does it have anything to teach us, we who live in a vastly different environment and sociopolitical context, where the notions of the sacred and the profane have been supplanted by new and more dynamic value systems? Relying not on a single perpetrator, it is set up through a system of negotiation and mutual acceptance of value systems. The heroic deed of the creator is neglected in favor of small-scale interventions by numerous authors. In a world characterized by mobility and globalization the task and method of forming communal unifiers and negotiation systems has to be reformulated. The traditional methods of the great interventions in the form of monuments do not apply anymore. The monumental, that is, petrifying single statements and value systems, are doomed by the migrational dynamics taking place and by the multiplicity of appropriations that the notion of space, embodied by the eruv, calls for.[8]

The Jewish Community Center in Munich as a Paradigm of "Jewish Architecture" in Germany

Abstracted from its religious context and analyzed in terms of an urban strategy, the eruv stands for a creation of a specific urban realm that nevertheless remains accessible to all groups of society and open to all uses. The two predominant urban characteristics of the eruv express themselves in the multiple layers of meaning and surplus of significance that are added to and augmented by the urban realm, as well as in the deployment of many small-scale interventions, which by being located at strategic points can affect large areas. The liveliness of a city-part and the changes in the activities that are carried out within that urban realm do not depend on a singular, visible and centralized intervention or building, but upon many small-scale changes, declarations, and shifts of signification, which are only noticeable to the informed and do not impinge on the unacquainted. The eruv as an urban space can be described as containing a minimum of "Jewishness" in a maximum of space. The eruv, both as a concept and as a pragmatic tool within everyday Jewish life, should be seen as the typical architectonic and urbanistic response to the condition of Diaspora. In Germany, representing one of the diasporic conditions "par excellence,"[9] one would have expected the existence of eruvin to be a recurring feature of urban life. However, no eruv currently exists in Germany. On the contrary, in Germany, the physical presence of Jewry—what I will call "architecture in a Jewish context"[10]— surprisingly follows an ideology that is remarkably opposed to that of the eruv, a presence that can be described as "anti-eruv", and maybe anti-diasporic. My claim is that the physical presence of German Jewry follows exactly the opposite strategy, and that this "anti-eruv" model has a specific political relevance, for the architects, the Jewish communities, as well as for the political establishment and society at large.

When the city of Munich started developing a concept for a new Jewish community center in 1999, it invited a number of scholars from all over the world to help formulate the program and guidelines for this daunting task. Even though earmarked solely as a symposium for the Jewish Museum, which was to be part of an entire complex of Jewish institutions included in the community center, the conference from its outset in fact discussed the fundamental guidelines of the Jewish community center at large.[11] Irrespective of the multiple voices that were heard (among them my own), a general strategy had been laid out from the beginning: the Jewish community center was to occupy the *St. Jakobs Platz*, one of the central places in the city, and being one of the few remaining open squares in the city center, close to the major shopping areas. The community center was to contain a synagogue, a Jewish elementary school, a Jewish school with evening courses for the general public, a multi-purpose hall, and connected spaces for community activities, i.e., the actual Jewish community center, a Jewish museum, a Jewish restaurant, a Jewish café (whatever that might be), and a Jewish bookshop. No other Jewish restaurants, schools, or cafés existed in Munich. This general strategy, which basically amounted to concentrating all physical Jewish presence into one city square, was developed, with a mutual understanding, by the local head of the Jewish community, Charlotte Knobloch, Munich's mayor, Christian Ude, and the then *Ministerpräsident* (governor) of Bavaria, Edmund Stoiber. A central structure, containing within a minimum of physical space a maximum of "Jewishness," represents the antipode to the diasporic strategy of the eruv. It is a prime example of what I call a concept of "architecture in a Jewish context" or "Architektur mit jüdischem Bezug," which in this specificity exists only in Germany.

Upon its completion, and in response to the inauguration held on November 9, 2006, the German media as well as the general public were almost unanimous in hailing the new Jewish community center of Munich as one of the best examples of recent sacral architecture.

When the journalist Ulf Meyer, of the *Neue Züricher Zeitung*,[12] began his very favorable description of the Jewish community center by stating that in Germany no other building typology had brought forward such interesting and innovative buildings as the typology of synagogues, he addressed a question that—unconsciously or not—went beyond the rather gloomy judgment that this comment cast on the general level of architectural culture in Germany. It seems remarkable that just a handful of buildings—of the most specific use and function—represent the engines for innovation in the profession of architecture. Why are synagogues, even more so than churches and museums, not to mention the seemingly banal typology of housing, at the forefront of architecture in this country? The answers to this question require considerations that extend into the social and political realm, stepping outside of the classical relationship of the architect as author, as well as his labor for the client and the brief. An explanation that only references the decision-making process of the architect or sees the architectural outcome exclusively as a product of the relationship between the architect and the client is too individualistic and apolitical, and would therefore conceal major issues in the sociopolitical spectrum. Especially in reference to the theme of the physical, and therefore visible Jewish presence, which is of vital political and social interest to Germany, any simplistic reference to the architect's preferences must be supplemented with another layer of explanation. Indeed, one can only adequately understand the specific expression of that architecture in a Jewish context in Germany when it is perceived in its collective, and thereby political, frame of reference.

What exactly is the specific character of those buildings located within a Jewish context in Germany? First and foremost it seems legitimate to state that a consistently high level of "good" architecture has been realized. Irrespective of the location and size, whether looking at the major buildings in a major city like Munich or at a small synagogue in a provincial city, whether it is an architecture following a classical or minimalist style or of sculptural expression, the spatial and conceptual quality that is reached in those buildings is exceptionally high. In particular, their concepts are rich in metaphors, unfold a plenitude of historical and thematic relationships, and often have a finely tuned and very well-formulated argument with respect to the buildings' shapes and their "morphogenesis." In this sense, they are examples of "proper" architecture following the requirements that are deemed necessary for architecture of quality. It has often been mentioned that Daniel Libeskind's Jewish Museum and Peter Eisenman's Holocaust Memorial, both in Berlin, are among the most interesting and outstanding examples of contemporary architecture anywhere in the world. The new synagogue in Munich has been described as being, "a masterpiece of contemporary sacral architecture."[13] Salomon Korn, himself an architect and member of the Central Council of Jews in Germany, writes that the building is a, "symbol of a sublimation of mankind, liberating itself from matter and moving from darkness towards the light, from immaturity to enlightenment and from barbarity to human kindness. (…) a building like a sparkling well of hope and confidence."[14]

Thus, the Jewish community center in Munich fulfills all requirements of a proper Jewish community center in Germany. The main body of the synagogue, a bit at odds with the slightly bucolic and rustic adjacent buildings, is positioned centrally on the previously open city square and references the Temple with its heavy and bold plinth, clad with Jerusalem limestone. This stone, however, citing the Holy City, has become the ubiquitous symbol of Judaism within architecture, and has by now become a nice and easy gesture, evoking with architectural tools a notion, origin, provenance, and feeling of warmth and decency, which taken together evade criticism and represent an architectural safety net of sorts. A glazed volume with a hexagonal (i.e., Star-of-David-patterned) structure rises over and above the plinth, housing the main praying hall itself, and is meant to symbolize the tent of the Tabernacle, being the precursor to the Temple. In the interior, the light falls through a metallic mesh, again referencing the tent and paramount to the formation of Judaism, and falls upon furnishings made of Lebanese cedar, one of the most recurring symbols used in the Bible and the Talmud and also one of the main building materials of Solomon's Temple. Thus, the full cornucopia of standard Jewish symbolism—the Star of David, the Temple, the biblical cedars and the cloth as well as the tent of the Tabernacle—is to be found architecturally and spatially expressed. The architects "Wandel Hoefer Lorch" even managed, through architecture, to show the genealogical relationship of the tent to the Temple, with the former physically contained by the latter. It is this kind of architecture, condensing every prevailing Jewish symbol into one building, almost oozing Judaism out of every pore, literally recreating elements of the Temple (so that people have actually started putting "kotel notes"[15] into the gaps between the stones of the plinth, just like the tradition at the Wailing Wall in Jerusalem), becoming unassailable in its "proper" use of Jewish tradition and multiple references, and concentrating all Jewish functions onto one city square, which represents the paradigm of "German Jewish Architecture." The conference that was held prior to the competition provided a platform for voicing other strategies for finding a physical manifestation for the Jewish community and its cultural, religious, and social activities in the city. Architects like Eyal Weizman suggested alternatives for the use of one of the last remaining open squares in Munich, through the infiltration of existing buildings or the dispersal of certain functions in the surrounding area, also questioning the use and occupation of this public square by a single group within society.[16] None of those suggestions, or others along similar lines, were picked up, however. The dominant use of this symbolism, the monofunctionality established on the previously open square, and the visible amassing of Jewish presence onto a single spot, so prominent in the case of Munich, but also prevalent in Germany as a whole, represent the exact opposite of a diasporic strategy of the eruv.[17]

Furthermore, Jewish architecture in Germany exhibits a certain tendency towards the breaking of rules. In various ways the buildings refute the generally accepted forms of planning and construction and acquire an anarchic quality. Not only is this "rebelliousness" present in the visible and formal, but it includes the institutional and bureaucratic aspects of construction as well. While the process of decision making and the planning stage went beyond any previously known and accepted scale in the case of the Holocaust Memorial, the design of the Jewish Museum in Berlin is literally pervaded by rule breaks and critiques of established conventions. Through the radicalism of its design and construction, architecture is meant to have a disturbing and disquieting power. By means of architecture, a critique of established values of German society is publicly expressed. In the case of Munich, even if formally rather austere and minimalist, the fact that a public square is occupied by the Jewish community center, represents a break of conventions of the highest dimension. The urban arrangement of those buildings does not follow the ever-present block structure of the city, which is otherwise held in high esteem. The buildings of the community center in Munich become a public reminder of the Jewish presence, and the Jewish history in Germany. Their visibility acts as a critical momentum within the urban fabric.

The Canadian architect and critic George Baird writes in a recent article on the lineage and function of criticality in architecture: "One of its most cogent and internally coherent renditions has been that of the practitioner—and no mean theorist himself—Peter Eisenman, accompanied by [Michael] Hays. Together, over the past two decades, these two have developed a position that has consistently focused intellectually on concepts of *resistance* and *negation*."[18] Baird describes criticality as the built form expressing a refusal of contemporary political establishment and the consumer society, and he traces back the conceptual genealogy to that of Jacques Derrida in the field of philosophy and to the Italian (Jewish) architectural historian Manfredo Tafuri in the field of architectural theory. It is this criticality, the attempt to establish architecture as a voice critical of society and the political system, which seems to be identifiable in 'Jewish' architecture in Germany today.

For sure, this architectural disrespectfulness is fought for very hard in the case of each and every one of those buildings. Before many of these buildings were able to be built in the most uncompromising way possible, it sometimes took years to overcome all of the institutional objections and resolve all the technical and legal difficulties that arose. The individual architect, though, is not the only author of this architecture. Since those buildings perform a vital political function in Germany, one has to observe and analyze them on a political level as well. As this anarchic quality can be observed in a multitude of buildings from various architects, it indicates that the interest in this disrespect of rules also arises from a different side.

No buildings in Germany other than synagogues, and other buildings with Jewish context, are inaugurated with greater expenditure and larger ceremonial events, and the amount of public response and media coverage. The significance given to this event, which seems almost of fundamental importance to the state, is astounding in view of its provinciality and practical irrelevance, when looking at the number of people who will use it actively. The Jewish community center in Munich was inaugurated in the presence of Horst Köhler, the German head of state, the governor of Bavaria, the federal minister of interior, the mayor of Munich, and several Bavarian state ministers and dignitaries. Virtually every local and regional newspaper published special issues or extra sections covering the building and its opening celebrations. Being located in the city center, the exhaustive security measures for the high-ranking guests brought public life to a standstill. On television, the evening news featured the inauguration as its first item. No one in Germany, whether locally present, or anywhere else in the federal republic, could escape taking notice of this opening of the Jewish community center in Munich.

Irrespective of the question of whether the synagogue is being actively used or whether it actually forms a functioning center of activity for the local community, its very existence serves as proof of a secured Jewish presence in Germany, which it symbolically and visibly expresses by merely being there. Therefore, an expressive building with radical architecture is desirable, as it cannot be ignored and will enter public consciousness. The Federal Republic of Germany needs the institutionalized experiment in the form of "wild" buildings and tolerated breaking of rules to liberate itself from the "unbearable burden" of the Holocaust. This overcoming of the Holocaust forms one of the basic narratives in the constitution of the German state, in which architecture in a Jewish context plays a substantial role today. Slavoj Zizek describes the fool, with a reference to Lacan, as the leftist intellectual, who, "in his hysterical satisfaction believes that through trickery he can steal from the 'master' a small piece of his (the master's) 'jouissance' (joy/ vividness)."[19] But, according to Zizek, this theft never questions the fundamental position and power of the ruling master. On the contrary, it only strengthens and stabilizes him within the existing order. This is exactly the role performed by Jewish architecture in Germany today. The architects attempt to create an architecture that is both "good" and rebellious, to position it as a critical voice within the general public, and to critically assess and question major aspects of German society. However, this superficially critical stance is tolerated and propagated by the institutions. Its density of symbolism and quality of architectural narratives is gladly taken up by the most conservative parts of the political spectrum. This architecture, in fact, reinforces the established society as it allows the official Germany to distinguish itself and present itself positively within the international arena. Counting on the naïveté and self-indulgence of the architects, those buildings achieve a good effect as a decorative embellishment. Here is the architect as the court's fool.

Eruv Urbanism

Ulf Meyer concludes his description of the new Jewish community center by stating that "Jewish life can now return to the heart of Munich."[20] Indeed, by placing the synagogue, the Jewish kindergarten, the Jewish school, the Jewish museum, a Jewish café, a Jewish restaurant, and the Jewish community functions in that central square of the historical city center of Munich, Jewish life will return to the heart of the city. But it will only take place there, and nowhere else. The concentration of all functions associated with Jewish life into one single square in the center of the city is the antithesis of the concept of the eruv. It represents the complete opposite of a spatial strategy that was developed in response to diasporic life. "Concentrated" community centers, of which Munich provides just one very telling example, but which exist in almost all German cities, have become the showcase of Jewish presence as their own illustrations: a pure representation of Jewish life to the non-Jewish public.

These examples of constructed and condensed Jewish presence in Germany, with their supposedly critical view, are received gratefully by the conservative voices within the Federal Republic of Germany. Contrary to criticizing the accepted points of view, this architecture proclaims solidarity with the conventional and conservative opinions, a closing of ranks. Just as the architect Rem Koolhaas once said, "[t]he problem with the prevailing discourse of architectural criticism, is [the] inability to recognize [that] there is in the deepest motivations of architecture something that cannot be critical,"[21] we might have to acknowledge that architectural (physical) form is either limited in its power to refuse a contemporary political establishment, or that it can be easily drawn into the very matrix of power relationships that it tries to negate or question through its own means. Instead of surrendering to this embrace, which has been termed as a shift from a "critical" to a "projective" approach, and barely but nonetheless obediently subscribing to a, "post-Utopian pragmatism so pervasive in leading Dutch architectural circles nowadays," as Baird describes it,[22] I would like to suggest a more urbanistic approach that Jewish communities can follow in order to establish a physical presence in German cities. In contrast to the spatial technique currently pursued by political actors as well as by most Jewish communities within Germany, the author would call for a twofold strategy based on the eruv as an urban model:

The construction of actual eruvin for the Orthodox Jewish communities represents the first part of how Jewish communities reformulate their physical presence in German cities. These eruvin allow Orthodox Jews a different kind of community life on the Sabbath, which can then also include the public realm. On a very practical and pragmatic level, it allows for larger freedom of movement of Orthodox Jews within the city, a visible presence of the community in streets within the area of each eruv, preferably covering the complete city. Apart from introducing a notion of "ordinariness" into the presence of Orthodox Jews, especially on the Sabbath, in German cities it also makes it easier for this group to live out their lives according to their own understandings.

Besides these direct and immediate consequences that eruvin in German cities would have, the construction and planning phase will inevitably be accompanied by a public debate. The proposal for an eruv in London during the 1990s triggered one of the longest, most public, and also one of the fiercest debates within all channels of media, and drew in a large part of society. Even though a certain part of the population was vehemently opposed to the construction of the eruv, the discussion as a whole can be seen to have played a very important role in the shaping of the (self-)understanding of the various Jewish communities and the non-Jewish population throughout and beyond London. When the proposal for the eruv in north London was published it was as if the floodgates of heaven had burst. Anti-eruv groups were formed, pro-eruv groups followed. Lawyers were hired, petitions were gathered, and letters were written in a long campaign that culminated in a demonstration in support of the eruv, held in front of Barnet Town Hall. The reality that the eruv had to deal with proved more complex than merely tuning a Talmudic concept to suit the material condition of the city. In the case of the north London eruv, different understandings of space and territory, ownership and meaning, made it difficult for the public to enter into a pluralism where objects and spaces could be subjected to more than one reading, as in the case where it was proposed that the walls of a church form part of the eruv's perimeter.

It is at this point that the creation of eruvin in German cities ceases to be only a service for a particular group of Orthodox Jews, which, we have to acknowledge, represents only a small share of the Jews living in Germany. A public debate on the construction of eruvin links seemingly quotidian aspects (e.g., planning applications for eruv poles, the declaration of garden fences or graffiti-sprayed walls to be part of the eruv boundary), with the role that Jews play in German society, i.e., questions that are at the very heart of the German state and its self-understanding. A debate on eruvin would allow the politics of everyday life to incorporate a subject, which so far has been relegated to the exclusive terrain of discussions in the house of parliament, or the feuilleton pages in broadsheet and quality newspapers. Furthermore it would make the public aware of the fact that Jews have arrived again in German society, also on a quotidian and everyday level, and are actively shaping it.

Beyond these issues, the debates on the eruv touch upon central questions for a contemporary society, first and foremost the relationship between religion, its symbolic operations, its modes of inclusion and exclusion, and the public realm. At a time when these questions are marked by a rather fierce debate of often simplistic notions of religious representation, the techniques of the eruv, with its symbolism relying on the all but invisible, and its method of declaration, which is always pluralistic and additive, could expand the range of views that these controversies expound. The construction of eruvin, and the debates thereby triggered, would reveal that there are alternative concepts of space in the urban realm and understandings of the urban fabric and the use of physical space that differ from those normally conceived of today. The eruv therefore, far from being a utilitarian instrument for a small group of Orthodox Jews in Germany, through its application serves a purpose in the larger public arena.

The second part of the call for an eruv-based approach to the ways in which Jewish communities formulate a different physical presence in German cities goes back to the very specific architecture of Jewish community centers in Germany. Instead of concentrating all Jewish institutions into one central community center, an architectural strategy following the ideology of the eruv would call for a distribution of Jewish facilities through the whole fabric of the city.

First and foremost, the eruv is the "urbanistic proof" within the vocabulary of Jewish thought, that the German concept of the "Einheitsgemeinde," the unified community meant to pull together all different strands of Judaism under one organizational umbrella as well as within one building, is neither practical nor in the interest of the communities themselves, nor is it inscribed within a religious tradition. On the contrary, it represents a very a-diasporic response to the diasporic condition. To have more than just one single central synagogue, serving a variety of different (or identical) Jewish groups within each city would correspond much better to the particular understanding of space as developed in the Talmudic tractate on the eruv.

Apart from the localization of synagogues within the urban fabric, an architectural strategy based on the eruv would lay importance on the mundane and the "banality" of the everyday. With very few exceptions, there are no Jewish bakeries, butchers, cafés, or bookstores in German cities. The visible and physical presence of those ordinary Jewish facilities could spread a minimum of "Jewishness" over a maximum of urban space. Obviously, being aware of the question of why one should have "Jewish" cafés at all (or what makes a café Jewish?) the simple existence of these mundane points of Jewish presence in German cities would sometimes serve a more important function in "normalizing" German-Jewish relationships than any large community center could. Jewish bakeries make the non-Jewish population aware of the fact that Jews exist not only in the shape of talk show interview-partners or visitors to synagogue inaugurations, but as participants in the everyday parts of life.

This call for an eruv-based strategy of the physical presence of Jews in Germany, a strategy that amounts to a distributed or dispersive approach playing an active role in the quotidian layers of public life, can at first glance easily be rebuffed on the grounds of "self-determination," as the Jewish communities themselves choose to follow a strategy of single, concentrated presence in the urban fabric, the Munich example being the best proof. As the interest in how Jewish presence is formulated in the physical realm reaches far beyond the internal hagglings of the communities and into the realm of state politics, an explanation, as mentioned above, cannot be reduced to a simple dialectic relationship.[23]

Perhaps more crucial is that the distributed strategy of the spatial presence of Jews in Germany could also be rejected on terms of security. At a time when political crime, especially right-wing offences, has reached all-time highs,[24] and when the threat of fundamentalist Islamic attacks—whether perceived or real—dominates many public debates, the urge to congregate in single places might be understandable, though wrong. Apart from the fact that the diasporic techniques of the eruv were precisely conceived and developed as a reaction to catastrophe and ultimate threat, the way in which German Jewry presents itself in public, how it approaches the general population, and how it plays an active role in the mundane aspects of urban life, has to be rethought. Community centers concentrating all parts of Jewish life within a single complex, and then necessitating police presence and very untalented and ineffective architectural solutions might not represent the best response. Security measures, which have become an omnipresent part of 'Jewish architecture,' often fulfill a more psychological function than actually being effective in defending against aggressive attacks. This both lulls the Jewish population into *feeling* secure and constructs adversaries as being invincible. But because it operates primarily on a psychological level, the security installations, visibly creating a split within society between the included and the excluded, also affect the population at large, who might develop feelings of doubt and rejection in view of the strong fortification that is apparently necessary to protect '*them*' from '*us*': "How bad must the non-Jews be that the Jews have to protect themselves so vehemently?" Right-wing or anti-Semitic crime is a problem in society that rather has to be addressed on a political level, and through social and communal processes that extend beyond the security installations of individual buildings.

Jewish bakeries do not lend themselves to a symbolic and publicly effective usage and are therefore resistant to the embrace of the political establishment and conservative state interests. In this resistance lies a much more fundamental critique and progressive stance than in any kind of radical, expressive, and rule-breaking architecture. After the fools, the time has come for the mundane.

Notes

1   'Eruv' is the Hebrew word for 'mixture' and denotes, amongst other things, the mixture of different kinds of spaces. It is defined in the identically named tractate of the Talmud. Jewish law defines four different kinds of spaces, with the private realm and the open, public realm being the most prominent. On Shabbat carrying objects from one type of space into another, being one of the 39 activities of work, is forbidden. The eruv of carrying defines an area by means of certain visual and physical boundaries, in which the differences between types of spaces are suspended, i.e. different spaces are mixed. Thus, carrying of objects becomes possible on Shabbat. The eruv is often used to facilitate movement and carrying of objects such as books from the private house to the synagogue, or in between private houses. Apart from this eruv of carrying, the Talmud also defines the eruv of cooking and the eruv of travelling, following similar logics.

2   Babylonian Talmud, Masechet Erubin.

3   For further reading on the issue of textual and symbolic interpretation of Jewish (rabbinic) writing and sequences of significations refer to Susan A. Handelman, *The Slayers of Moses: The Emergence of Rabbinic Interpretations in Modern Literary Theory* (Albany: State University of New York Press, 1982). Its specific nature in the writings of Emmanuel Levinas is expounded in Susan A. Handelman, *Fragments of Redemption: Jewish Thought and Literary Theory in Benjamin, Scholem and Levinas* (Bloomington: Indiana University Press, 1991). Also refer to Gillian Rose, "Architecture to Philosophy: the Post-modern Complicity," in id., *Judaism and Modernity: Philosophical Essays* (Oxford: Blackwell, 1993) for a more spatial and architectural discussion of symbolism and the chains of significations in Jewish thought.

4   Edmond Jabès, *The Book of Questions* (Hannover, NH: Wesleyan University Press; 1991), Vol. 1.

5   In this context, though outside of a specific religious realm, the work of Michel de Certeau is crucial. Especially his book *The Practice of Everyday Life*, trans. Steven Rendall (Berkeley: University of California Press, 1984) shows the way in which people appropriate seemingly mundane components of the public sphere, such as language, urban elements, or material commodities, and how these techniques of appropriation, and their corresponding evaluation, are structurally different from state actors, the media, or other influential institutions.

6   Emmanuel Levinas, "Assimilation Today," in id., *Difficult Freedom: Essays on Judaism* (Baltimore: Johns Hopkins University Press, 1990), 257.

7   Jacques Derrida writes on this notion in his essay "Edmond Jabès and the Question of the Book": "When a Jew or a poet proclaims the Site, he is not declaring war. (…) The site is not the empirical and national Here of a territory. It is immemorial, and thus also a future." In *Writing and Difference* (London: Routledge, 1978), 66. For further reading on a dynamic

notion of Jewish thought and its opposition to a Greek tradition connected with stasis, refer to Thorleif Boman, *Hebrew Thought Compared with Greek* (London: Norton, 1970).

8   For further reading on issues of Diaspora in a contemporary context, refer to Jonathan Boyarin and Daniel Boyarin, *Powers of Diaspora: Two Essays on the Relevance of Jewish Culture* (Minneapolis: University of Minnesota Press, 2002). For a more general debate on the issues of Diaspora and migration, refer to: *Theorizing Diaspora: A Reader*, eds. Jana Evans Braziel and Anita Mannur (Oxford: Blackwell Publishing, 2003). For further reading on issues related to the eruv refer to: "Jewish Conceptions and Practices of Space," *Jewish Social Studies* 11, no. 3 (2005).

9   With the risk of being too speculative or falling into clichés, two main reasons can be stated for the German Jewry representing a paradigm of Diaspora: First of all, the Jewish research and center of study in the SHUM communities of Speyer, Worms and Mainz, which dramatically transformed Judaism at its time, especially through Rashi and Gershom ben Judah, who was also named 'Light of Diaspora'. Secondly, the Jewish 'Bildungsbürgertum' (trans: 'educated bourgeoisie') was prominent in Germany, especially during the 19th century, modernizing and transforming the recognition of a Jewish self-identity, as well as the perception in the society at large.

10 To speak of a "Jewish" architecture is obviously very problematic, not only as it leaves undefined whether Jewish users, Jewish architects, or buildings with a Jewish theme are referred to. As this analysis concerns the physical presence of Jewish communities in German cities, a rather comprehensive notion of this type of architecture will be used in the following observations.

11 This is documented, among others, by the fact that of all the architects invited to that conference, none had yet built a museum, but all had been involved in the building or the design of a Jewish community center.

12 Ulf Meyer, "Eine Festung aus Stein und Glas: Einweihung von Münchens aufsehenerregender neuer Synagoge," *Neue Zürcher Zeitung*, November 10, 2006.

13 Ibid.

14 Salomon Korn, "Funkelnder Quell der Zuversicht," *Süddeutsche Zeitung*, November 9, 2006.

15 Kotel notes are papers of little prayers and wishes that are placed in between the stones of the Western Wall (the Kotel) in Jerusalem.

16 See the Proceedings of the Conference "Ein Jüdisches Museum für München," ed. Kulturreferat der Landeshauptstadt München, 2000, 96.

17 In line with this "centralization" of all Jewish presence is Munich's stance on the "Stolpersteine," or "stumbling-cobblestones." Small copper plaques are inserted into the pavement, with a simple name and date inscribed, to commemorate the places of habitation of Jews who were killed in the Holocaust. They have been installed in most German cities, thus paying a homage to the "ordinary" Jewish citizens and their annihilation from "ordinary" areas of residence throughout those cities. Munich is the only larger German city that has adamantly refused to install these little mundane memorials, which were developed by the artist Gunter Demnig. The head of the Jewish community, Charlotte Knobloch, and mayor Christian Ude, feared an "inflation of sites of memory." ("Neue Diskussion über die Stolpersteine," *Süddeutsche Zeitung*, June 13, 2004).

18 George Baird, "'Criticality' and Its Discontents," in *Harvard Design Magazine* 21 (2004/2005), http://www.gsd.harvard.edu/research/publications/hdm//back/21_baird.pdf (accessed August 8, 2007).

19 Slavoj Zizek, *The Plague of Fantasies* (London: Verso, 1997), 47.

20 Meyer, "Eine Festung aus Stein und Glas," translation by the author.

21   Rem Koolhaas, quoted by Beth Kapusta in: *The Canadian Architect Magazine* 39 (August 1994), 10.

22   Baird, "'Criticality' and Its Discontents."

23   Berlin, the German city that perhaps comes closest to the idea of a distributed Jewish presence, shows an example where the concept of the "Einheitsgemeinde" has in practice been outdone by the multiple heterogeneous actors within the Jewish community, or rather, communities. Even though sharing some of the characteristics of this distributed presence, the three "big" institutions connected to a Jewish presence and the Holocaust—the Jewish Museum, the Holocaust Memorial and the Topography of Terror—plus the Synagogue Oranienburger Strasse, outshine much of the Jewish presence that occurs on an everyday level in the city.

24   For reports on the criminal statistics of 2006 see the report of the German Ministry of the Interior (Bundesinnenministerium) from March 30, 2007 and various reports in the general media, for example: "18,000 Mal Angst," *Süddeutsche Zeitung*, March 28, 2007.

3 From State-Imposed Urban Planning to Israeli Diasporic Place

The Case of Netivot and the Grave of Baba Sali

Haim Yacobi

The entrance facade of the Baba Sali Yeshivah, Netivot, 2006

Space, Place and the Third Space

To a visitor arriving in Netivot, an Israeli Development Town located in the peripheral Negev region, the city looks like many other development towns in Israel that conform with the modernist form of space. The housing blocks and the semi-detached houses are in accord with the road system that marks the planned neighborhood units. This urban scheme, designed by Tzion Hashimshoni,[1] is based on modernistic planning principles such as zoning (i.e., the divided location of urban functions); open, public spaces linked to pedestrian paths; and efficient road systems that link the different zones.[2] Nevertheless, this schematic urban morphology is visually and spatially disrupted by indications of a different layer of urban life and experience that reflect diverse perceptions of what constitutes a city—an approach that highlights the way in which, "people are never passive recipients of external initiatives, but rather always struggle within their own immediate contexts of constraints and opportunities to produce meaningful life with their own particular values and goals."[3]

Following this observation, this article critically examines the Western-Modern orientation of Israeli space production vis-à-vis its diverse population, which in many aspects represents material culture that does not comply with the national supremacy. Specifically, this article will focus on the case of Netivot, a peripheral development town that offers an alternative experience of sense of place, linked to the diaspora and *Mizrahi*[4] identity that subverts the Israeli hegemonic production of space by creating a hybrid place.

A vast amount of literature discusses the notion of the terms space, place and sense of place. Yet in the scope of this introduction, let me highlight only the analytical distinction between "place" and "space" that emerged during the 1970s, when a qualitative shift in the field of geography paved the road to the development of social and cultural geography. Yi-Fu Tuan,[5] for instance, identified space as a general term in opposition to place that was defined as material. This distinction also appeared in the definition of absolute space as a container of material objects, in opposition to relational space that was defined as perceived and socially produced, emphasizing the phenomenological dimension, claiming that place is not an abstract but an experienced phenomenon linked to a process that involves the perception of objects and activities that are used as sources of personal and collective identities.[6]

Space and place became fundamental terms in the field of architectural theory and criticism. Christian Norberg Schultz followed this line of argument, as described above, and claimed that space is nothing but the relationship between objects. On the other hand, he argued that place is a defined built or natural space that has meaning, which stems from personal and collective memories as well as from identity.[7] Indeed, space will transform into place only when we are identified with and define ourselves through it. This work was viewed by many as a critique of the modernist movement in architecture, claiming that it had produced spaces but not places.

Simultaneous with the emergence of the phenomenological perspective, a new generation of geographers and urban sociologists pointed to the capitalist system as a social structure that may serve as a key to understanding the organization of space. For them, spatial practices such as those of planning and architecture, which were seen by the phenomenologists as agents for the production of places, were viewed as tools in the service of capitalism, which aim to balance private and collective capital, and thus hold potential for social oppression.[8] Indeed, this school of thought was significant in revealing interrelations between society, space, culture, and economy.

Yet, such a Marxist point of view demonstrates a lack of understanding of the everyday practices of the users and their struggle to transform space into place. Here, the work of Henri Lefebvre, who aspired to integrate theories and abstract thought with practice and the tangible daily urban experience, is significant. For Henri Lefebvre,[9] space is a social product, and thus a "sense of place"—though he does not explicitly use this term—cannot be seen solely as a reflection of either experience or knowledge. Rather, it is the juxtaposition of three interrelated dimensions: perceived space, conceived space, and lived space.[10] This approach enables us, analytically, to examine, on the one hand, the way in which space is appropriated by those in power who are motivated to reinforce the hegemonic narrative by using and implementing specific settlement structures or certain architectural styles. On the other hand, and most importantly, Lefebvre also refers to the users, to their everyday lives, and to their ability to produce a counterhegemonic meaning of place.

Beyond the general discussion presented above, the very particularity of the Israeli spatial reality calls for a localization of such theories towards the meaning of the built environment. The "Israeli place," as I will elaborate, is the product of a contested sociohistorical process, characterized by the motivation for controlling national space and framing it in a total manner. Such a decisive approach generates counterproducts that are also spatially expressed. The methodological roots of my claim originate from the tendency of urban research in Israel to focus on formal processes of space production, dictated from above and burned onto the collective mind by means of plans, thus reproducing the perception of what a place is and which sites do not warrant being called places. Connected to this debate is the centrality of the argument that the production of Israeli-Zionist space can be understood along three axes: the denial of the Orient, the rejection of the bourgeois, and the invalidation of diaspora culture.[11] However, vis-à-vis the short theoretical notes that open this section, let me propose that such an argument is partial since it refers to place production from above, ignoring the fact that vast parts of the built environment in Israel do not comply with standard regulations (legally as well as architecturally) and thus they penetrate into the spatial order created by the national culture and by so doing produce hybrid places.

Even at this stage it is important to clarify that the notion of hybridity accordingly is not a third concept that relieves the tension between cultures, hence resulting in the recognition of the subordinate culture by the hegemonic one.[12] Rather, it is formulated within a third space—a discursive junction in which the sovereign and the colonial subject are not exclusive alternatives, and the construction of their identities involves "mutual contamination."[13] During this process, which involves mutual reproduction and imitation within the intervening, namely third space, the colonial power also produces its alien. Therefore, claims Homi Bhabha, the third space is potentially a site of resistance, undermining the polar perception that poses identities as opposite, authentic, ethnically and racially essentialist entities, and hence can be perceived as a site of struggle and negotiation.[14] The significance of Bhabha's argument is its recognition that power relations are the basis for the production of subaltern culture, and it proposes a wide sociological understanding of the range between the top-down power and the voice of the subaltern subject. This insight, I would suggest, is an appropriate vehicle for examining space production in social and cultural theory in general and in the Israeli case in particular.

At the core of criticism against Bhabha's third space conceptualization, emphasizing the discursive aspect lays the material question. This critical tone appears both in relation to the distinction between politics and discourse and in the call for the examination of hybrid spaces within postcolonial contexts of specific geography, history, and economics. Postcolonial studies have focused on textual and literary studies, being only vaguely concerned with what happened. In the context of these criticisms, it is necessary to engage in material practices, in the actual spaces and the real politics, which have increasingly, if belatedly, brought recent as well as earlier studies of colonial urbanism and architecture into the debate, that have been largely ignored by the literary discourses within the postcolonial context.[15] Indeed, it is important to ground the formulation of the third space in meaningful practices; hybrid places are the result of interactions that are located in concrete, differing positions of power, which must, nevertheless, cohabit.[16]

This article joins that call; it aims to acknowledge the centrality of practices that are being conducted in the third space as a tangible site where diaspora place is produced within a national-sovereign space. To put it differently, the discussion concerning the Israeli place, on which I will focus, allows for recognizing the significance of practices occurring within the third place, not merely as a metaphor, but also as a concrete site in which material practices, producing the physical space, are activated. My argument follows Amnon Raz-Krakotzkin's suggestion that disavowing the diaspora past in the Israeli context is part of the implementation of the regime of modernity. It should be recognized that a strict denial of the diaspora exists owing to the formulation of cultural identity in terms of disavowal, and more importantly, the fact that repudiating the diaspora means repudiating the Jewish memory.[17]

Jewish Place, Modernity and Sovereignty

The general debate concerning the Israeli production of space-place focuses on the "local" versus the "other," ignoring the dynamic nature of identities. The same applies to the examination of the role of architecture and planning within the Zionist enterprise, which has so far been focused on the Jewish-Arab or Israeli-Palestinian issue. This is connected to a common point of departure "1948" and to the discussion of the appropriate spatial form of habitat for the Jewish people in their motherland. Yet, it is important to follow this debate from an earlier period—the pre-state period—as I will discuss in the following section by means of an analysis of the architectural discourse in the 1930s as presented in the first issues of a journal named *Habinyan* (The Building).

> These issues of Habinyan provided a forum for serious discussions concerning the commitments of architects and planners to defining an appropriate form of Jewish habitat and its relevance to the construction of identity and the attempt to prove territorialization. In its first issue, there is a clear expression of the tension between the western approach to planning and the local geographical and economic conditions in Eretz Israel: In the course of our adaptation to the conditions of the Land we learned (…) that neither American nor European models of development, even the most progressive, are appropriate for our capabilities (…) since in the future they will cause an increase in public expenditures.[18]

For the architect Julius Posner, it is significant to adopt a neutral modernist attitude. Thus, he argues that "lately Jews have taken part in the development of European taste." This is expressed by the fact that the Jewish people, "are distancing themselves from traditional forms, they are learning to appreciate cleanliness and simplicity, and are thus liberating their homes from the memories of the past." This liberation from the past has a considerable impact upon the presence of Jews in the Land of Israel, not just as the denial of the diaspora, but also, as the denial of the Orient that is presented by means of Oriental morphology:

> First of all, people are no longer captivated by the Oriental appearance. Anyway, we have relinquished the Oriental character created from constructing domes and arcades. This reaction is necessary as well as suitable to the real demands of Jewish taste.[19]

The third issue of the journal, which deals with villages in the Land of Israel, draws attention to the dichotomous attitude towards the Oriental-Arab landscape. In the opening essay, Posner categorizes settlements and cites their disadvantages and merits. He suggests that the village in the Land of Israel, "is ancient and has hardly changed." Thus, he asks, "what can we learn from such ancient experiences? Probably we can learn from their economy, their social relations, their collective agricultural manners (…). Some people claim that the home in the Arab village protects one from the climate better than our homes in the *moshavot*."[20]

Beyond such an Orientalist approach, Posner continues to argue that the journal equally avoids romantic superlatives concerning the wholeness of the Arab agricultural villages, stating that "we would not say that we must build so traditionally and we would also say it is prohibited to build so badly and oddly. The Arab village is not a model for replication by us (…)."[21] This approach indicates how central architectural discourse and space production are to identity. More specifically, it reveals the duality in relation to the Oriental landscape: it is on the one hand an authentic object of desire that might inspire the shape of habitat of the Jewish people, and on the other hand it is the signifier of the underdeveloped Oriental-Arab society. Yet the spatial implementation of such an approach was limited. As I will discuss in the following section, it was only when the geopolitical conditions had changed and the Israeli state was established that a modernist paradigm in planning and architecture became central to the production of the new Jewish place.

But which landscape was supposed to replace the built environment that was marked as Oriental? The answer to this question was obvious at the time and can be related to the modernization project that provided justification for the rejection of an Oriental past and present as I have discussed above. In addition, it should be related to the manner in which power relations enable the implementation of a plan aimed at providing the means to create social transformation. Realizing this plan requires the extensive involvement of the state and a centralistic planning approach to enable the fulfillment of a vision that provides "an opportunity to rewrite the national history."[22]

Modernity and urbanism in this sense are not part of an uncontrolled evolutionary process. Rather, as sociological and political processes, they crawl along and in most cases erupt via their various agents—settlement, nationalism, immigration, professional experts, and capital—guaranteeing a change in society and consciousness that will eventually lead to an inevitably better future.[23] Indeed, modernity as a social project includes a doctrine of progress accompanied by the creation of a new subject, the agent of modernity, who is freed from the bonds of tradition in order to fulfill her/ himself as an individual. Seeing modernity as a neutral process strengthens the dominance of western culture and transforms it into the sole default option to which to aspire in order to justify being termed 'modern', and this is a necessary condition in order to benefit from the distribution of rights and goods.

Let me illustrate the above argument while using one of Israel's iconic architectural objects—the *shikun* (the Hebrew word for tenement housing block), whose political and architectural meaning has been the subject of many papers,[24] indicative of its dominance in the Israeli landscape of development. Apart from being a manifestation of a certain school of planning, it is also linked to an ideology that perceived the formation of modern space as a means of constructing a sense of collective belonging. The *shikun* in its modernistic form assumed a double function in the Israeli context: it reflected sovereignty over national territory, and at the same time served as an incentive to economic, social, and identity production and reproduction.[25]

Indeed, the construction of the *shikunim* during the 1950s was considered revolutionary. The project, which was part of a comprehensive national plan for spatial development in Israel conducted by Arieh Sharon, head of the Planning Division of the Prime Minister's Office, presumed to provide housing for a population that doubled in size during the first decade of the state.[26] The plan, entitled "Physical Planning in Israel," reflected the centralistic statehood that characterized the Israeli regime in the 1950s. The Sharon Plan defined three dimensions of spatial design (and in my opinion, a pedagogic objective as well)—land, people, and time—as a basis for a professional, physical construction plan. These imaginary concepts facilitated the formation of the new national space:

> This assorted immigrants' ingathering will become uniformly consolidated only if supported by comfortable physical, social, and economic conditions (…). A social composition and a planning framework should be provided in order to facilitate assimilation and stimulate the process of integrating different types of settlers (…) into one unified creative whole.[27]

This approach enhanced the importance of the home as a vehicle for the creation of a collective sense of identity and belonging, as a means of transforming immigrants into locals; or in spatial terms, to produce place in the new territory. Golda Meir confirms this claim by stating that

> (…) inadequate accommodations are evident everywhere around the globe. In Sweden, no Swedish-born individual whose ancestors resided there will cease to be Swedish just because he has no home. Here, however, this is severely problematic. The housing problem is highly significant, and it will determine whether that same family that emigrated with its children, foreign and unacquainted with the language, the conditions, and often also the goals—it will determine whether these family members will become Israeli or remain foreign, albeit holding Israeli citizenship.[28]

The necessity for domesticating the immigrants' culture coincided with the modernist approach to planning that took it upon itself to design the housing unit. This fact had social implications that came to bear on the everyday use of private space since it aimed at liberating the family from its traditional domestic perceptions. This planning and architectural paradigm dovetailed with the objectives of the Israeli regime in the 1950s that aimed to transform the immigrants through a process of de-Arabization.[29] Architectural modernism can therefore be contained within the parcel of national belonging under the guise of civil and secular culture—terms that according to Homi Bhabha were exploited to draw people into the human community, but at the same time were used to exclude them from it as "others."[30] Modernism clearly reflects the double mechanism that produced the new habitat in Israel—an approach of efficiency, order, and planning on one hand and the application of ethno-national logic which replaces what has been considered as underdeveloped on the other.

It seems that Amos Oz' description encapsulates the transformation of the *shikun* from a pedagogical architectural object:

> The large distance between the buildings, planned by the architect, make the shabbiness more marked than it would be if the buildings were close together—a Mediterranean town, house touching house, spaces of more human proportions. Were these neglected lots intentional, in the planners imagination perhaps, meant to be vegetable gardens, small orchards, sheep pens, and chicken coops: a North African *nahalal* on the rocky slopes of Judea? What did the town planner know or want to know about the lives, the customs, the heart's desires of the immigrants who were settled here? Was he aware of, or partner to, the philosophy prevailing in the fifties that we must change these people immediately—remake them completely—at all cost?[31]

As noted by Oz, North Africa has been one of the main origins of immigration to Israel since the establishment of the state. The new national project referred to the Oriental immigrant culture as an object that demands special treatment by pedagogic westernization and modernization aimed at reshaping the immigrant's everyday life. This was the contribution of the architectural practice and discourse to affixing antinomies such as west/ east, first world/ third world, modernity/ backwardness and sovereignty/ diaspora.

Images of the tenement housing block, as a signifier of the Mizrahim, definitely appear in several representations that deal with the Mizrahi culture, political activism, and protest.[32] Over the years, the users have transformed their housing environment. These additional constructions, which are not the product of professional logic and esthetics, undermine the power of national logic, supported by professional knowledge. In other words, the modification of the housing environment is a counter act of place determination that goes beyond the inhabitants' motivation to improve their physical quality of life;

A typical street placard with a logo of the Rabbi Yisrael Abuhatzeira Institutions, Netivot, 2006

rather, it is a manifestation of past Mizrahi cultural affiliations—a debate to be discussed in the following section.

The Baba Sali gravesite, Netivot, 2006

Netivot as a Diasporic Place?

> In the year 1956 the first settlers arrived at Netivot from the Maghreb countries. The Olim [new immigrants] were loaded on trucks and taken in the middle of the night to the place, the object of their yearning. Many of them believed that they were taken to Jerusalem, but under cover of darkness they were transported to the town of Netivot.[33]

The above quotation narrates in a nutshell the re-territorialization of Israel and the attempts to stabilize its sovereignty by establishment of the new development towns.[34] As part of the physical plan for Israel discussed in the previous section, the town of Netivot was established in 1956 as a regional center for the northwestern Negev agricultural settlements, and the first wave of Netivot's inhabitants were primarily Jewish emigrants from North Africa. Several reports since the establishment of the city point to its economic underdevelopment, attributing this to its ethno-demographic composition.[35] Even more recent data from the Central Bureau of Statistics[36] indicates that the city is ranked at socioeconomic level 3 (out of 10). In the year 2000, Netivot was officially declared a city by the Ministry of Interior and its population, as of 2004, stood at 26,000 inhabitants,[37] seventy percent Mizrahim and twenty-five percent Russian immigrants.[38]

The main entrance to Netivot is Jerusalem Street, with other streets branching out from it and bearing the names of Jewish-Moroccan saints and rabbis. At the main streets' corners, on top of the official street signs, an additional placard is placed containing an image of Rabbi Yisrael Abuhatzeira, known as the Baba Sali ('Praying Father' in Moroccan Arabic). Born in 1890 in Morocco, he immigrated to Israel in the 1950s, and several years later settled in Netivot.

As a consequence of his religious spiritual, and, political influence, and the subsequent creation of such informal shrines, these constructions have gained political importance at the municipal as well as the national level.[39] Netivot also begun to attract Jews who have returned to their religious roots.

The Baba Sali died in 1984. His funeral in Netivot's cemetery drew an estimated 100,000 people. The influence of the Baba Sali has grown since and the city has become a renowned focus of pilgrimage for the Moroccan Jewish community in Israel as well as from abroad, i.e., for Moroccan Jews residing in France or Canada. His gravesite in Netivot has become a popular pilgrimage site in Israel, especially on the date commemorating his death. It is important to mention Oren Kosansky's remark that the Jewish pilgrimage shares similarities with saint veneration as practiced by both Muslims and Jews in Morocco.[40] Though several anthropologists have written extensively about this phenomenon, exploring its cultural, social, and political dimensions,[41] no special attention has been given to its spatial influence nor to its contribution to the creation of Netivot as a place—a void that the following section aims to fill. If the visitor to the city were to follow the signs along Abuhatzeira Street, s/he would begin to recognize a different architectural expression of the buildings—contradicting the modernist space and commemorating the past of the Jewish community in the diaspora. This issue was raised in an interview with the representative of the Baba Sali Institutions[42] who claimed that the use of such an architectural style that "purposely does not fit the Netivot cityscape (…) is the appropriate way to commemorate the *Tzadik*[43] (…).The buildings in Morocco in the *Tafilalt* region (an oasis in the Moroccan Sahara) are similar. We replicated them here in Netivot, in order to symbolize the past."

Down the road, Abuhatzeira Street leads to the edge of the city, where the modernist housing blocks mark the end of the urban constructed area. The back of these buildings faces a neglected open space, which according to the planning regulations detaches the city from its cemetery. Nonetheless, Netivot's cemetery is not a dead place—the Baba Sali burial site has become a focal point of religious, spiritual, and social encounters, especially at the time of the *Hillulah,* or celebration day. From the architectural point of view, the place constitutes an attempt to establish an icon that commemorates not only the Baba Sali legend but the memory of the Jewish community in Morocco as well. This notion was raised by several people during the last *Hillulah* when I asked them what the significance of Netivot is for them. A man in his fifties told me that he was a child when his family immigrated to Israel from Morocco: "I do not remember myself what it was like there, but we come here every year with my mother (…). She has told us that it is exactly the same. I feel as though I were there."[44] Let me suggest that here lies the notion of diaspora experience for those who are here but still attached to there.

Interestingly enough, the modification of the cemetery into a pilgrimage site has occurred through official planning procedures. The authorized new urban scheme that has been authorized enables the modification of land use from a cemetery into a pilgrimage site: "the objective of plan No. 103\03\22 is (…) altering the existing land use from public, open space into a burial plot of 4,339 square meters (…)."[45] Moreover, the modified urban scheme of the cemetery acknowledges the pilgrims' needs according to their tradition and allocates space for the construction of a "feast shelter" for use by the pilgrims, the establishment of a structure for commercial activity, and the construction of three rest units:

> There is a custom among some ethnic groups that the terminally ill seek healing at saints' graves by praying and seclusion, as well as by adjacent sleeping accommodations. The purpose of the rest units is to enable these people to realize their wishes under the same roof [as the other activities].[46]

Spatially speaking, on the day of the *Hillulah*, the neglected space between the edge of the city and the cemetery is transformed into a meeting place of the pilgrims. Thousands of people visit the Baba Sali burial site and a lively market of religious goods, food and clothing serves the crowds.

The extensive city life takes place in a public space, which is actually a parking lot, while the modernist *shikunim* that house many of the Baba Sali's community members serves as the backdrop. The modernist urban order is further transformed by means of the cemetery. The new religious and educational institutions that have been established by the Baba Sali Foundation are designed with reference to the "old-new" architecture and are used as landmarks on the urban scale (see opening page of this article).

The effect of the diaspora-Mizrahi religious notion of the city is acknowledged by the municipality of Netivot, which participates (in terms of budget) in the *Hillulah* events and acknowledges the contribution of the institutions to the city, stating that "On its 50th anniversary, all Netivot's inhabitants appreciate the contribution of the Baba Sali to the development and progress of the city. The municipality is committed to act, by all means, in order to commemorate the Baba Sali legend and to support its institutions."[47] The reconstruction of Netivot's Morrocan-like sense of place coincides with Oren Kosansky's observation of the considerable Jewish element in the Moroccan city of Fez in the *mellah*[48] (the segregated Jewish quarters in Morocco), and of the city's Jewish cemetery. Architecturally speaking, he suggests that there is a specific Jewish architecture expressed in the *mellah*: the main road is lined with exceptionally tall buildings bearing a distinct decor that marks their front elevations. Also, he observed that the balconies extend beyond the lanes below—an element not to be found elsewhere in the old city of Fez.[49] Moreover, the interim space created in Netivot has also constructed a virtual network of places. Praying to and imploring the Baba Sali is possible via several internet sites, which also transmit images of the burial site and recitations of prayers twenty-four hours a day.[50] I would suggest that Netivot appears to offer some insight into the potential of diaspora communities within a forceful national context to express, and often glorify, their ties to their Morrocan homeland.

Let me elaborate on the relevance of the notion of diaspora to our case. The concept of diaspora has traditionally referred to cases of communities living outside of their homeland. The term is used extensively when referring to emigrants, expellees, alien residents, and ethnic minorities. Beyond the different definitions, there are shared characteristics of the notion of diaspora as referring to a given social group that has dispersed from its territory to a different, foreign region. In this process, the specific group constructs its collective memory concerning its origin, location, history, and culture. More importantly, Safran argues, the diaspora group perceives itself as an excluded social entity that cannot be fully integrated into the host society.[51]

The above definition stems from the very specific case of the exile of the Jews from the Holy Land and their dispersion throughout several areas of the world.[52] Such an approach must be seen as an ideal since a comparison to other cases demonstrates that most diasporas do not match this definition. In contemporary postcolonial literature, there is a wider understanding of the term that broadly refers to it in relation to displacement, dislocation, and reformation of the "double consciousness" of being "inside and outside."[53] Indeed, diaspora discourse, "is loose in the world, for reasons having to do with decolonization, increased immigration, global communications, and transport—a whole range of phenomena that encourage multi-locale attachments, dwelling, and traveling within and across nations."[54]

The notion of homeland is an inherent component and used as a *raison d'etre* for the production of space (as presented in the previous sections), which is related to the ideological and political circumstances that caused these people to immigrate to Israel, while at the same time, Israeli-Zionist ideology denies the notion of the diaspora past, geographies and culture of these immigrants. More specifically, as already suggested by Levy, though Jews perceive the Land of Israel as the core of their collective history, they furthermore, "conceive of Morocco as a symbolic center; a homeland for those who remained behind as well as for those who migrated."[55] Likewise, as in other cases, the role of religion in the diaspora experience and place making is central; places accumulate meaning beyond their function for religious practices and thus gain value and become social, cultural and political signifiers of diaspora identity.[56]

The Third Space: Beyond Recognition?

This essay examined the transformations in the discussion of the Oriental nature of the Israeli built environment, supporting the claim that identity—as a political and cultural construct—is related to the formulation of new time and space created by communal imagination processes that intertwine past, present, and future. This process is a manifestation of hegemonic culture, which frames the place while intervening and generating spatial transformation, using space production as an instrument for their realization. Thus is formed the pedagogic landscape, the spatial fabric that teaches us about our past and our identity, and within which the built environment assumes its structured symbolic significance, being justified as a representative of the collective desire and thought.

The discussion points out the role of the built environment in the production of Jewish place in the old-new space, and, as indicated, this is the site of ongoing struggles, in which top-down power creates counterreactions that do not adhere to the desire to modernize/westernize space. Indeed, within the Israeli space, as a product of counter-acts, the modernist housing machine became a hybrid twice. First, it was transformed from a site that expressed a unified modernist national identity into a site that symbolized the excluded *Mizrahi* population. Second, it was transformed from a site based on the notion of progressive modernity into a site of alternative modernity, where individual place production violates the top-down initiatives.

Indeed, space is not a static container of social relations; people create alternative local narratives that do not necessarily reflect the rationale of the nation or of capital, nor the social hierarchy or the power relations that create them.[57] The counterproduction of urban order, I would propose, is a direct result of the Zionist ideology based on the denial of a diaspora past. However, this conclusion does not aim to idealize or essentialize the Arab character of Jews in the diaspora. Following Raz-Krakotzkin, historically there had been tension among the Jews as a minority in their Arab countries of origin. However, this tension was not defined as a cultural gap to be overcome. It is the Eurocentric model of the Israeli national project that contains the East-West dichotomy as an objective category of modernity and space ordering that leads to the conclusion that being included in the Israeli collective is both the tangible and symbolic act of Jews foregoing Arab culture and place construction.[58]

Theoretically, I have indicated the relevance of postcolonial theory to the understanding of the Israeli space/place production.[59] First, the postcolonial body of knowledge critically examines the social structures that result from ideologies of domination stemming from colonial histories.[60] Secondly, postcolonial criticism has enabled an analysis of the ways in which subaltern cultures are shaped while internalizing hegemonic culture. Thirdly, a significant issue in the postcolonial theory is hybridity, which is linked to the notion of diaspora. Here, allow me to rephrase the notion of diaspora as it is exposed by the case study. The concept of diaspora indeed incorporates the transnational experience of those who returned home according to Zionist ideology, a situation producing negotiable, multidirectional ideas and spatio-cultural urban topographies.

Through these lenses, this paper rethinks the traditional view of architecture that assumes the national category as the natural realm for space production—a perspective that places a high priority on official planning and architectural practices as apparatuses for the production of national sovereignty, an issue discussed in the first section. Yet, it also suggests that the multiple loyalties of people is simultaneously molded in different spaces/places, locating themselves between here and there, within sovereign, state boundaries, and at the same time in their diaspora experience that produces a hybrid place. This term is not a fixed topographical site of negotiation between different locales (i.e., societies, cultures), but rather a zone of deterritorialization, which in turn produces identity.

Hybridity as a site of negotiation was not confined merely to the tenement housing blocks environment. The new social, economic, and political structures as indicated in Netivot enabled the shifting of the excluded imagined place and desires to be included in the official mapping of the city, and the extensive infiltration of this architecture into other spheres to become visible. In fact, the modernist model that seeks to level the range of identities has become an indication of a multicultural option that grows from the bottom up and enables a discussion of Netivot as a project of alternative modernity—a concept focusing on the significance of modernity in daily life among societies and spaces that are not part of the "first world." This type of modernity rejects the bourgeois ethos of modernity, and, instead, seeks recognition of the fact that different modernization projects have not produced uniform results.[61] At the basis of this cultural theory lies recognition for the many expressions of modernity. The capitalist economy, technology, and bureaucratic organization of the state are inherent elements of modernity, but they lead to different types of modernity that diverge from the binary view of modernity versus traditionalism.

In this context, the peripheral city of Netivot can be seen as a third space, where the recognition of planning authorities enabled the expression of communal architecture that is not subjected to the hegemonic narrative. A similar argument is presented by Ben Ari and Bilu,[62] who suggest that the emergence of sacred sites of Jewish saints in Israeli development towns is not just rooted among diaspora North Africans, but instead, strengthens people's sense of belonging to their places:

By constructing these sites people in development towns come to terms with their peripheral status in Israel. This phenomenon is related to what maybe termed an internal Israeli cultural debate centering on its identity as a "Middle Eastern" society; to the extent which Israel shares with its Arab neighbors a set of cultural concepts and guidelines by which public life is carried out.[63]

The question that remains is to what extent it is possible to view the third space as an arena of subversive struggle and negotiation, as suggested by Homi Bhabha. Let me suggest that though the third space can be seen as an element that challenges the hegemonic perception of space; it does not transform it strategically. If we return to the peripheral characteristics of Netivot (in terms of socioeconomic, class, and ethnic stratification that I have cited above),[64] this recognition cannot come to replace or be separated from distributive questions.[65] Rather, it should not draw attention away from distributive issues, as then the city would fall into the trap of perpetuating the hierarchy as dictated by the state's spatial ordering.

Notes

1   David Zaslevsky, *Seker pituach be-netivot* [A survey of Netivot's development] (Jerusalem: Ministry of Housing, 1969).

2   For a detailed discussion see: Zvi Efrat, *Ha-proyekt ha-israeli: Bniya ve-adrikhalut 1948–1973* [The Israeli project: Building and architecture 1948–1973] (Tel Aviv: Tel Aviv Museum of Art, 2004), 807–24.

3   Goh Beng-Lan, *Modern Dreams: An Inquiery into Power, Cultural Production and the Cityscape in Contemporary Urban Penang*, Malaysia (Ithaca, NY: Cornell: Southeast Asia Program, 2002), 202.

4   Mizrahi Jews (plural: *mizrahim*), are those who come from Arab and Muslim countries.

5   Yi-Fu Tuan, *Space and Place: The Perspective of Existence* (Minneapolis: Minneapolis University Press, 1977).

6   Ali Madanipour, *Design of Urban Space: An Inquiry into a Socio-spatial Process* (Chichester: John Wiley and Sons, 1996). Edward Relph, *Place and Placelessness* (London: Pion, 1976).

7   Christian Norberg-Schultz, *Genius Loci: Towards a Phenomenology of Architecture* (New York: Rizzoli, 1979).

8   Manuel Castells, *City, Class and Power* (London: Macmillan, 1978).

9   Henri Lefebvre, *The Production of Space* (Oxford: Blackwell, 1991).

10 Ibid.

11 Alona Nitzan-Shiftan, "Batim mulbanim" [Whitened houses], *Teoria u-vikoret* 16 (2000).

12 Homi Bhabha, *The Location of Culture* (London: Routledge, 1994), 113–14.

13 Ibid.

14 Homi Bhabha, "The Third Space: Interview with Homi Bhabha," in *Identity, Community, Culture, Difference*, ed. Rutherford Jonathan (London: Lawrence and Wishart, 1990), 211.

15 D.A. King, "Actually Existing Postcolonialism: Colonial Urbanism and Architecture after the Postcolonial Turn," in: R. Bishop, J. Phillips (eds.), *Postcolonial Urbanism* (London: Routledge, 2003), 167–83; here 167.

16 Nezar AlSayyad, "Hybrid Culture/Hybrid Urbanism: Pandora's Box of the 'Third Place,'" in id., ed., *Hybrid Urbanism: On the Identity Discourse and the Built Environment* (Westport, Conn.: Praeger, 2001).

17 Amnon Raz-Krakotzkin, "Galut betokh ribonut: Likrat bikoret shlilat ha-galut be-tarbut ha-israelit" [Exile within sovereignty: Towards a critic of the 'negation of exile' in Israel culture], *Teoria u-vikoret* 4 (1993): 113.

18 Avraham Schiler, "Land Development Problems for Housing," *Habinyan* (1937), 28–9 trans. by the author.

19 Julius Posner, "Ha-kfar be-eretz-israel" [The village in Eretz Israel], *Habinyan* (1938), trans. by the author.

20 Ibid.

21 Ibid.

22 James Holston, *The Modernist City: An Anthropological Critique of Brasilia* (Chicago: University of Chicago Press, 1989), 5.

23 Charles Taylor, "Two Theories of Modernity," *Public Culture* 11, no. 1 (1999): 153–74.

24 See for example: Hadas Shadar, "Between East and West: Immigrants, Critical Regionalism and Public Housing," *The Journal of Architecture* 9 (2004); Rachel Kallus and Hubert Law-Yone, "National Home/Personal Home: Public Housing and the Shaping of National Space in Israel," *European Planning Studies* 10, no. 6 (2002): 765–79.

25 Kallus and Law-Yone, "National Home/ Personal Home."

26 For a detailed study see: Smadar Sharon, "Livnot u-lehibanot ba": Tikhnun ha-merchav ha-leumi ba-shanim ha-rishonot le-medinat israel" ["To built and be built": Planning the national space in the formative years of Israel] (Thesis submitted at Tel Aviv University, 2004).

27 Arye Sharon, *Tikhnun phisi be-israel* [Physical planning in Israel] (Jerusalem: Government Printing Press and Survey of Israel Press, 1951), trans. by the author.

28 Golda Meir as cited in David Zaslevsky, *Diur le-olim: Bniya, tikhnun u-pituach* [Housing for immigrants: construction, planning and development] (Tel Aviv, 1954) trans. by the author.

29 Yehuda Shenhav, *The Arab Jews: A Postcolonial Reading of Nationalism, Religion, and Ethnicity* (Stanford: Stanford University Press, 2006).

30 Bhabha, "The Third Space: Interview with Homi Bhabha," 211.

31 Amos Oz, *In the Land of Israel* (Tel Aviv: Am Oved, 1983), 28–9.

32 Haim Yacobi, "Ha-makom ha-shlishi: Arkhitectura, le'umiut u-postcolonialism" [The Third Place: Architecture, nationalism and postcolonialism], *Teoria u-vikoret* (2007 forthcoming).

33 Netivot Municipality website: http://www.netivot.muni.il.

34 Erez Tzfadia and Oren Yiftachel, "Between Urban and National: Political Mobilization among Mizrahim in Israel's 'Development Towns,'" *Cities* 21, no. 1 (2004): www.geog.bgu. ac.il/members/yiftachel/new_papers_eng/Cities.pdf.

35 Zaslevsky, *Seker pituach be-netivot*; id., *Diur Le-Olim*.

36 Central Bureau of Statistics, ed., *Characterization and Ranking of Local Authorities According to the Population's Socio-Economic Level in 2001* (Jerusalem: Central Bureau of Statistics, 2004).

37 Ibid.

38 According to the available data 22 percent of the population were born in Africa while 68 percent were born in Israel, mainly a second and third generation of the Mizrahi newcomers to the city. See http://www.netivot.muni.il/#Top (accessed September 27, 2007).

39 Eyal Ben-Ari and Yoram Bilu, "Saint's Sanctuaries in Israeli Development Towns," in: *Grasping Land: Space and Place in Contemporary Israeli Discourse and Experience*, eds. Ben Ari Eyal and Bilu Yoram (Albany: State University of New York Press, 1997), 61–83.

40 Oren Kosansky, "All Dear Unto God: Saints, Pilgrimage, and Textual Practice in Jewish Morocco" (PhD Diss., University of Michigan, 2003), 553.

41   Yoram Bilu and Eyal Ben-Ari, "The Making of Modern Saints: Manufactured Charisma and the Abu-Hatseiras of Israel," *American Ethnologist* 19, no. 4 (1992): 672–687.

42   Interview with the representative of the Baba Sali Institutions in Netivot, September 25, 2006; trans. by the author.

43   *Tzadik* is a righteous person.

44   Informal interview, January 23, 2007; trans. by the author.

45   Urban scheme No. 103\03\22; trans. by the author.

46   Ibid.

47   Yehiel Zohar, mayor of Netivot, in a brochure published by the Baba Sali Foundation (2006).

48   On the development of the *mellah*, see Susan Miller's article in this volume.

49   Oren Kosansky, "Reading Jewish Fez: On the Cultural Identity of a Moroccan City," *The Journal of International Institutes* (8 March, 2001): http://www.umich.edu/~iinet/journal/vol8no3/kosansky.html (accessed July 31, 2007).

50   See, for example "POIP Live Videostreaming," http://www.po-ip.co.il (accessed September 25, 2007).

51   William Safran, "The Jewish Diaspora in a Comparative and Theoretical Perspective," *Israel Studies* 10, no. 1 (2005): 36–60; also online: http://muse.jhu.edu/.

52   Ibid.

53   André Levy, "Center and Diaspora: Jews in Late-Twentieth-Century Morocco," *City and Society* 13, no. 2 (2001).

54   Clifford James, *Routes: Travel and Translation in the Late Twentieth Century* (Harvard University Press, 1997), 249.

55   Levy, "Center and Diaspora," 245.

56   John Fenton, *Transplanting Religious Traditions: Asian Indians in America*. (New York: Praeger, 1988).

57   James Holston, *The Modernist City: An Anthropological Critique of Brasilia* (Chicago: University of Chicago Press 1989), 31-4.

58   Raz-Krakotzkin, "Galut betokh ribonut," 126.

59   In the scope of this paper, I did not address the relevance of postcolonial discussion to the understanding of social and political structures in Israel (e.g., in *Teoria u-vikoret*, 20 and 26). However, one topic that postcolonial discussion has not included is the position of space designing practices, such as urban planning and architecture, which challenges this paper.

60   Jane M. Jacobs, *Edge of Empire* (London and New York: Routledge 1996).

61   Arjun Appadurai, *Modernity at Large: Cultural Dimensions of Globalization* (University of Minnesota Press, 1996).

62   Ben-Ari and Bilu, "Saint's Sanctuaries in Israeli Development Towns," 61.

63   Ibid.

64   New research findings point to the way in which Netivot's image among the Israeli public has been improved. One explanation for this is the transformation of the city into a spiritual node that exposes it to the public. Furthermore, in comparison to other development towns, Netivot's economy is improving, independent of state subsidies or intiatives (*Ha'aretz*, January 26, 2007).

65   Nancy Fraser, "Social Justice in the Age of Identity Politics: Redistribution, Recognition, and Participation," in *Redistribution or Recognition? A Political-philosophical Exchange*, eds. Nancy Fraser and Axel Honneth (London: Verso, 2003).

Part II
Jewish Quarters

4 Ghetto Gardens

Life in the Midst of Death

Kenneth Helphand

In the spring of 1942, a few young children in the orphanage of the Lodz Ghetto, about seventy-five miles southwest of Warsaw, turned a piece of forlorn ground into a garden.[1] The ground was hard, and "whatever their little hands could get hold of they used; spoons, forks, sticks and so on." As the garden grew, the children were awed by the first shoots, but the plants, like the children, looked malnourished. As their teacher, Sheva Glas-Wiener, recounts in her memoir, every day the children waited "to see the fruit of their labor. Every morning I would see them looking at the sky and at their little garden waiting for the beets to grow larger and to ripen but they waited in vain."[2]

When the children of the Lodz Ghetto learned they were to be deported, a spontaneous fury seized them. They went to their gardens and, in a burst of anger, trampled the few beds of pathetic beets. With the toes of their shoes they kicked every clod of earth and, in a rage of frustration, tore the plants out by the roots.

"Nothing will grow after we have gone! Nothing will bloom in this garden!" a girl about ten years old screamed in rage.

"Nothing will grow! Nothing will bloom!" the others repeated and trod everything into the ground even more passionately.

This protest burst forth from the depths of their souls and they knew no other way to express their sense of betrayal and anger. In the moment of their pain, they had no one to cry out to, so they turned on the earth itself, for it had failed them. They had loved it, nurtured it, and watched over it, and it had not heeded them or responded to their loving care.[3]

Gardens in the Ghetto? It seems a preposterous yet amazing proposition. How were they possible? There were gardens in the major Ghettos of Warsaw, Lodz, Kovno, and Vilna. Gardens, like other aspects of life that could be considered "normal"— going to school, strolling in the park, attending a concert, praying, taking a walk, or having enough for dinner— were all present in the ghetto, but only through extraordinary effort. Though short-lived, like the ghettos themselves and their prisoners, ghetto gardens were mechanisms of resistance against the horrific conditions under which people lived. They were acts of hope and defiance. By attempting to create conditions for their survival by creating the gardens, ghetto residents also provided work, relief, solace, and food for themselves.

Gardens created in the ghettos are examples of what I have termed 'defiant gardens', gardens created under extreme or difficult environmental, social, political, economic, or cultural conditions or situations.[4] They all represent acts of adaptation to their challenging circumstance, but they can also be viewed from other perspectives, as sites of assertion and affirmation. This certainly holds true for gardens in the ghettos.

George Eisen's Children and Play in the Holocaust: Games among the Shadows offers instructive parallels. Play, like the garden, isn't always taken seriously. The fact that pleasure, joy, abandon, innocence, and freedom often accompany play disguises its more profound aspects. Eisen discovered that children's play in ghettos, concentration camps, and wartime hiding places was an "enterprise of survival, a defense of sanity and a demonstration of psychological defiance."[5]

Gardens are subject to a similar phenomenon. Ironically, the fact that we associate the garden with the archetypal ideal of paradise and the nostalgic allure of the pastoral may account for our marginalization of garden making, where we see gardens as a luxury, frill, pastime, or leisure time activity, and not as an essential component of culture and human existence. And the beauty of the garden may also mask its deeper messages.

The Eden of Genesis is the archetypal garden, the prototype of Western garden design: in the tale of Eden we have condensed and concentrated the central images and ideas of gardens and begin to recognize the complexity of the garden idea. Recall the attributes of Eden. The world begins with a garden. It is the territory and microcosm of creation, where all the world's species are contained. It is where humans talk with God. It is sacred; a place of innocence, peacefulness, perfection, and complete harmony—although of course it is also flawed and contains the seeds of its own destruction.

Adam is the steward tending the Garden of Eden, but after the Fall he is cursed and it becomes his lot to work the earth and toil in the garden. Presumably labor in Eden was not "work." Indeed, the Bible tells us that on the sixth day of the creation, "The Lord God planted a garden in Eden, in the east, and placed there the man whom He had formed. And from the ground the Lord God caused to grow every tree that was pleasing to the sight and good for food" (Genesis 2:9). Remarkably, in this first commentary on creation, "pleasing to the sight" ("good to look at" in many translations) is cited before food, which is necessary for physical sustenance. Some religious scholars note that this pairing—and its order—is significant. Perhaps the aesthetic is as essential a need as food.

Gardens offer a promise of beauty where there is none, hope instead of despair, optimism versus pessimism, and finally, life in the face of death. If the Edenic ideal of paradise is found at one extreme, its opposite is found in the landscape of hell. Perhaps the garden exists on a higher plain, perhaps at its highest, when hell is near.

Setting Apart the Ghettos: The Historical Context

As soon as the Nazis took power in Germany, their persecution of Jews pervaded all aspects of life. Even before they were restricted to living in ghettos, Jews were forbidden to enter and use public open spaces.

The fate of those imprisoned in ghettos is known: a large proportion died from disease and starvation, and those who survived were shipped to extermination camps, where most of them perished. But when the ghettos were first established, the ultimate fate of those imprisoned there was still unknown. Indeed, ghetto inhabitants had no reason to suspect they would not survive: in 1930s Europe there was no precedent for modern, industrialized mass murder, and over the course of their long history, Jews had already endured much persecution. So the early, natural response of ghetto residents was to form organizations to try to maintain as much of the continuity of life and society as was possible under Nazi rule. Among the various organizations that arose in the ghetto were those that would come to create gardens.

While our knowledge of life in the ghettos owes a debt to diaries and memoirs, the material provided by official archivists and historians is particularly valuable. In the Warsaw Ghetto, resident and historian Emanuel Ringelblum organized a staff of dozens of workers to create a ghetto archive, which was named Oneg Shabbat, and included accounts of street traffic, religious life, humor, and information about the school system in the ghetto.[6] When it gradually became clear to residents that they would not survive, Ringelblum wrote, "the keeping of the records became meaningful as a gesture for posterity—a pure historical act," with the hope that "the future would avenge what the present could not prevent."[7] In the Lodz Ghetto, which the Germans renamed Litzmannstadt, a ghetto archives office was created. Oskar Rosenfeld, Oskar Singer, Josef Zelkowicz, and others worked as official ghetto chroniclers. The collection is named for Nachman Zonabend, who worked in the archives and retrieved material he had hidden after the war.

Warsaw was the largest Jewish city in Europe, with more than a quarter of a million Jews, about one-third of the city's population. In October 1940 the Warsaw Ghetto was established. About 1.3 square miles (3.36 km^2) or about 100 square city blocks, the Ghetto, encircled with a wall, became a compressed city of almost half a million people. For most people the wall was the boundary between death and life. It is the extreme opposite of a garden wall that is intended to circumscribe a paradisical and idealized environment. The ghetto wall was an inversion; the other side held at least a potential of life, all that gardens signify. A view out over the wall was a view to hope.

An alternative life existed parallel with the omnipresent horrors. Organizations and other aspects of prewar Jewish society reconstituted themselves in the ghettos, and a new organizational structure also emerged; both old and new organizations struggled against German rule in an effort to ameliorate conditions in the ghettos and help people survive. A long, Eastern European Jewish tradition of community self-help was already in place, with specific organizations able to respond to the most practical and critical necessities of food, clothing, and shelter, as well as the desire for education, arts, and culture.

Not everything, however, was about physical survival. Attention was also given to the continuing survival of culture and social life. YIKOR, The Yiddish Cultural organization, organized scientific and scholarly assemblies as well as literary and artistic events. Mobile libraries were formed, there were numerous coffeehouses, theaters holding performances in Polish and Yiddish, and a newspaper. It was not all high culture. There were brothels, taverns, restaurants, nightclubs, and as Ringelblum notes, sixty-one "night spots in the Warsaw Ghetto."[8] All of these organizations and their activities were directed toward the survival of the community and in resistance to German policies.

Toporol: Creating Gardens in the Ghettos

One organization that continued its prewar mission and activities was the TOPOROL (Towarzystwo Popierania Rolnictwa/The Society to Encourage Agriculture among Jews), which since its founding in 1933 had trained Jewish agricultural workers in Poland. Hehalutz, the umbrella organization of Zionist youth groups, had established hachshara centers (training farms) to prepare their members for making aliya (immigration to Palestine).

Inside the Warsaw and Lodz Ghettos, Toporol assumed responsibility for garden creation. The Ringelblum archive, in a report on Jewish social self-help, includes a "Report on the Activities of Toporol" in the Warsaw Ghetto from December 1, 1940, to May 31, 1941.[9] Our limited knowledge of the Toporol and its activities is supplemented and described in more personal terms in diaries and memoirs. It is important to bear in mind that, however impressive their activities, like virtually all ghetto actions of survival and resistance, while the efforts were valiant, the results were meager. Perhaps that makes Toporol gardens even more inspiring.

Gardens straddled the territory between the practical and the aesthetic. Given the horrific conditions of the ghettos of Eastern Europe, a garden or park might be seen as a luxury, but the desire for some contact with the natural world and a quest for the restoration of some semblance of normalcy represented by the common landscapes of garden and park was critical. A garden was a minimal form of modest recompense. Its creation was one of many mechanisms directed at individual and community survival.

Toporol did everything from acquiring land, seeds, equipment, and funds to mobilizing and training workers. The network created gardens in every imaginable spot of land, even on balconies. They gave away seeds, planted vegetables and trees, and made small parks.

Cabbage field on the site of Skra, the former sports stadium in Warsaw (which later became the site of mass graves), sometime before 1941

This community organization's bylaws included the following goal: "We are to instill in the children an aesthetic appreciation of their surroundings (…) direct their attention to growing plants that might bring them closer to nature and provide them with aesthetic experiences."[10] Toporol, assumed responsibility for its own gardens and also trained workers who created gardens for other public institutions. They reconstructed a glass house and glass-sheltered seed beds.

The planting of vegetables is to be expected, but they also planted flowers! Working with housing committees Toporol planted gardens in over 200 courtyards, the center of outdoor life for most residents. Most courtyards established children's corners as play areas. Halina Birenbaum, who was eleven in 1939, and a survivor of Warsaw and Auschwitz, recalled: "The yard on Muranowska Street was my world of games and daydreams. In the ghetto these yards served children as garden, school, reading room, and playing field."[11] In early 1940, Gustawa Wilner worked with the building committee at her apartment to set up a children's garden she describes as "without flowers or even a single blade of grass."[12] Ironically, this was on Ogrodowa Street, which means Garden Street in Polish.

The attempts to grow food were valiant, but despite the efforts of Toporol, the need far outstripped any ability to supply and could only stave off starvation for short periods but however modest the contribution, the impact was felt nonetheless. Helena Szereszewska remembered that "in the first days of April 1941, the day before the first Passover meal in the ghetto we were sent a bundle of red radishes wrapped up in paper and some leaves of lettuce so young it hadn't formed heads yet. 'It's from Toporol,' we said happily."[13] The red radishes were grown in Toporol's cemetery gardens. At the Seder they ate hard-boiled eggs and radishes dipped in salt water. The Warsaw Ghetto Uprising began on Passover two years later.

Workers in a plant nursery, Glubokoye Ghetto (Belarus), between 1941–1942

Lodz

The conditions in Lodz Ghetto, Europe's second largest, were different. The Marysin area of the Ghetto was semi-rural and became the site of most of the garden creation in Lodz.

A gardening consulting service was established along with a lecture series that dealt with the practicalities of preparing the soil, weed removal, hoeing, and the use of edible weeds. They published the mimeographed notes of the lectures along with professional instructions for growing vegetables. Oscar Singer found that "Every inch of soil has been planted with vegetables."[14] But "that only in the rarest of cases is the ground thoroughly prepared" for cultivation. People who were "reduced to skin and bones" did not have the energy to water in the morning and evening, there were no tools, and a watering can was "almost a luxury item."[15] "The person outside the ghetto will never understand that the kinds of things that a gardener takes for granted, which otherwise have hardly anything to do with money, were almost insoluble in the ghetto." His report continues: "The Ghetto [Lodz], with its approximately 85,000 inhabitants, is probably the only city in the world without any, or almost any, flowers."[16]

In the summer of 1944, weeks before the Nazis would liquidate the Ghetto, starvation was rampant in Lodz and mobs from the city descended on the Marysin gardens. Roman Kent and his fellow gardeners gathered together to protect their crops, but to no avail:

"During the rampage, there were hundreds of people, and whatever they saw growing was hurriedly pulled out of the ground. Within two hours, it was all over. There was no evidence remaining that anything had even been planted."[17] Kent stood sobbing in the denuded field he had worked so arduously. But, ever resourceful, he and his brother returned the next day to harvest the potatoes buried in earth. Kent says simply that the gardens were a "lifeline to life," and essential to his family's survival.[18]

There is a common fascination with gardens created in unexpected, unconventional, or imaginative locations. It is rare, though, that a garden can be the source of levity, but there is a report on what some called the Eighth Wonder of the World, the mobile garden:

> A worker in the tailoring plant who had a small garden last year but lost it because of the new 'agriculture policy' quickly resolved to fill an old baby carriage with soil and plant a few onion bulbs. He now pushes his garden to the workshop every day and parks it in a sunny spot in the courtyard; from the window he can easily guard his 'garden plot'. The mobile dzialka (plot) is quite a conversation piece, and the shrewd farmer has people laughing with him, not at him.[19]

About 204,000 Jews had lived in the Lodz Ghetto. About 10,000 survived.

Blessings of the Earth

The boundaries of the Warsaw Ghetto were deliberately drawn to exclude any parks—the only green area inside the walls was the cemetery. Jan Jagielski, the photo archivist at the Jewish Historical Institute in Warsaw told me that people also had "fantasy gardens." Ghetto residents went to upper floors or onto roofs to look out over the green areas adjacent to the Ghetto: Krasinski Park and the Saski Gardens. Both of these borrowed gardens could also be seen from the bridge connecting the two main sections of the Ghetto.[20] What did the view mean? It could be gratifying, but perhaps it was also a reminder of their earlier lives—and in the Ghetto's early period, even a promise that the residents would survive this horrible experience. In his poem, "A Window on That Side," W. Szlengel writes about having a borrowed view of greenery:

> I have a window on that side
> An impudent Jewish window
> Onto lovely Krasinski park
> With autumn leaves in the rain…
> In the evening, the dark violet
> Branches make a bow
> The Aryan trees look straight into
> That Jewish window of mine.[21]

Nature close at hand was forbidden to ghetto inhabitants. They were taunted by a nearby nature that they could not touch. Only fragments of the natural world were available in the ghetto. There were the teasing glimpses over walls or through barbed wire, the mixed messages of the sun and season, the rare trees, and the few gardens and fields. The sky was seen only in slivers and strips. From the courtyard in Warsaw where she lived, Janina David saw "only a small square of sky" significant enough to become the title of her memoir. Ringelblum found not a single tree in the Warsaw Ghetto, while in Vilna there was one grand tree around which "the youth of the ghetto took their walks."[22]

What did these gardens mean? They were about spiritual resistance, life, hope, work, and beauty. There were also innumerable acts of what has been termed "spiritual resistance." These actions were also translated into the religious precept of Kiddush Hahayyim. In 1940 in Warsaw Rabbi Isaac Nissenbaum said,

> It is time for Kiddush Hahayyim, the sanctification of life, and not Kiddush Hashem, the holiness of martyrdom. In the past the enemies of the Jews sought the soul of the Jew, and so it was proper for the Jew to sanctify the name of God by sacrificing his body in martyrdom, in that manner preserving what the enemy sought to take from him. But now it is the body of the Jew that the oppressor demands. For this reason it is up to the Jew to defend his body, to preserve his life.[23]

Garden creation was only one of many such actions. It could serve as both a psychological protection from harmful forces, but also as an assertive action in an attempt to preserve life and dignity. Eisen says that play served as a "protective cloak—a spiritual shelter from which the wounds of the ghetto would not seem as appalling."[24] To some degree, gardens could make that real.

The slogan of Jewish communal activity in the Warsaw Ghetto was, "Live in dignity and die in dignity!" How does one sustain or even enhance one's dignity and remain fully human? One way is through cultural continuity, honoring and even reasserting the habits and traditions that have served you well. The communal institutions and spirit of the Jewish community were aids to survival, requiring great collective resolution and courage. Stanislaw Adler used a flowering metaphor to find that "in spite of all, the ghetto had produced some remarkable standards of conduct, and the cultural life among the doomed masses has not died, but rather blossomed."[25]

Some forms of resistance were "permitted" and others were "clandestine." Obviously gardening fell into the permitted category, for secret gardens were virtually impossible. A garden could be a place of dignity and hope. Garden work is purposeful action, occuring over time and carrying with it the expectation of tangible results. It is inherently optimistic, done in the spirit of hope, anticipating a harvest in which you will be able to reap the fruits of your labor.

Among the papers in the Zonabend archive is "Blessings of the Earth," probably written by Josef Zelkowicz, one of the chroniclers of life in Lodz. It is a remarkable document that takes the point of view that life in the ghetto taught Jews things both good and bad. Gardens are definitely on the positive side of the ledger. Zelkowicz's reasoning goes beyond the gardens as a source of food. Zelkowicz describes Jews as historically alienated from nature and even willfully ignoring its beauties, a people who have extolled a "head" life at the expense of a "hand" life. He wrote that "Those who had never seen a plant in their lives are today gardeners." The ghetto inhabitants gardened with the zeal of converts. "Taking a stroll through the ghetto fields at 5 p.m., when the work in the factories is over and the population is already in the fields, your impression is that the population of the ghetto has lost its face and is nothing more than backsides, sticking out from amid green leaves and thick stalks."[26]

Zelkowicz then constructs a fascinating and almost humorous dialogue between "head" life and "hand" life. He says that traditionally peasants never bother to ask why, since they work based on their experience and tradition, but these neophyte gardeners need to know why, and ask incessant questions:

> Why is it better to water cucumbers when it is sunny and between 10:00 and 3:00 in the afternoon?
> Why does one have to tie garlic shoots?
> Why have you placed stakes near your tomatoes?[27]

Of the new Jewish gardeners he speculates:

> Maybe he demonstrates with his bowed posture that field work does not require any head, but only healthy hands and strong legs, but finally, he can't shed his own skin that easily. In the end, the character of a people changes only through evolution—The Jew who has always lived a head-life can't suddenly overnight go over entirely to a 'hand' or 'leg' life.

Even when it comes to farmwork, he needs his head; he needs to understand to work; he must know the why.[28]

Nature is alive and as human beings we identify with its vitality. The desire to connect with something living is profound. Genia Silkes, in the Warsaw Ghetto, told of a young girl whose sister was dying and just wanted to hold something green. Risking her life she snuck under the wall to the Aryan side and fetched a single leaf. She placed the leaf in a glass by her sister's bed, so she could see that bit of green before she died.[29]

When Warsaw native George Topas was fourteen, he entered agricultural school near Wlodzimierz in preparation for emigration to Palestine. Three years later he became a Toporol instructor, for which he was paid a loaf of bread a week. He supervised a group of teenage students in clearing a site on Gesia Street and was amazed, given the level of their starvation, at the success of their efforts. Although pleased when the site began to produce, Topas thought that the harvest "was symbolic more than real." At the cemetery, where crops were being raised, he marked the "incredible contrast" of students nurturing tomato plants to "preserve life" and corpses being "dumped into mass graves."[30]

Hope

Gardens grow, they take time to conceive, start and develop. Since time immemorial cultures have wondered at the biological miracle of a seed, which, planted in the ground emerges out of the darkness of the earth into the light and life. Hope is embodied in this transformation. Hope is not naïve optimism. Hope takes persistence, resourcefulness, and courage.

On December 8, 2003, in the YIVO library in New York, I sat listening to Esther Mishkin, a short woman with a brilliant smile. She vividly recalled events of sixty years earlier. Her father, Reuben Yitchak, a rabbi and shochet, planted a garden at 16 Parvenų Street in the Kovno Ghetto. Yitchak planted cucumbers, tomatoes, and potatoes. His daughter could not recall where he got seeds for the garden. "Did you work in the garden?" I asked. "Sometimes," she said, but added that her father "adopted the whole thing." She recalls observing her father watching the garden grow. She remembers "him sitting in the garden, [with] his vegetables; it became part of him." The output was not substantial—he grew only food for their table, but it meant much more. She said it was important that "something was growing there." About her father, she said that watching the garden grow "gives him a feeling that something is growing, that we can survive somehow."[31] His wish was fulfilled only for his daughter: the remainder of her family died in the Holocaust.

Halina Birenbaum was a teenager when she was sent from the Warsaw Ghetto to the Majdanek concentration camp, where she worked as a slave laborer. Even there the power of contrast and the imagery of a garden contributed to a sense of hope.

> Sometimes we worked in fields only a stone's throw from a village hut – armed SS sentries cut us off from freedom. Then I used to look with envy and longing at children playing freely in the peasant's gardens, at the hens pecking the dust, the people bustling about (...). I just could not comprehend that there still existed another world in which people were allowed to move around open spaces not cordoned off with barbed wire and in which children played! (...) The existence of life outside the camp, the sight of the bright sky and green fields alleviated the tragedy of the camp (...). Here in the open fields, closer to people's houses, it was easier to hope.[32]

In landscape terms, hope has often been embodied in ideas of the pastoral. The pastoral in its literary and garden forms is a retreat into the past, into nostalgia and often into the country, physically or symbolically. There is a danger of the garden becoming a trite palliative, a mask, and that working in the garden reflects an abdication of involvement and responsibility. At its extreme, the site of the garden seems a place of cowardice. One of the paradoxes of the pastoral is that it is not always based on an urge to get away from something. The garden may represent a desire to have a different kind of involvement. The pastoral can be conceived of as a movement toward an ideal. The peace that the garden may offer is not merely the absence of conflict found in quiet and tranquility but is a positive assertive state. The garden is not only a calm retreat and welcome respite, but also the assertion of a proposed condition; a model of a way of being.

The pastoral vision always implies a contrast. The power of the imagery and perhaps the depths of its meaning derive in part from the dimensions of the contrast. At the Majdanek extermination camp, Halina Birenbaum was assigned to weeding the grass between the two rows of electrified barbed wire that divided the women's and the men's camps. Just touching the wire meant death, and some hurled themselves onto the wire in order to end their suffering. As the weeders worked between the wires, the German guards left them alone. In this gap Birenbaum found that she "could sit and rest, picking at the weeds and grass (...). I preferred this work to any other. Here I had the peace I longed for."[33] This is a horrifying pastoral vision of a bizarre respite, weeding green grass hemmed in by a deadly electrical current and the prospect of death.

Hope is not just an emotion but also a force. Janusz Korczak, the director of the orphanage in the Warsaw ghetto, called hope the "crutch of illusion essential to the survival of human dignity."[34] On the one hand illusion could be a necessary support; on the other hand it could be a willful deception that functioned as a defense mechanism. That the hope was baseless for all but a few survivors does not diminish the force of the emotion or desire, and what it may have meant to people in the moment. Korczak's final diary entry of August 4, 1942, reads: "I watered the flowers, the poor orphanage plants, the plants of the Jewish orphanage. The parched soil breathed with relief."[35] Korczak accompanied his charges into the gas chambers.

There is a distinction between the garden as a refuge and as a respite. The garden as refuge suggests that it can function as a sanctuary, a place protected from outside forces. A respite is only a temporary material or psychological sanctuary, for one will soon be at the mercy of external conditions that lay siege to one's shelter. The ghetto gardens offered places of sensory difference, of quiet, shelter, and elements of the natural world. It offered opportunities for calm, a change of mood, even a temporary forgetfulness about one's conditions. In the ghetto, gardens were only brief respites, but that does not lessen their significance for those moments. Gardens conformed to the expected cycle of seasons and growth and life; a garden was a demonstration of life as ordered, not a world turned upside down.

Work and Beauty

Garden is a verb as well as a noun, an activity as well as a place. Gardening takes mental as well as physical effort. The work has satisfaction and rewards attuned to the processes of garden creation: the conception, marshaling and gathering of materials, propagating, planting, crafting, monitoring, maintaining, harvesting, and the whole panoply of garden uses, from aesthetic satisfaction to a setting for leisure. Work can be productive, but it is also a way to keep one's mind and body occupied, fostering a sense of dignity, self respect, identity and relief from boredom and restlessness—it can be something "to do." For the depressed, it can be a "natural" antidepressant.

Roman Kent wrote about the arduous and time-consuming labor in the gardens in the Lodz Ghetto, and the gift this work offered residents:

> Slowly but surely, we learned how to toil and utilize every inch of the land more and more efficiently (…). To my surprise, in spite of the heavy load I had to carry, I became very fond of working in the field. Even though it was only temporary relief, for the time that I was there I could forget about the daily problems of ghetto life we were forced to endure and our unfortunate situation in general.[36]

Gardens are beautiful. Gardens counterbalance the ugliness of war, but clearly the weight is not equal. The destructive power of war and its wanton killing, death, and destruction can only be temporarily and modestly restrained. The garden's calm and security are ultimately only a respite from war's terror. But the garden can function as a reminder of a peaceful prewar period and make it possible to look ahead to a longed-for postwar time, when domesticity and simple pleasures will prevail and the "ugliness of war" will be replaced by the beauty of creation.

Stafania Ross, wife of the photographer Henryk Ross, in a flower garden of the Lodz Ghetto, sometime between 1941–1944

Even in the most extreme situations, people not only see beauty; they seek it and will create art that celebrates beauty and life. In writing about Theresienstadt and its artists, Milan Kundera asks, "What was art for them? A way of claiming the full array of emotions, of ideas, of sensations, so that life should not be reduced to a single dimension of horror."[37] Oskar Singer lauded the work of Lodz Ghetto gardeners. "If yields should be recorded under these circumstances," he said, "then one must testify (or certify or recognize) that the small farmers in the ghetto are great poets."[38]

In Warsaw, Mary Berg wrote what is a common garden sentiment, but its meaning was poignantly dramatized: "The greens and the sun remind us of the beauty of nature that we are forbidden to enjoy. A little garden like ours is therefore very dear to us. The spring this year is extraordinary. A lilac bush under our window is in full bloom."[39]

Gerda Klein was eighteen in 1942, and for the next three years she would be shuffled from work camp to work camp. Her last camp was Grünberg, a place Klein describes as: "cruelty set against a background of beauty. The gentle vineyard-covered hills silhouetted against the sapphire sky seemed to mock us." Each morning, two thousand girls were marched to the factory. In the factory courtyard were star-shaped flower beds, exquisite rose gardens, and beds of tulips, each lending a touch of "beauty against the grim reality." Klein craved something beautiful. "Day after day," she says, she "had to resist the desire to run out of line and touch those beautiful blossoms." One morning on their daily march, a crocus had broken through the concrete and was sprouting. The mass of girls parted, and silently, "hundreds of feet shuffled around it" to avoid trampling the flower.[40]

In Search of the Warsaw Ghetto

What is heroic? Actions that transcend the normal range, that confound our expectations, that take courage, when people overcome danger and fear. These were heroic gardens and gardeners. These gardens astonish us by their presence and their persistence. They seem out of place, yet paradoxically, upon careful examination these gardens reveal themselves to be supremely adapted to the specifics of their condition.

The gardens speak to us. Many of us have a friend or relative whose company we appreciate or find congenial, who then responds with unexpected aid or depth of compassion in a crisis. Gardens are like that: comfortable companions whose capacities remain dormant, awaiting a crisis to release their potential.

These ghetto gardens were all short-lived, but their brief duration is not at all proportional to their meaning. In fact, their brief life spans may accentuate their significance. That gardens rarely outlive their makers (at least in their original form) is just one of the reasons change is at the essence of garden meaning and experience.

Gardens require tending, and without that attention they begin their slow return from a place of culture back to nature, whose rules, cycles, and imperatives then become paramount. Sometimes the restoration of a natural state is rapid, with a single season's growth obscuring the marks of people on the land. War's effect on humans also varied in the duration of its impact: for some individuals the wartime events were forgotten or repressed, never to emerge consciously; for others, thoughts of their time of distress and horror never left them, and they remained traumatized for a lifetime, the consequences even affecting their children. The second generation of Holocaust survivors also suffers in ways different from their parents.

What is left of the Ghetto today? It was essential for me to visit the sites of the making of these gardens. I felt that I could not understand their meaning or their creation without being in the same place, looking at the same landscape, touching the same material.

The land remembers, but the return of nature after war blurred and eventually obliterated the death and destruction that occurred there. Wars leave ghost marks on the land. The actual experience of the land, its fields and hills, its dimension, and the resonance of its spirit can bring visitors closer to a realization of what events were like in this place. This happened "here" is an extraordinarily powerful emotional catalyst.

Together with my wife, I visited the Warsaw Ghetto in the spring of 2004. I was born three years after the Warsaw Ghetto uprising was quelled and the Ghetto razed. I knew the history and the site from reading, maps, and photographs. I knew that nothing remained, but I still looked, in vain for something, a fragment, a remnant. It takes an act of imagination to replace the scenes of blocks of modern housing and new streets with the crowded streets and courtyards of the Ghetto, the few people who now walk the streets with the crowds who, even in winter once filled the streets. I know that I am stepping on ground where thousands died, that the sidewalk I am walking on is where people stepped over starving children and around dead bodies. I know that here soldiers rounded up families. I know that death was rampant. I know that the sewers underneath the street were passages for smugglers and escape routes for the few. I know that the full spectrum of human emotions happened right here: unthinkable horror, acts of unimaginable courage, the loss of all faith, as well as moments of divine presence. I know that few survived and that fate was random. I know that on this ground a few people planted seeds to grow food, to foster hope, to keep busy, and to make something beautiful.

I picked up a stone in the Warsaw Ghetto. I don't know if it is a piece of rubble from the Ghetto that was obliterated in the spring of 1943 or of more recent vintage, but it is of that place. I now place it on our Seder plate at Passover every year as a reminder of the Warsaw Ghetto uprising that began on the first night of Passover in 1943, and of our visit in 2004.

Notes

1   The material in "Ghetto Gardens" is adapted from Kenneth Helphand, *Defiant Gardens: Making Gardens in Wartime* (San Antonio: Trinity University Press, 2006).

2   Sheva Glas-Wiener, *Children of the Ghetto* (Fitzroy, Australia: Globe Press, 1983 [1974 in Yiddish]), 160–70.

3   Ibid.

4   In my book *Defiant Gardens* I have also explored gardens created by soldiers of World War I on the Western front, by Allied POW in both World Wars, and by detainees in the Japanese American Internment Camps between 1942–1945.

5   George Eisen, *Children and Play in the Holocaust: Games among the Shadows* (Amherst: University of Massachusetts Press, 1988), 8.

6   On Emanuel Ringelblum and his important role for the Polish-Jewish Landkentenish movement in the Interwar period, see Samuel Kassow's article in this volume.

7   Emanuel Ringelblum, *Notes from the Warsaw Ghetto: The Journal of Emmanuel Ringelblum*, ed. Jacob Sloan (1958; repr., New York: Schocken, 1974), xvi.

8   Ibid., 146.
9   Emanuel Ringelblum Collection, ARI/ PH/ 5-3-3 in Joseph Kermish, ed., *To Live with Honor and Die with Honor!*, trans. M.Z. Prives et al. (Jerusalem: Yad Vashem, 1986), 519–29.
10 Eisen, *Children and Play in the Holocaust*, 39.
11 Halina Birenbaum, *Hope Is the Last to Die: A Coming of Age Under Nazi Terror* (1967; repr., Armonk, New York: M. E. Sharpe, 1996), 11.
12 Gustawa Wilner in Michal Grynberg, ed., *Words to Outlive Us: Eyewitness Accounts from the Warsaw Ghetto* (New York: Henry Holt, 2002), 472.
13 Helena Szereszewska, *Memoirs From Occupied Warsaw, 1940–1945* (Portland, Oregon: Valentine Mitchell, 1997), 51.
14 Oscar Singer, "Something about Horticulture in the Ghetto," May, 30, 1942, trans. Roberta Newman, Ghetto Fighter's House Archive (unpaginated).
15 Ibid.
16 Ibid.
17 Roman Kent, interview with the author, April 22, 2004.
18 Ibid.
19 Lucjan Dobroszycki, ed., *The Chronicle of the Lodz Ghetto, 1941–1944* (New Haven: Yale University Press, 1984), 491.
20 Jan Jagielski, interview by the author, March 16, 2004.
21 See Barbara Engelking, *Holocaust and Memory: The Experience of the Holocaust and Its Consequences: An Investigation Based on Personal Narratives*, ed. Gunnar S. Paulsson (New York: Leicester Univ. Press, 2001), 42.
22 Ringelblum, *Notes from the Warsaw Ghetto*, 259.
23 Joseph Rudavsky, *To Live with Hope, to Die with Dignity* (Lanham, MD.: University Press of America, 1987), 5.
24 Eisen, *Children and Play in the Holocaust*, 42.
25 Stanislaw Adler, *In the Warsaw Ghetto 1940–43: An Account of a Witness; The Memoirs of Stanislaw Adler* (Jerusalem: Yad Vashem, 1982), 255; see also Rudavsky, *To Live with Hope*, 60; Chaim A. Kaplan, *Scroll of Agony: The Warsaw Diary of Chaim A. Kaplan, ed. Abraham I. Katsh* (New York: Macmillan, 1965), 244.
26 Josef Zelkowicz, "Blessings of the Earth," Zonabend Collection, YIVO RG 896.
27 Ibid.
28 Ibid.
29 Eisen, *Children and Play in the Holocaust*, 83.
30 George Topas, *The Iron Furnace: A Holocaust Survivor's Story* (Lexington: University Press of Kentucky, 1990) 50.
31 Esther Mishkin, interview with the author, December 8, 2003.
32 Birenbaum, *Hope Is the Last to Die*, 88.
33 Ibid.
34 Janusz Korczak, *Ghetto Diary* (1978; repr., New Haven Yale University Press, 2003), 34.
35 Ibid., 113.
36 Roman Kent, "Ghetto Years," unpublished manuscript, courtesy of the author.
37 Milan Kundera, "Such Was Their Wager," *New Yorker*, May 10, 1999, 94–6; here 95.
38 Singer, "Something about Horticulture in the Ghetto."
39 *Warsaw Ghetto: A Diary by Mary Berg*, ed. S.L. Shneiderman (New York: L.B. Fischer, 1945), 144.
40 Gerda Klein, interview with the author, July 12, 2004. Gerda Weissman Klein, *All But My Life* (New York: Hill & Wang, 1957), 167, 171.

5 The Mellah of Fez

Reflections on the Spatial Turn in Moroccan Jewish History

Susan Gilson Miller

28. - Le Mellah, vue prise du sommet du ravin Oued Zitoun

View of the Fez mellah from the South (historical postcard), c. 1912

Al-Hassan al-Wazzan, or Leo Africanus, as he is known in the West, was an intrepid Muslim merchant of Granada who traveled widely throughout Northern and sub-Saharan Africa in the early years of the sixteenth century. Around 1520, on one of his far-ranging circuits, he was captured by pirates off the coast of Tunis, and delivered to Pope Leo X as a valuable prize. Baptized and catechized, he spent eight years in Roman captivity, during which time he wrote a detailed account of his African journeys. His book became a sixteenth century best seller, and was soon translated into several European languages, including the decorous English of the Pory edition published in London in 1600.[1]

One of the chapters in his *History and Description of Africa* is about Fâz, at that time commercial and intellectual capital of the Morocco. A cosmopolite who was not easily impressed, Leo Africanus was nevertheless struck by the beauty of Fez, and especially that part of the city where the Sultan's palace was located, known as Fâs al-Jadîd, or New Fez. He noted that Fâs al-Jadîd

> ...is now inhabited by Jews, [who] first dwelt in Old Fez, but upon the death of a certain king Abusabid [Abu Sa`id Uthman III, ruled 800–23/ 1398–1421] caused them to remove to new Fez, and by that means doubled their yearly tribute. They therefore even till this day do occupy a long street in the said new city, wherein they have their shops and synagogues, and their number is marvelously increased ever since they were driven out of Spain.[2]

Leo Africanus was clearly amazed by the prosperity of the Jewish quarter and its proximity to the seat of royal power. His testimony reaffirms that in the early sixteenth century, the city of Fez was home to a growing Jewish population, many of whom were *expulsados* from Iberia who found there a safe haven from the terrors of a renascent Christendom.

The Jewish Quarter in Moroccan History

Leo's text is the departure point for a discussion of space as a conceptual framework for understanding Jewish-Muslim relations in Morocco. The Jewish quarter that Leo Africanus so admired was not only exceptional for its beauty, but also because it stood as a prototype for Jewish quarters elsewhere in Morocco. These quarters, or *mellah-s*, as they are known in Arabic, followed the Fez model closely; they were usually separated from the rest of the city by gates and surrounding walls, and they were usually located close to the royal compound. Although Jews had been a component of the Moroccan social fabric for millennia—no one knows precisely when they arrived, but some have argued that they first came with the Phoenicians — until the fifteenth century Jews were never compelled to live apart. At that time, and for reasons that are not precisely clear, the *mellah* of Fez was created on the site of a salty marsh; hence the toponym *mellah*, derived from the Arabic root meaning "salt."[3] Subsequent *mellah-s* were built in other Moroccan towns, and although the founding circumstances and positioning of each Jewish quarter differed from place to place, curiously, the name *mellah* was applied to all. These enclaves were created by order of the ruling sultan; once built, Jews were compelled to live in them. Some moved reluctantly, while others went more willingly, believing that the high walls and the nearby palace guard offered some guarantee of protection.[4]

Over time, the *mellah* came to represent the complex reality of Jewish minority existence in Morocco. Each enclave differed in its details, but the common thread running through them all was their quality of encompassing the totality of elements needed for Jewish life in the Muslim city—houses, schools, workshops, and ritual space. But the *mellah* was more than living room; it was also the setting for enacting the social practices, ceremonial performances, desires and memories that inscribed a specific Jewish identity onto a Muslim context. In this sense, the *mellah* was less a "container" for Jewish life, and more a "stage" on which historical processes relating to both the inner and outer worlds of Jewish experience were worked out. In pre-modern Morocco, *mellah* and Jew were co-terminal; in the popular imagination, the *mellah* was synonymous with Jews, Jewish life, and Jewish history.[5]

Despite these patently positive attributes, history has treated the concept of the *mellah* unevenly. In the intervening years between Leo Africanus' warm appraisal and the mid-twentieth century, the *mellah* somehow acquired a dark and pernicious reputation. It lost its aura of being protected space, and became, in the minds of many, a place of suffering, misfortune, and degradation. Europeans are partly to blame; visitors who came to Morocco noticed the excessive overcrowding, the strange costumes and customs of the Jews, the streets strewn with refuse, and were revolted by them. Vicious anti-Semitic stereotypes in European travel literature cemented the evil impression. The description by popular French travel writer Pierre Loti, who visited the Fez *mellah* in 1889, is not unusual:

> In front of the [Jewish] quarter is the public dumping ground for dead animals, a courtesy prepared especially for them; to get there, you have to pass among dead horses, dead dogs, various carcasses rotting in the sun, emitting an indescribable odor, for [the Jews] have no right to remove them—a great concert of jackals takes place each night under their walls—in the narrow streets, too narrow to pass, they do not have the right to remove the rubbish they throw from the houses; over time, the bones, the peels, and other refuse pile up, until some Arab official takes it upon himself to have it all swept away in return for a large sum of money—in this dark and fetid quarter, there is a peculiar moldy stench, and the faces of the inhabitants have a sickly pallor...[6]

The disjuncture between these two sets of images—the positive one of Leo

Africanus, and the repulsive one of Pierre Loti—poses a challenge to the historian. Where does the truth reside? Was the *mellah* a place of Jewish sanctuary, a safe harbor in a Muslim sea, or was it a dark and morbid prison? More to the point, to what extent can Jewish space become a metaphor for understanding the tenor of Muslim-Jewish relations? It is important to keep in mind that the *mellah* was not only a real place; over time, it was transformed into a compressed metonymic device for capturing the fullest meaning of the minority condition. If one assumes that the central problem of diasporic Jewish life was that of assuring communal survival, then we may ask the further question: How does a passionate identification with place sustain the collective Jewish consciousness and fortify it for a life of exile?

Here we shall examine some of these questions by looking at the case of Fez and its Jewish quarter as an exemplar of the *mellah* phenomenon. However, prior to that, we should consider three issues regarding historiography and the difficulty of working in the charged environment that surrounds this topic.

The first issue is that for historians of Morocco, whether they are Jewish or Muslim, the very word *mellah* is imbued with all the tension and emotion that surround questions of Muslim-Jewish relations more generally. The *mellah* was not a neutral place that just happened to be inhabited by Jews; rather, *mellah* space was inflected with the complex qualities of *dhimmitude* [i.e. the condition of religious and social subordination] that defined Jews in the Muslim sphere. In an earlier generation, studies of this topic carried out by non-Jewish scholars in the West were based largely on the legal literature that dealt with laws and regulations that demoted Jews to positions of inferiority. For these historians, the notion of structural segregation could hardly be attacked, since it was part of sacred law.

But for Jewish historians, the problem was quite different; for them, the separate Jewish quarter was a place of categorical exclusion, where Jews lived apart due to attitudes of religious hatred. For the Jewish historians, the theme of persecution was foregrounded, good times were assigned to brief "golden" moments, and bad times were the norm, especially in a place like Morocco, where the strict Maliki legal tradition (*madhhab*) was enforced.[7] In this historiography of oppression, the underlying leitmotiv of inter-communal relations was conflict and violence, and the question of minority space as a socially constructed phenomenon had no place.[8] Nor was the *mellah* particularly interesting in itself; rather, it was simply the backdrop to a history of loss. Here we confront a longstanding problem in Jewish historiography, in which suffering produced an attitude of exceptionalism, allowing historians to write Jewish histories that were distinct from those of the non-Jewish environments that surrounded them. Closer to our own time, historian Mark Cohen aptly labeled this tendency as "the neo-lachrymose conception of Jewish-Arab history," and gave examples of the genre.[9] That is to say, the biases of Jewish historians meshed with stereotypes present within the popular culture, each reinforcing the other in a dark vision of negativity.

A second point is that traditionally, historians have often located Moroccan Jews on the periphery of power, by depicting them as weak, subordinate, and contingent—as social marginals with little to contribute to the wider society. But more recent research shows this was absolutely not the case. In fact, both individual Jews and Jews as communities were regarded as an asset, a reservoir of skills, capital, and expertise. In Morocco, Jews constituted a separate corporate entity, similar to other corporate groups in society, such as the religious nobility [shurafâ] the brotherhoods [turûq], and the learned class [ulamâ]; like them, Jews played a key role in enhancing and extending the power of the state.[10] At different times and places, the Sultan mobilized Jews as auxiliaries to his rule, just as he mobilized other foundational groups. Jews in turn manipulated and negotiated their relationships with authority to win advantages and concessions. Indeed, our ideas about the relationship between Jews and the Moroccan state must be recast to introduce the notion of Jewish agency and even Jewish power. Pre-modern Moroccan Jewry are not, of course, unique in this respect; David Biale has made the same point about pre-modern Jewish communities in Europe, stressing the capacity of rabbis and other communal leaders to respond to crisis by organizing for collective action.[11]

Yet the forms in which hidden sources of power are revealed are not always clear, and may be buried in innuendo, or occulted within the micro-event, only to be made transparent by "reading between the lines." In order to reposition individual histories within larger and more interconnected themes, we must rely on our interpretive creativity. It is here that the spatial turn is most helpful, for as we shall see, a reading of space clarifies the sometimes partial and even mysterious intimations of the text, driving home points that might otherwise easily be missed. The methodologies of histoire croisée employed by contemporary German and French historians to correct the deficiencies of Euro-centric readings of the past, the thrust of the subaltern critique emanating from the history of the Indian subcontinent, the rejection of hyper-nationalism and the production of its "grand narratives," are all critical approaches that can contribute to the making of more balanced Jewish histories.[12]

The third and final point is the need to understand the specificity of the Moroccan Jewish experience and how it differed from Jewish life in Europe. In pre-modern Europe, the long-term position of Jews was always a contingent one, dependent on changing political, economic and ideological factors; Jews might be tolerated in one set of conditions, but considered completely noxious in others. The inevitable periodic expulsions and banishments that European Jews suffered were aimed at *uprooting and even wiping out* the Jewish presence; this was hardly ever the case in Morocco, where Moroccan Jews (who also lived lives fraught with contingency) were considered a perennial and structural component of society. The permanent and durable nature of the Jewish presence in Morocco is the fixed background to our discussion of *mellah*.

The Geography of the *Mellah* of Fez

Using the example of Fez as a case study that will make some of these themes more specific, we begin with a map. The *mellah* of Fez is one of the four cities of Fez, an urban conglomerate that is actually a string of settlements. The oldest part of Fez, called Fâs al-Bâli, or Old Fez, dates from the ninth century; it is the commercial and religious center. In the thirteenth century a new town was built alongside Old Fez to house the sultan and his court; it was called New Fez, or Fâs al-Jadîd. The two together form a dense concentration of architectural splendor designated by UNESCO as a World Heritage site in 1980. A third element in this string of settlements is the *mellah*. It sprang up in the shadow of Fâs al-Jadîd in the fifteenth century.[13] Finally, the fourth city of Fez is the New Town, or *ville nouvelle*, built by the French after 1912 at some distance from the Old City.

The *mellah* covers an area of about five hectares, or about twelve and a half acres. It is completely walled, with a gate at both ends. A main commercial street runs on an east-west axis, and branching off it are side streets that contain the residential quarters. The morphology of the quarter has changed only slightly over time. If you visit the Fez *mellah* today, you will easily detect its original form. In its heyday, at the end of the nineteenth century, it was home to about 6,000 Jews, while the Muslim population of New and Old Fez together was about ten times that; a ratio of one Jew for every ten Muslims, keeping in mind the steep fluctuations caused by natural and man-made disasters.[14]

The Social Construction of *Mellah* Space: The Chronicle "Divre Ha-yamim"

The Jewish sources give the most complete picture of what life was like inside the pre-modern Fez *mellah*, and how it was conceived of in the minds of its inhabitants. The most important source for this history is a chronicle entitled *Divre Ha-yamim*, a history of Jewish Fez compiled over the course of three centuries by scholar-rabbis of the Ibn Danan family. The first texts in the chronicle date from the fifteenth century, and later texts bring the story up to the late nineteenth century. Written in Hebrew and the local dialect of Judeo-Arabic, the entire work was translated to French, edited and chronologically rearranged by the French scholar Georges Vajda in 1948. The Diaspora Research Institute at Tel Aviv University published a Hebrew version in 1993, and recently, an Arabic version of the text was edited by the Moroccan scholar `Abd al-`Aziz Shahbar. I have used all three texts in my analysis, reading them side by side in the case of problems or obscurities.[15]

What are the main concerns of the writers of the chronicle, what events do they record, how do they construct an idea of the community? How is the *mellah* rendered up as material and conceptual space?

Vajda, the French translator, says that the chronicle relates to "the humble realities of everyday life" but in fact, its concerns are much wider.[16] Much of it deals with calamities and misfortunes that befell the community. These misfortunes varied from light to heavy, from short-term to extended, from individual to collective. They took the form of acts of violence, exactions, epidemics, droughts, and famines. Their number and variety was such that this work could easily be read as a chronicle of woe—another example of that "lachrymose" genre. The various writers who record these events seem to have regarded calamity as collective punishment for unnamed sins; not only did bad things happen, but they could happen at any time. Remarkably, the omnipresent threat of calamity did not propagate a defeatist attitude. Close attention to the text shifts our understanding away from reading it as a self-referential lament, to understanding it as a collective history in which actors saw themselves as instruments of a divine intervention in human affairs. Our analysis begins by identifying the themes around which the various entries in the chronicle aggregate. First among them, in terms of both quantity and sheer dramatic effect, are acts of violence against the Jews of Fez.

Who were the perpetrators of this violence? Different categories of perpetrators emerge, depending upon their physical proximity to the *mellah*: 1) neighbors of the *mellah* living in the nearby palace quarter or Fâs al-Jadîd, 2) the people of the more distant city of Old Fez 3) tribesmen from the surrounding countryside, 4) rebellious tribesmen and soldiers, and finally, the ruler himself and the people around him. Was it random violence, occurring because the authorities turned a blind eye, or deliberate violence, encouraged and abetted by the state? By far, the acts of violence perpetrated against the Jews as reported in *Divre Ha-yamim* were the work of tribesman on the rampage or rebellious soldiers who took advantage of a failing state, a period of chaos, or a moment of internecine warfare to attack the Jews. Rare, indeed, were moments of state-sponsored violence. Most acts of aggression had as their purpose indiscriminate pillaging; a chance to get rich. Wealth seemed to be inextricably linked in the minds of Muslims with the Jewish population. The amount of goods robbed, commandeered, and simply removed from the *mellah* over the years is simply staggering. The stolen wealth consisted of precious metals, clothing, weapons, spices, silk, emergency stocks of food, jewelry, and cash; the bleeding of the *mellah* of its accumulated riches over the centuries was unremitting.[17]

The built environment of the *mellah* figures prominently in these tales of pillage. The rage of interlopers to find hidden treasure was such that they often tore open walls, dug trenches under people's houses, and "found treasure belonging to ancestors" that had been long forgotten.[18] The *mellah* was viewed from the outside as an overflowing but poorly guarded treasure house, a ripe plum hanging tenuously from the branch of a tree, just waiting to be plucked. In a society in which wealth was unevenly distributed, where capital was mainly held in the form of goods, where survival often depended on access to stockpiled food, the *mellah* was simply too tempting a target to be ignored.

The attitude of Muslim authorities that Jews were a bottomless source of wealth is also evident in terms of the various taxes, gifts, supernumerary payments, and extortions imposed on the Jews, a second important theme highlighted in the chronicle. How did Jews react to this ceaseless pressure to yield up their cumulative wealth? In the chronicle, the legal poll tax (*jizya*) is only mentioned twice, and in a manner that indicates it was grudgingly accepted as part of the human condition. The payments that caused real anxiety were those imposed irregularly, at the whim of the sultan or the local governor. They were a major source of distress because the community had to organize itself on each occasion to raise the money. These moments were occasions for the leaders of the community to exercise their authority by managing the distribution of the burden. They were also occasions for elaborating the social hierarchy, giving high status to those most capable of meeting the sudden demand, and low status to those who were not. In other words, this sort of crisis was an instrument for sorting out the collectivity by etching onto the social fabric the relative position of each member of it, depending on his capacity to pay. Ironically, through this system of periodic extraction, it was the outer society, the Muslim sphere, that determined the contours of the inner, Jewish one.[19]

A third important theme covered in the chronicle is the torments of nature such as droughts, famines, disease, and earthquakes visited on the community. The people of Fez, Jews included, lived in a precarious ecological environment, not only because of the constant warring and competition for resources, but also because nature itself often conspired in making life difficult. For example, the danger of famine was ever-present. In the period 1603 to 1606, food disappeared altogether and "people hunted through the garbage to find a morsel to eat."[20] Food security was an ongoing concern, starvation an ever-present threat, and the Jews of Fez were proactive as a community in meeting it. They stockpiled food, and the *mellah* house was specifically designed to accommodate the need for food storage. In 1680, the chronicler proudly recorded that "these days, Jews are hardly bothered by famine; because there are enough rich people in the *mellah* whose houses are stockpiled with all sorts of food."[21] Research into the structure of the *mellah* house conducted during field work in Fez in 1999 revealed commodious spaces built into the underground level of the more wealthy Jewish houses. These spaces were used to store food, and they were absent in Muslim houses of a similar typology.

Plague and other diseases were frequent visitors to the Jews of Fez, causing people to flee, while earthquakes reshaped the urban fabric by causing houses to fall like stacks of cards.[22] The chronicle suggests that the *mellah* was built and rebuilt repeatedly over the centuries because of damage from earthquakes. Moreover, natural disasters tested the moral fiber of the community. During the famine of 1603–1606, more than 2000 Jews converted to Islam and another 3000 died of hunger. Some victims committed suicide by drowning or cutting their own throats.[23] These same catastrophes were visited on their Muslim neighbors as well, but the manner in which they are discussed in the chronicle offers subtle insight into the tenor of inter-communal relations. Rather than creating empathy between the two groups, this sort of calamity placed them in contention, for chroniclers would measure the severity of a crisis in terms of the Jewish capacity for survival. For example, the chronicler Shaul Ben David Serero proudly asserts that while many Muslims perished in the great famine of 1613, not a single Jew died, a statement that resonates with the timbre of the inherent competition that prevailed between the two groups.[24]

A fourth topic that recurs often in the chronicle is public performances that took place inside the *mellah* meant to reinforce communal solidarity. In this sphere, the rabbis reigned supreme. The entry for the years 1583/84 describes a six-month long drought so severe that the rabbis called for days of fasting. Prayers were held in the synagogue, all business came to a halt, and the inhabitants of the *mellah* poured out into the streets in a show of public lamentation.[25] On another occasion, the chronicle describes a kind of communal confession, in which individual Jews publicly asked for pardon from one another for sins committed as long as thirty years before. This ritual of mass penitence and reconciliation, not unlike the offices of Yom Kippur, took place in the synagogue under the guidance of the rabbis; its purpose was to inspire a sense of collective restitution following a time of inner divisiveness and turmoil.[26]

Rituals of exclusion occasioned another kind of public performance. Moral breaches were punished by a group ceremony involving the whole community, where the theme was the solemn abasement of the wrongdoer and the purification of the body social through his separation and banishment.

Ironically, the misery of the guilty one became an occasion for renewal for the rest of the community. The case of Gedeon Kohen is instructive: In the year 1616, Kohen, a member of the priestly class, committed adultery with his "cursed girlfriend (*arura hevrato*), the daughter of Moshe Ibn Azulay." Rabbi Samuel Ibn Danan publicly pronounced the ban of *herem*, or excommunication, on Kohen for his moral depravity, "in the streets of the *mellah*, on the tombs of the saints, and in the synagogue, to the sounds of the shofar" in a clear demonstration of rabbinic power.[27] The rabbinic impulse was to translate events within the *mellah*, both good and bad, into social dramas that became thaumaturgic displays of their own moral authority.

The *mellah* was also the setting for joyous events in which Jews and Muslims sometimes celebrated together. This same chronicler, Rabbi Samuel tells us an amusing and quite preposterous story that is redolent with meaning. In the year 1699, according to the chronicle, the famously cruel Sultan Mawlay Isma'il threw four Christian captives accused of thievery into a pit and called in the lions. The Sultan climbed onto a rock to enjoy the spectacle with his soldiers and entourage, and began throwing stones at the captives. Meanwhile, one of the captives spoke to the lion in some strange language, ordering him to jump onto the rock and attack the Sultan. The lion did as he was told, seized the Sultan and began to maul him, biting his leather vest but not injuring his flesh. One of the soldiers grabbed his gun and shot the lion between the eyes, the lion fell back into the pit, while the Sultan tumbled off the rock unhurt. The chronicler goes on to tell us that the miraculous delivery of the Sultan became the pretext for a kind of Jewish-Muslim bacchanalia that ran through the streets of the *mellah* and beyond:

> Jews and Muslims had a great celebration together; banquets and festivities were set up in each town. The rabbis asked everyone to close their shops and wear their best clothing. The roofs and windows of the *mellah* were adorned with flags. Muslims came into the *mellah* to drink *mahiya*, [a strong variety of brandy], and Jews entered the palace of the Sultan, into the houses of the nobility, and into the mosques and madrasa-s wearing their shoes, while no one said anything. The Jews even robbed the shops of the Muslims in Old Fez and no one chastised them, because this event was a great miracle…[28]

Did this remarkable event really happen, or was it a hallucination? It is not reported elsewhere, neither in European nor Moroccan Muslim sources. In this Jewish version, the intimations of cultural inversion are unmistakable. Transgressions took place in the form of border-crossing and the breaking of moral and religious taboos, such as Jews wearing shoes in mosque, Muslims getting drunk inside the *mellah*, Jews robbing Muslim shops and going unpunished. These images of a world turned upside down, in which the weak suddenly become powerful, and Jews and Muslims are suddenly equal, remind us of the Purim ritual that is also a ceremony of inversion.[29]

Perhaps this celebration of the imagination—for it is difficult to see here much semblance of the truth—was a Jewish joke meant to both entertain and mock the indignities of everyday life that the minority had to endure. This story that evokes the boundaries that separated Muslims and Jews, it speculates on how dangerous it was to ignore them, and it demonstrates how fragile they really were.[30]

In the final analysis, the intense identification of Jewish personhood with *mellah* space that emerges from the Ibn Danan chronicle makes it difficult, if not impossible, to imagine the unfolding of the Jewish history of Fez in any place other than the *mellah*. Over time, the physical attributes of the *mellah* became less salient to its Jewish inhabitants than its moral ones. While the *mellah* space expanded and contracted over the centuries with each turn of events, those changes were minor compared to the deep attachment of its Jewish inhabitants toward their surroundings. The protective ramparts came to represent security, even though they were periodically breached; the family house, or *dâr*, came to represent the continuity of the bloodline, passing from one generation to the other; the synagogues were rostrums for enacting rituals of communal solidarity; the cemetery was not simply a place of the dead, but also a shrine where one could talk directly to God. The *mellah* became a micro-cosmic Jewish state in exile, and within its walls, Jews exercised the sovereignty denied to them in the world-at-large. *Mellah* space increasingly lost its material existence and became a function of other processes and scenarios by which Jewish life in Fez was constructed, observed, imagined, and remembered.[31]

To demonstrate this point, one final story. An interlude reported by the chroniclers that took place at the end of the eighteenth century underscores this intimate connection between Jewish self-identity and Jewish space. In April 1790, the brutal and infamous Sultan Yazid, [who reigned 1790–1792] invaded the *mellah* and ordered all the Jews to leave; none of the usual bribes could deter him from his plan. Immediately, soldiers came to enforce this unprecedented order, which had no parallel in the annals of post-medieval Moroccan history.[32] It was a day of suffocating heat, yet the Jews were forced to march out of the *mellah* barefoot, carrying a few household effects to their new home in the *qasba* [citadel] of Shrarda. They were required to leave behind "enormous amounts" of food, including the stores of raisins used to make *mahiya*, the powerful fig brandy that was a Jewish specialty, thus allowing the Sultan to take over the monopoly of this lucrative industry. Inside their new home the Jews set up tents where they were besieged by "rats, lice, snakes scorpions and flies." Meanwhile, "delicate" Jewish women, unused to the task, were forced to draw water from the communal well. These accumulated insults caused the chronicler to lament that "Fez has lost all its beauty."[33]

The disaster did not end with that. According to the chronicle, all the synagogues of the *mellah* were destroyed, and as a final blow, the Sultan ordered the Jews to dig up the bones of their ancestors from their ancient cemetery and remove them to another resting place. He then used the abandoned tombstones to construct a mosque in the *mellah* on the site of the most ancient synagogue, and renamed the *mellah* "Kabir," [the Great One] forbidding anyone "under pain of sanctions, to call it thereafter the *mellah*."[34] The effects of this displacement on the Jews of Fez were devastating; they not only lost their homes, but also their sense of who they were. "All the Hebrews [sic] became the object of hatred and ignominy," notes the chronicle.[35] Not long after, Sultan Yazid left town and charged an unnamed *sharîf* [descendent of the Prophet] with continuing the torments in his absence. Finally, after twenty-two months of exile, the chronicle tells us that a courageous *qâ`id*, or military officer, approached the Sultan and asked him: "Sire, who is this petty *sharîf* who wants to exterminate the entire Jewish community?…What crime have they committed to be so put to death?" This question was replete with meaning, for the *qâ`id* understood better than the Sultan that it was madness to destroy the Jews, and that expelling them from the *mellah* was self-defeating. Not long after, Yazid was mortally wounded before the gates of Marrakesh, the Jews were permitted to return home "in great joy and jubilation", the offending mosque was dismantled, and the *mellah* was restored as exclusively Jewish space.[36]

Was the Jewish passion for their own quarter unique to them, or was it a cultural value that Jews shared with Muslims? Attachment to place, and especially one's home town, was a common feature of North African culture and a clear marker of identity. Family names were often derived from one's city, and the name "al-Fâsî," or a person of Fez, is both a Jewish and Muslim surname. Moreover, there was special pride associated with coming from the city of Fez, the center of Maghribi and Moroccan learning, whether one was a Jew or an Arab. In Jérôme and Jean Tharaud's portrait of Fez, *Fès ou les bourgeois de l'Islam*, a Muslim father dismisses the proposal of an Algerian suitor by saying that he would prefer to give away his daughter to "a Jew, for the Jew was at least a Fâsî, and his parentage and origins were known."[37] Jewish and Muslim mentalities overlapped in many places, but especially in the domain of identification with their home town, where matters of lineage were transparent and the quality of one's personal reputation was common knowledge, regardless of religious affiliation.

The Unmaking of Place

How did the *mellah* die, if it is indeed dead? The causes were complex and cumulative over time. Usually this process has been framed in terms of a sweeping "modernity" that came out of the West and drove Jews along some pre-plotted evolutionary path, making *mellah* existence obsolete.[38] But in fact, the primary causes of the abandonment of the *mellah* were largely internal and tied to its physical degradation, the disruption of its economy, and the political fragmentation of its population. These transformations were not merely the result of the encounter with the West. They were due to other processes not directly related to Europe and connected to regional developments such as the disruption of trade routes, the migration of the rural population to the cities, and the breakdown of centralized rule.[39]

The disengagement of the Jews of Fez from their *mellah* was a trend that gathered momentum over time: it began in the late nineteenth century, when Jews began to leave, first in a trickle, then in a flood. A crucial turning point came in 1912 when Fez was besieged during the French invasion of Morocco, and many buildings in the *mellah* were destroyed, first by pillagers, then by French artillery seeking to drive them out. This catastrophic event caused a psychological break that stripped the *mellah* of its qualities of refuge and revealed it in a new light, as a walled prison locking Jews in and depriving them of the freedom of movement necessary to join an evolving urban culture in contact with the wider world.

In the early years of the French Protectorate, a French military officer administered the *mellah* of Fez, but the real master of affairs was the Resident-General of Morocco, General Louis-Hubert Lyautey.[40] Conservative and a pro-monarchist, Lyautey was a staunch believer in rationality and order; he was appalled by the chaotic situation he found in Fez, much of it due to the French intervention. In the case of the Jews, establishing order meant returning to the traditional institutions that had stabilized *mellah* existence over the centuries. For some time, the *mellah* had been the scene of a bitter battle between younger, Western educated Jews and the older, entrenched elites. In an effort to "reconstitute" the *mellah* as it once was, Lyautey supported the old elites and reintroduced them into the positions of authority.[41] This was a grievous error that hastened the *mellah*'s decline. Disaffected younger people, seeking opportunity in a booming colonial economy and eager to trade their old corporate status for a new civil one, began to move outward in droves, depriving the *mellah* of its most vital resource. By the 1930s, the *mellah* had become a squalid rooming house reserved only for the poor. Even though Jewish institutions continued to function in the *mellah*, their purpose now was mainly charitable. Cut off from the rest of the city by bureaucratic barriers, deprived of its most upwardly mobile elements, and riven by factional disputes, *mellah* society entered its last phase.[42]

The nail in the coffin, so to speak, were the ordinances promulgated during World War II, when France was torn between pro- and anti-fascist forces. Pro-Fascists wanted to implant in Morocco the racist policies being enacted in France and elsewhere in Europe, while anti-Fascists tried to prevent them. A recumbent Sultanate anesthetized by the colonial regime now rose up and emerged as a potent political force; one of its stated goals was to protect the security of Moroccan Jewry. In August 1941, the Vichy government ordered all Jews living in the European quarters of Moroccan cities to prepare to leave their homes and return to their *mellah*-s. Plans were laid to extend the *mellah*-s to accommodate them. But the Allied landings of 1942 stymied these plans, and all thoughts of segregation were swiftly cancelled.[43]

The racist policies of Vichy were the penultimate act that turned the *mellah*-s into places of abandonment; largely because of them, the Jewish understanding of *mellah* space as positive and life-sustaining gave way to one that viewed the *mellah* as a place of stagnation and death. By mid-century, a dark shadow hung over Morocco's *mellah*-s. Other forces came into play as well, such as a strident nationalism that deemphasized the role of minorities and ignored the *mellah* as a stabilizing social phenomenon. Disengaged from their defining space, scattered throughout the social fabric, atomized internally into opposing factions, and devalued by the independence movement, Moroccan Jews were now positioned for displacement. In these conditions, the great exodus began. It is not surprising that by the mid-nineteen sixties, most Jews of Morocco were gone, dispersed to the countries of the West and Israel.

A *Mellah* Without Jews

What remains of the *mellah* of Fez? A few street names, one or two restored synagogues, an impeccably maintained cemetery. The splendid interiors of once great family houses are now subdivided into smaller units to accommodate newcomers from the countryside. A sheet hung across the balcony fashions a flimsy privacy. The Fez *mellah* today is above all home to those rural migrants seeking to gain a tenuous foothold in the city.[44] Some inhabitants recall former Jewish neighbors, but most do not. The majority are under the age of thirty-five, and for them, Jews are simply hearsay. Paradoxically, the name *mellah* persists, despite efforts of the local Muslim authorities to stamp it out by absorbing *mellah* space administratively into neighboring Fâs al-Jadîd. But a stubborn and inexplicable resistance by the people of Fez seems to have stalled this attempt, and the name lives on as an inadvertent witness to those who once lived there. A *mellah* without Jews is now a reality, making its claim on the popular imagination.

The question remains: how much attention will be given to this artifact of the Moroccan past by those responsible for writing the history of Morocco? To what extent will the *mellah*, and the history it represents, be integrated into new narratives? Today Muslim historians of Morocco are calling for a reevaluation of the received texts, and for a more generous spirit of inclusiveness.[45] Will Jews, Berbers, dissidents, and other cultural outsiders finally be recognized and given space on the national intellectual agenda, as some in the historical profession have promised? Will political circumstances allow for such a reevaluation? Moroccan Jews from Israel and the Diaspora return to their former home with regularity, seeking answers to questions of origins and identity.[46] Jewish scholars of a new generation show a poignant curiosity to examine ancestral roots. But will this trend help to propagate a revitalized historiography of Moroccan Jewry that is more balanced and faithful to the past? If it does, we may see a renascent *mellah* in all its contemporary representations—as icon, figure of speech, place of memory, and vernacular monument—assuming its rightful place in the unfolding story of Morocco.

Notes

1   A new critical study of Leo Africanus' life and work is Natalie Zemon Davis, *Trickster Travels: A Sixteenth-century Muslim between Worlds* (New York: Hill and Wang, 2006). See also the luminous scholarly work on this subject by Oumelbanine Zhiri, *L'Afrique au miroir de l'Europe: fortunes de Jean Léon l'Africain à la Renaissance* (Genève: Libr. Droz, 1991), and *Les Sillages de Jean Léon l'Africain, du XVIe au XXe siècle* (Casablanca, Maroc: Wallada, 1995). The Pory text is available in a modern reprint edition: *The History and Description of Africa and of the Notable Things therein Contained, written by al-Hassan ibn Mohammed al-Wezaz, al-Fasi, a Moor, baptized as Giovanni Leone, but better known as Leo Africanus.* Translated into English in the year 1600 by John Pory, and now edited, with an introduction and notes, by Robert Brown (New York, B. Franklin [1963]) and also on line at the EEBO website.

2   Leo Africanus, *The History and Description of Africa*, 3 vols. (New York: Burt Franklin, 1967), 2, 476–77. The Jews of Fez—artisans, merchants, jewelry makers, vendors of luxury goods, weavers of cloth, purveyors of goods and services to the nearby palace as well as to the rest of Fez—occupied a vital niche in the local and regional economy. Some Jews may have lived in the *mellah* in the late 1300s, but the big population shift took place in 1437, not long before the visit of Leo. A comprehensive post-medieval history of the Jews of Fez, rich in documentation but constrained by its Orientalist perspective, is Jane Gerber, *Jewish Society in Fez 1450–1700: Studies in Communal and Economic Life* (Leiden: 1980).

3   On the founding of the Fez *mellah*, see Mercedes García-Arenal,"The Revolution of Fâs in 869/ 1465 and the Death of Sultan `Abd al-Haqq al-Marînî," *BSOAS* 41, part 1 (1978): 43–5, and by the same author "Les Bildiyyîn de Fès, un groupe de néo-musulmans d'origine juive," *Studia Islamica* LXVI (1987): 115–16. On the formation of Moroccan *mellahs* more generally, see: Daniel J. Schroeter, "The Jewish Quarter and the Moroccan City," in *New Horizons in Sephardic Studies*, ed. Y. Stillman and G. Zucker (Binghamton, NY: SUNY Press, 1993), 67–81; H. Z. Hirschberg, "The Jewish Quarter in Muslim and Berber Areas," *Judaism* 17 (1969): 405–21; Harvey Goldberg, "The Mellahs of Southern Morocco," *The Maghreb Review* 8 (1983): 61–9; Kenneth Brown, "Mellah and Madina: A Moroccan City and its Jewish Quarter (Salé, ca. 1880–1930)," in *Studies in Judaism and Islam*, ed. S. Morag (Jerusalem: 1981), 253–81; Susan Gilson Miller, "Apportioning Sacred Space in a Moroccan City: The Case of Tangier, 1860–1912," *City & Society* 13, no. 1 (2001): 57–83; A. Khimlishi, "Ḥawl mas'alat al-maḥaṭṭat bil-mudun al-maghribiyya,'" *Majallat Dâr an-Niyâba*, part

one, 14 (Spring 1987): 21–8; and part two, 19/ 20 (Fall 1988): 30–41; and more recently, the excellent book length study by Emily Gottreich, *The Mellah of Marrakesh: Jewish and Muslim Space in Morocco's Red City* (Bloomington: Indiana University Press, 2007).

4   The status of Jews in Islam, like that of Christians, was that of *dhimmi-s*, or "protected" people, subordinate to Muslims. Jews in Morocco rarely suffered the outrageous persecutions perpetrated against their co-religionists in the European context. Since the coming of Islam to Morocco in the 8th century, only one major genocide is recorded during the 12th century, under the Almohad dynasty. Jacqueline Hadziiossif, "Les conversions des juifs à l'islam et au christianisme en Méditerranée XIe–XVe siècles," in *Mutations d'identités en Méditerranée: Moyen Age et époque contemporaine* ed. Christiane Veauvy, Henri Bresc, and Eliane Dupuy ([Saint-Denis]: Bouchène, 2000), 159–73.

5   A more comprehensive definition of the term *mellah* is found in the *Encyclopedia of Islam*, 2nd edition, *s.v.* "Mallâh," authored by Haim Zafrani, who devoted his lifelong scholarly efforts to studying the literature of the Jews of Morocco. Also, by the same author, *Mille ans de vie juive au Maroc: histoire et culture, religion et magie* (Paris: G.-P. Maisonneuve & Larose, 1983). On Moroccan Jewry in general, see André Chouraqi, *Between East and West: A History of the Jews of North Africa* (Philadelphia: Jewish Publication Society, 1972). H.Z. Hirschberg, *A History of the Jews in North Africa*, 2 vols. (Leiden: E.J. Brill, 1974, 1981), and most recently, Mohamed Kenbib, *Juifs et Musulmans au Maroc, 1859–1948* (Rabat: Université Mohammed V, 1994).

6   Pierre Loti, *Au Maroc* (Paris: Boîte à documents, 1988), 206.

7   A technical term often translated as "school of law"; in particular, one of the four legal systems recognized as orthodox by Sunni Muslims. *Encyclopaedia of Islam*, Online Edition, *s.v.* "Madhhab".

8   David Nirenberg, *Communities of Violence: Persecution of Minorities in the Middle Ages* (Princeton, NJ: Princeton University Press 1996), see especially the Introduction. Nirenberg's approach to majority-minority relations, though dealing with another time, place and principals, has much informed my own.

9   Mark R. Cohen, *Under Crescent and Cross: The Jews in the Middle Ages* (Princeton: Princeton University Press, 1994) and the review by D. Lasker in the *Jewish Quarterly Review 88*, 1–2 (July-October, 1997): 76–8.

10 On corporate groups in Moroccan society, see Mohamed El Mansour, "Saints and Sultans: Religious Authority and Temporal Power in Precolonial Morocco," in *Popular Movements and Democratization in the Islamic World*, ed. Masatoshi Kisaichi (London, New York: Routledge, 2006), 1–32.

11 David Biale, *Power & Powerlessness in Jewish History* (New York: Schocken Books, 1986), Chapter 2.

12 Michael Werner and Bénédicte Zimmerman, "Beyond comparison: histoire croisée and the challenge of reflexivity," *History & Theory*, 45, no. 1 (Feb 2006): 30–50; Dipesh Chakrabarty, *Provincializing Europe: Postcolonial Tthought and Historical Difference* (Princeton, N.J.: Princeton University Press, 2000); and by the same author, *Habitations of Modernity: Essays in the Wake of Subaltern Studies* (Chicago: Chicago University Press, 2002) ; Prasanjit Duara, *Rescuing History from the Nation: Questioning Narratives of Modern China* (Chicago:University of Chicago Press, 1995).

13 Ordered to move there by the Sultan, many complied; but others refused and were forced to convert to Islam. A number of old Fâsî Muslim families still trace their ancestry to those Jewish converts of the fifteenth century and carry distinctive names such as al-Kûhin, Bannânî, Jassûs, Ibn Shaqrûn, and so on. H.Z. Hirschberg, *History of the Jews*, vol. 1, 191.

14 Susan Gilson Miller, Attilio Petruccioli and Mauro Bertagnin, "Inscribing Minority Space in the Islamic City: The Jewish Quarter of Fez (1438–1912)," *Journal of the Society of Architectural Historians* 60 (2001): 310–27.

15 Georges Vajda, "Un Recueil de textes historiques judéo-marocains," *Hespéris* 35–36 (1948–1949): 35 (1948): 311–58 and 36 (9): 139–88; A. Shahbar, ed., *al-Kitâb al-tawârîkh aw tarîkh fâs* (Tetuan: Association Tetuan Asmir, 2002); Meir Benayahu, ed., *History of Fez: Misfortunes and Events of Moroccan Jewry as Recorded by Ibn Danan's Family and Descendents;* in Hebrew, *Divre Ha-Yamim shel Fas* (Tel Aviv: Diaspora Research Institute 1993).

16 Vajda, 35, 313.

17 *Ibid.*, 36, 150.

18 *Ibid.*, 36, 168.

19 The role of the *nagid*, or communal spokesperson, was critical to the process of tax collection. He was chosen from among the wealthiest families, and the function, like that of rabbi, seemed to remain in certain families over the centuries. Gerber, *Jewish Society in Fez*, 86–94.

20 Vajda, 35, 326.

21 *Ibid.*, 36, 142. On the architecture of the *mellah* house, see Miller et al., *Inscribing Minority Space*, 315–16

22 Vajda, 35, 346–47.

23 *Ibid.*, 35, 326–27.

24 *Ibid.*, 35, 337.

25 *Ibid.*, 35, 324.

26 *Ibid.*, 35, 334.

27 *Ibid.*, 35, 340; for the Hebrew, see Benayahu, *Divre*, 66.

28 *Ibid.*, 35, 147.

29 It was not uncommon to celebrate such "miraculous" rescues through the celebration of a Purim ritual. Susan Gilson Miller, "Crisis and Community: The People of Tangier and the French Bombardment of 1844," *Middle Eastern Studies* 27 (1991): 583–96, and references cited.

30 On Jewish converts in the Muslim West, see Mercedes García-Arenal, "Jewish Converts to Islam in the Muslim West," in *Dhimmis and Others: Jews and Christians and the World of Classical Islam*, ed. Uri Rubin and David Wasserstein (Winona Lake, Indiana: Eisenbraus, 1997), 227–48.

31 See Denis Cosgrove, "Landscape and Landschaft," GHI Bulletin 35 (Fall 2004), 57.

32 Mawlay Yazid was particularly violent toward the Jews who had played a leading role in Morocco's economic opening to Europe orchestrated by his father, Sultan Sidi Muhammad b.`Abd Allah (reigned 1757–1790). Muhammad El Mansour, *Morocco in the Reign of Mawlay Sulayman* (Wisbech, Cambridgeshire: Middle East and North African Studies Press Limited, 1990), 18.

33 Fez is known for its bountiful water supply, and most houses were equipped with running water. The need to draw water from a common well was an activity associated in the minds of Fâsî-s with rural living. Vajda, 36, 170. Y.D. Semach, "Une Chronique juive de Fès: Le "Yahas Fès" de Ribbi Abner Hassarfaty," *Hespéris* 19 (1934): 88.

34 Vajda, 36, 172.

35 Ibid., 36, 173.

36 This story is the compilation of several different texts. See Vajda, 36, 166, notes 3, a, b, c. See also the oral version in judeo-arabic collected by L. Brunot and E. Malka, *Textes judéo-arabes de Fès: Textes, transcription, traduction annotée* (Rabat: Ecole du Livre, 1939), 199.

37 Jérôme and Jean Tharaud, *Fès ou les bourgeois de l'Islam* (Rabat: Marsam, n.d.), 146.

38   Daniel J. Schroeter, "Orientalism and the Jews of the Mediterranean," *Journal of Mediterranean Studies* 4 (1994): 183–96.

39   There is no one authoritative account of Moroccan history in this period. The best source is the collective effort by Jean Brignon et al., *Histoire du Maroc* (Paris: Hatier, 1967). See also Edmund Burke III, *Prelude to Protectorate in Morocco: Precolonial Protest and Resistance, 1860–1912* (Chicago: University of Chicago Press, 1976).

40   Lyautey served in Morocco from 1912 to 1925, laying the groundwork for the modern state. Daniel Rivet, *Lyautey et l'Institution du protectorat français au Maroc*, 3 vols (Paris: L'Harmattan, 1988).

41   Archives nationales, Rabat; Archives du Protectorat, [hereafter, AN/ AP] Municipalités, Arrêtés municipaux, Medjles israélite de Fés, A 1712. 25 December 1913. Gouraud to Lyautey. AN/ AP, Municipalités, Arrêtés municipaux, Medjlis Israélite du Fés, carton A 1712, Lyautey.

42   AN/ AP, Municipalités, Arrêtés municipaux, Medjlis Israélite du Fés, Extrait du register des Procés verbaux des délibérations du Medjless El Baladi Israélite de la ville de Fès..., séance du 15 février 1926, carton A 1712. AN/ AP, Bureau du contrôle administrative, carton A 1046; see also Schroeter, "Orientalism," 187–91.

43   Michel Abitbol, *The Jews of North Africa during the Second World War*, transl. C.T. Zentelis. (Detroit: Wayne State University Press, 1989).

44   Hassan Radouine, "Conservation-Based Cultural, Environmental and Economic Development: The Case of the Walled City of Fez," in *The Human Sustainable City: Challenges and Perspectives*, ed. L. F. Girard (Aldershot, Hants: Ashgate, 2003), 457–77.

45   Interview with Prof. Jamaa Baida, Secretary-General of the Moroccan Association for Historical Research published in *Le Matin* (Casablanca) 2/ 3 December 2006, entitled "Le judaïsme marocain est partie intégrante de l'histoire du pays."

46   André Levy, "To Morocco and Back: Tourism and Pilgrimage among Moroccan-born Israelis," in *Grasping land: Space and Place in Contemporary Israeli Discourse and Experience*, ed. Eyal Ben-Ari and Yoram Bilu (Albany, NY: SUNY Press, 1997), 25–46.

6 Religious Microspaces in a Suburban Environment

The Orthodox Jews of Thornhill, Ontario

Etan Diamond

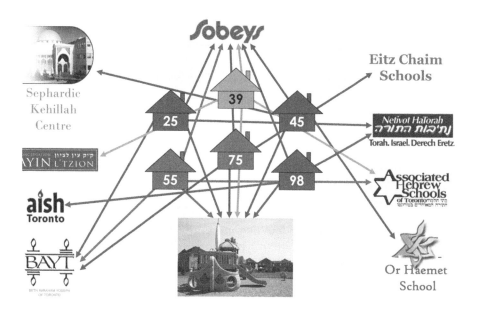

Overlapping religious microspaces of Ramblewood Lane, Thornhill

It is a typical Shabbat morning on Ramblewood Lane, a quiet residential street in the suburban community of Thornhill, just north of Toronto.[1] A family with three young children leaves home around nine o'clock for the twenty-minute walk to their synagogue, the Beth Avraham Yoseph of Toronto (BAYT) congregation. As they proceed down Ramblewood, they meet one of their neighbors, pushing his son in a carriage, as he walks to a different shul, Congregation Ayin L'Tzion. Then, they meet a third neighbor, who is headed to yet another shul, Aish HaTorah. At the end of the block, a path takes the group into Downham Green Park, a large park that sits at the center of the neighborhood. As they enter the park, the walking group splits. The young family heads to the right to the BAYT, and the others head to the left to Aish HaTorah and Ayin L'Tzion.

On the other side of the park, the family continues east on Milner Gate to the traffic light at Bathurst Street. At the intersection, they meet a father and son who are walking to the Chabad Lubavitch shul on the other side of the street. They also pass more shul-goers heading south on Bathurst to the Sephardic Kehillah Center.[2] When they reach York Hill Boulevard, they can see the BAYT building in the distance. They also see many other families, some walking to the BAYT and some walking home, having already attended the BAYT's early *hashkama minyan*.[3]

This scene happens in Thornhill, Ontario, but it could just as easily occur in Silver Spring, Maryland, Skokie, Illinois, North Miami Beach, Florida, or any number of other places across North America where Orthodox Jewish communities are found. In these suburban neighborhoods, the scene repeats itself each Saturday morning: families and individuals walking this way and that, to and from their particular synagogues, sharing the sidewalks with one another, and transforming their automobile-dominated neighborhoods into miniature "walking cities." But underneath this weekly ritual lies a more surprising notion—that these shared experiences are not really so shared at all. In fact, what looks like a single "suburban Orthodox Jewish community" is in fact a much more complex agglomeration of many communities, each with its own institutional affiliations and corresponding spatial arrangements.[4]

As this essay explores, the institutional nature of Jewish life, when combined with the patterns of suburbia, creates a situation where multiple and overlapping spaces exist. These "religious microspaces" vary down to the family and even individual level, and when uncovered, reveal a web of religious identities and meanings. In the following pages, we explore this web as experienced in one suburb, Thornhill, a community tucked in the southeast corner of the city of Vaughan, just over the northern boundary with Toronto, Ontario. This exploration is an extension of my earlier work on Orthodox Jewish suburbanization in Toronto.[5] In that work, I argued that the experience of Orthodox Jews in post-World War II suburbia represented a challenge to the traditional assumptions about both suburbia and Judaism. The observant Jewish community's geographic mobility in relation to the metropolitan periphery went hand in hand with a socioeconomic mobility — two trends that went against the standard postwar story of suburbanization, upward mobility, and religious laxness. In that initial research, the movement out of Toronto and into Thornhill in the 1980s and 1990s represented the end of the story, a culmination of four decades of northward expansion along Bathurst Street. Now, more than a quarter-century after the initial settlement of Thornhill's Orthodox Jewish community, it is useful to return to this Jewish suburban space and reexamine its growth.

More than just a continuation of the story, however, this essay offers an opportunity to reflect on the geographical dimension to traditional Jewish religious life. To an outsider, Thornhill's Orthodox Jews appear as one homogeneous mass; the experiences of one family are seemingly no different from the next. To an insider, however, the day-to-day patterns of religious life — the day schools that our children attend, the stores where we buy kosher food, and the synagogues we attend — are sufficiently different to translate into different spatial patterns. In other words, the religious institutional choices made by the Orthodox Jews of Thornhill mean that we experience our suburban environment in different ways from one another. Thus, through this study, we can begin to see the interplay between religion and space in the modern metropolitan environment.[6]

Creating a Suburban Jewish Space

In 1981, Toronto real estate developer Joseph Tanenbaum announced plans to create a new subdivision in Thornhill, a rural village just beyond the northern border of the city. Tanenbaum's plan, known as Spring Farm, including the normal components of a suburban subdivision: over 2000 homes, a shopping plaza, schools, and parks. But whereas other new suburban developments stopped there, Tanenbaum went one step further, and planned for a synagogue-community center that would act as the anchor for the entire new community.[7]

Tanenbaum's plans might have seemed rather audacious to an outside observer. After all, at the time, Thornhill was a sleepy village, composed of a few streets with homes from the early-to-mid-twentieth century and a handful of local "Main Street"-style businesses. Most of the area was devoted to farmland and the Canadian National rail tracks, which served as a buffer against Toronto's northward expansion. Jewishly, Thornhill was even less developed, with no Jewish institutions or Jewish residents. The closest synagogue was almost 2 miles to the south, and Jewish schools, kosher grocery stores, and other communal buildings were even further away. If Jews were to move to Thornhill, the popular opinion said, the neighborhood would only attract less-observant Jews who were not as tied to the Jewish religious infrastructure, rather than Orthodox Jews who relied on nearby community institutions.

A quarter-century later, Tanenbaum's audacity was trumped by his perceptiveness. With more than 20,000 Jews living in well-kept middle-class and upper-middle class subdivisions—about one-third of the total population in the area—Thornhill is home to a dynamic mix of Ashkenazic and Sephardic Jews, Russian, South African, and Israeli Jews, along with those families who have lived in Toronto for several decades. There are Orthodox, Conservative, Reform, and completely unaffiliated Jews. And there are the institutions: more than a dozen Orthodox synagogues, four Jewish day schools, numerous kosher restaurants and food stores, Judaica and Jewish book stores, and two Jewish social service agencies.[8]

Something seemed to have gone right.

In retrospect, given the history of Jewish suburbanization patterns in Toronto, Joseph Tanenbaum's gamble was perhaps not a large gamble at all, but rather an astute understanding of the geography of the local Jewish community. For the three decades prior to the settlement of Thornhill, Jews had been steadily moving northward along Bathurst Street. Starting in the early 1950s, the first pioneers left the downtown neighborhoods for newer suburban subdivisions north on Bathurst Street. New synagogues began to dot the landscape, followed by day schools and other Jewish-oriented institutions and businesses, including kosher butchers, bakeries, and restaurants. By the end of the 1960s, suburban development had reached the northern boundary of North York at Steeles Avenue.[9] At that point, instead of continuing their march north, Jews moved eastward, to neighborhoods along Bayview and Leslie streets. A couple of synagogues opened in northeast neighborhoods as did a branch of the Associated Hebrew Day School, and, for a while, it seemed as if these areas would be the new geographic center for the Jewish community.

One key factor inhibited further growth in the northeast neighborhood, however: the distance from Bathurst Street. Because of the historic ties to Bathurst Street and the general north-south geographical orientation of the Jewish community, sustainable growth to the east seemed to stand outside the mental maps of most in the Jewish community. It was simply not in the daily paths of most Jews to travel in an east-west direction for their Jewish religious needs. When Tanenbaum began his project in Thornhill, then, it was as if someone had released a pressure valve that allowed the flow of Jews to continue in their "natural" path northward. The concurrent result was the rather swift emptying out of the northeast neighborhoods, as families realized that the Jewish future was in Thornhill and not in the Bayview-Leslie area.

The Early Years of Jewish Thornhill: An Uncomplicated Spatial Pattern

For the first decade or so of Jewish settlement in Thornhill, the limited institutional presence made the community's spatial patterns relatively uncomplicated. The BAYT synagogue had not yet built its own building and, instead, met in a Tanenbaum-built house on Bevshire Court. The synagogue held only a single service on Shabbat morning, meaning that everyone who came to the shul prayed together. A Chabad Lubavitch center had opened on Chabad Gate near the southern edge of the neighborhood, but that shul was targeting a different membership than the BAYT and no other modern Orthodox synagogue options were available. No day schools had been built in the area yet, and the Spring Farm Marketplace that would later house a grocery store and other kosher businesses had not yet opened. For the first wave of families who had made the move up north, there was a "special feeling that the members formed a large family" who had to support one another. It was very much of a "pioneering community."[10]

The pioneering atmosphere extended beyond the Jewish community. Clark Avenue, today the busy east-west artery through the heart of Jewish Thornhill, was not completed initially. Arnold Avenue, now home to million-dollar mansions, was still populated by smaller farm houses. "You could hear the chickens on Arnold as you walked down the middle of Clark Avenue," recalled one of the earliest suburban pioneers.[11] What would become the Promenade, a major regional shopping center just west of the Spring Farm subdivision, remained undeveloped farmland.

By the late 1980s, the simple religious geography of Thornhill's Orthodox Jews had begun to change as new institutions appeared on the scene. In 1988, the BAYT moved into its new building on the south side of Clark, opposite the original house on Bevshire. The 90,000 square foot building was one of North America's largest synagogue structures, despite having been scaled back because of financial difficulties and the recessions of the mid-1980s. The new building, with two social halls, an educational wing, a *mikveh* (ritual bath), and a large *beit midrash* (study hall), represented a sharp transition from the earlier pioneering period. Not only was the building bigger, but the membership had grown to over 400 families.[12] An early morning *hashkama minyan* had been formed, and there was interest in starting a separate minyan for younger children. The BAYT building also included space for Ohr Sameach, an organization dedicated to outreach to unaffiliated Jews, and a separate minyan for this group was held on Shabbat mornings as well. With the relocation, the experience of everyone knowing each other, and everyone sharing the same spaces, disappeared.

Around the same time that the BAYT building opened, the Spring Farm shopping plaza opened further east on Clark Avenue, with a grocery store and over a dozen other businesses. Soon, the Spring Farm branch of the Eitz Chaim Day School opened. Aish HaTorah began to hold Shabbat services, first in the Westminster Public School, then in the Promenade Shopping Mall, and finally in its own building on Clark Avenue. By the mid-1990s, a branch of the Associated Hebrew Day School opened on Atkinson Avenue, and a new congregation began to meet in the building. Soon thereafter, Netivot HaTorah Day School opened its building, also on Atkinson, and a breakaway congregation from the Chabad Lubavitch shul began to meet in that school. Then, in 2003, a group broke away from the BAYT and formed yet another shul, which also met in Netivot. This wave of institutional expansion had important ramifications for the religious geography of the community.

Thornhill from the East

Spaces that Segregate, Spaces that Unify

In looking at the ways that the local institutional spaces shape the geographical patterns of Thornhill's Orthodox Jews, we can identity two opposite trends: there are religious microspaces that serve a segregating, or compartmentalizing, function, and others that serve a congregating, or unifying, function.[13] Segregating spaces divide the Orthodox Jewish community according to different religious and institutional affiliations. This institutional compartmentalization impacts the geographical experiences of the community by structuring the places within Thornhill that people frequent, the paths we take to get there, and the people we interact with in those spaces. On the other side of the ledger are the congregating, or unifying, spaces within Thornhill. These are places that transcend religious affiliation and bring the entire Thornhill Jewish community together.

The segregating and congregating spaces also have a temporal dimension in terms of religious time. The restrictions on vehicular travel on the Sabbath and Jewish holidays serve as a stark reminder that the Orthodox Jewish community is not fully enveloped by the suburban environment. Yet, even on the Sabbath and holidays, even during times when suburban mobility and consumerism are put aside, there are local spaces that divide the community and other local spaces that unify the community.

The interaction of the two spatial types—segregating and congregating spaces—and the two temporal dimensions—secular and sacred—create a grid that can categorize the different religious microspaces discussed (see Table 6.1). The following discussion explores each of the four microspace types.

Table 6.1   *Religious microspaces in a suburban environment*

	Segregating/ compartmentalizing spaces	*Congregating/ unifying spaces*
Secular suburban time	Day schools	Kosher stores
Sacred time	Synagogues	Parks/ playgrounds

Religious Segregation in Secular Time: The Jewish Day School

During the week, the day school serves as the clearest institutional marker of religious spatial segregation. Where we send our children to school in Thornhill shapes our geographical and religious identities. Orthodox Jews in Thornhill have both local and non-local options for their children's schooling. Locally, we can choose a religious Zionist school (Netivot HaTorah), a Zionist-but-less-religious-school (Associated), a religious-but-less-Zionist (Eitz Chaim), or a religious Sephardic school (Or Haemet). Outside Thornhill are other day schools, including Lubavitch and Haredi yeshivas, and even a school dedicated to special needs children. Thornhill has no local Jewish high schools, but again, residents can choose from across the religious spectrum, from the Haredi to religious Zionist to non-denominational, elsewhere in Toronto.

In choosing the schools for our children, Thornhill's Orthodox Jews make choices that structure their spatial patterns and give meaning to our religious identities in a segregating fashion. The daily routines of carpool drop-off and pick-up, as well as evening activities such as curriculum night, parent-teacher conferences, and school plays, bring together other parents and families with similar religious orientations. The experience of carpooling is a particularly relevant one for our discussion because it brings into play both the suburban (driving) dimension and the religious (segregative) dimension. Carpooling on a regular basis means driving the same paths, at the same times of day, with the same people. Certain roads become familiar, to the exclusion of others. When traffic back-ups or other factors force one to take a different route, those new routes take on a sense of the unfamiliar. Furthermore, because different families on the same street send their children to different schools, our sense of the local spaces differs. A family driving their children to the southern Eitz Chaim branch in Toronto might care a lot whether traffic on Bathurst Street is backed up, because that is their primary route, whereas a neighbor driving her children to Netivot HaTorah cares not at all about Bathurst traffic because she merely crosses the street rather than drives on it.

Religious Segregation in Sacred Time

The spatial segregating function of the Jewish day schools during the week has an analog on the Sabbath in the synagogues. Religious spatial distinctions exist both in the relation of one synagogue to another and even within a single synagogue. As this essay's opening vignette makes clear, you might meet your neighbor on the way to synagogue, but if you are both going to different synagogues, you will experience the same spaces differently.

Synagogues function as segregated microspace in several ways. First, they serve as identification badges, as labels to identify and characterize synagogue members. When your neighbor goes into Aish HaTorah but you continue onto the BAYT, you will likely see that neighbor as fitting into a certain category of observant Jew, with the associated characreristics that come with that particular congregation. Aish HaTorah is the "outreach" shul, Westmount is the "South African" shul, the southern Chabad shul is for long-time Lubavitch families, whereas the northern Chabad shul on Flamingo is more of an outreach shul with very few "real Chabadniks." Ayin L'Tzion is for young families who might otherwise have attended the BAYT but who found it too large or impersonal. The BAYT is for people who want to be in the "in crowd." And so on. It is irrelevant whether any of these stereotypes actually hold up to scrutiny (they usually don't). What is important is that their persistence creates sharp divisions across Thornhill's Jews with clear spatial implications. Because people's synagogue habits remain fairly rigid, we will rarely venture into a different shul unless specifically invited to a *simcha* such as a bar mitzvah, and, even when we do, we continue to see the other congregational spaces as "other" and "foreign."

The religious congregations also contribute to a sense of segregation by shaping the ways Thornhill's Orthodox Jews residents think about our particular streets. Overall, Thornhill is rightly characterized as a middle-to-upper middle class suburb dominated by detached, single-family homes.[14] *Where* those homes are, however, matters. For BAYT members, for example, proximity to the synagogue is an important distinction. Streets such as Bevshire and Tangreen are closest to the shul and were among the first streets built as part of the Spring Farm development. They have wider lots and larger homes than streets such as Thornbury or Jessica Gardens, which are on the outer edges of the development. More generally, streets on the east side of Bathurst—the side of the BAYT—have a higher cachet than do streets on the west side, and streets west of New Westminster are seen as "beyond the pale," too far to be even considered part of the community. Note that these nuanced socio-geographic variations in the religious geography of Thornhill are largely meaningless to outsiders. Local residents understand the religious/social implications of living "west of Bathurst" or "north of Centre," whereas outsiders see the entire area as "Thornhill."

Note that the perceptions of "near" and "far" in religious-geographical terms are relative. What a BAYT member considers to be on the edge of the community is quite different from what a member of Aish HaTorah thinks; each has a different religious central point of reference. Two neighbors living side-by-side on, say, Janesville at the northern end of Thornhill will see their houses differently depending on which congregation they belong to. The one who attends the Associated shul sees himself as sitting near the heart of his community, while the BAYT member is on his community's far northern edge.

The perceptions are also quite fluid and subject to external factors, such as economics. For example, neighborhoods on the west side of Bathurst were long perceived as being "far" from the BAYT. Yet, as housing prices in Thornhill increased in the late 1990s and early 2000s, families found that the western neighborhoods offered better value than those streets closer to the shul—sometimes as much as a $100,000 difference between comparable houses. As more families moved to streets such as Ramblewood, Tansley, Colvin, and Millcroft, these streets have become less "far" than they once were. (Of course, those streets were never far at all to members of Aish HaTorah, which is on the west side of Bathurst!)

Then there are streets that serve as social markers regardless of the congregational affiliation. One block north of Spring Gate is Arnold Avenue, home to some of the most expensive addresses in Thornhill. Once filled with one-story ranch houses that dated to the 1940s and 1950s, many of Arnold's houses are replacement mansions, with half-acre lots and million dollar homes. Further to the south, at the corner of Hilda and York Hill, sits Theodore Place, a single circular street built in the 1990s, with very large homes that fill almost their entire lot line. The grandness of the homes is exaggerated by the above-grade basements, which raise the first-floor entrances almost to the second story. Yet, whereas Arnold's mansions communicate a slightly older, more established sense of wealth and prestige, Theodore represents nouveau riche, a kind of aspirational, somewhat more flashy wealth. The physical differences between the two streetscapes capture these distinctions. Arnold's homes are sprawling and set back from the street (some are even gated), while Theodore's homes are built almost up to the lot line and seem to overwhelm the street itself. Yet filled as they are with Orthodox Jewish families who belong to various synagogues, Arnold and Theodore segregate based on price and housing style more than they do based on geographic proximity to any particular synagogue location.

Religious microspaces can divide even within a single synagogue. One of the most common examples in North American Orthodox Jewish synagogues is the fault line between the service in the main sanctuary and the *hashkama minyan*. Although different reasons have been given as to the original popularity of the *hashkama minyan*, in the past two or three decades the early Shabbat service has become associated with "serious" worshippers who prefer a quicker service without the frills of a main sanctuary service.[15] Often a *hashkama minyan* provides tables for daveners[16] to sit at, rather than just chairs, to provide a more "yeshiva-style" environment that encourages studying. This is contrasted with the services in the main sanctuary, which might have pews or fixed seating without space for extra books for learning and which is typically filled with sounds of talking rather than praying. In many synagogues, the religious division between main and *hashkama* service has a spatial dimension, with many *hashkama* services meeting somewhere "downstairs" in an auxiliary social hall or the weekday chapel. Again, it is irrelevant whether one service is actually louder or more "serious" than another, and one can always find examples of talkative *hashkama minyanim* and quiet main services. The fact remains that these distinctions have an impact on people's spatial and social behavior within the synagogue.

Given the family orientation of most suburban Orthodox Jewish synagogues, it is not surprising that spatial segregation exists along age lines as well, particularly in terms of separating out children into their own activities. Often called "programs" or "groups," the typical Shabbat morning schedule for children includes structured, supervised activities starting around 10 a.m., usually with a shortened davening, some games, and a snack, and lasting until the end of the main service (usually about two more hours). The fact that most services begin at 9 a.m. and groups do not start for another hour means that children are expected either to come to shul later or, more often, hang out in the hallways until groups start. Though the official goal of groups is to provide structured programming for children, the primary purpose in many shuls is to keep the kids occupied so they do not disturb the main services or run in the hallways. The irony, of course, is that the absence of children does not keep the services from being noisy (adults do the talking instead) or the hallways empty (those who are not talking inside the sanctuary congregate outside). In addition, when children are segregated into their own activities, they miss out on the chance to be socialized into the acceptable synagogue behaviors, thus leading people to want the children segregated because they are too noisy and ill-behaved.[17]

Thornhill's best example of how religious microspaces exist within a single congregation is found in the BAYT. It should be noted that the acronym "BAYT" has a place-based connotation, since it is pronounced like the Hebrew word for "home" (*bayit*). Yet, the segregative nature of the shul's microspaces can make it seem far from home-like. Consider that, on a normal Sabbath, someone attending the BAYT has seven different services to choose from, each with its particular characteristics. Thus, one could pray at:

- an early *hashkama minyan*, starting at 7:30 a.m. Meets in the downstairs beit midrash with tables. Has a reputation as a no nonsense service for "serious" daveners.
- the main service at 8:45 a.m. Reputation as drawn-out, noisy service with lots of distractions. People attend here because they like to hear the rabbi's sermon.
- a *beit midrash minyan*, starting at 9 a.m. Meets downstairs in a social hall, also with tables. Has reputation as a no nonsense service for "serious" daveners, but with a later starting time than the *hashkama minyan*.
- a "family" minyan, oriented to students in Grades 4 to 8, starting at 9 a.m. Meets in the main floor social hall. Stereotyped as being a Netivot HaTorah minyan because almost all of the children attend that school. Also reputed to be almost as noisy as the main service.
- a "young adults" minyan, starting at 9:15 a.m. Meets in a downstairs beit midrash, formerly used by the Ohr Sameach outreach organization. Started out as a service for university students, but as they grew older, they kept the service going. Reputed to be quick, no frills service.
- a high school yeshiva minyan, starting at 9 a.m. Meets in the shul library, downstairs. Reputed to be a serious "yeshiva-type" minyan for boys only.
- a high school minyan, starting at 9:15 a.m. Meets in the "youth minyan room" downstairs. Nominally sponsored by NCSY (National Council for Synagogue Youth), an Orthodox Jewish youth group organization targeting high school students from less observant backgrounds.

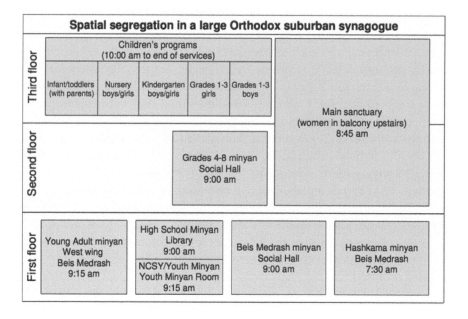

Spatial Segregation at the BAYT synagogue in Thornhill, 2006

Nowhere are the above descriptions accepted as official policy of the shul and all of the services are open to anyone—and plenty of BAYT members do circulate among the services. Nonetheless, the perceptions within the shul of the different services are real and there are in fact many in the congregation who almost never attend a service other than their own. These devotees are often convinced that theirs is the "best" service and that the others are somehow inauthentic/ less serious/ more noisy/ more drawn out than theirs. Only when there is a congregational celebration, such as a bar mitzvah or a guest speaker, will these members leave the familiar surroundings of their regular service and venture to a different part of the synagogue.

These divisions not only have social implications (one might think of the BAYT as seven different shuls under one umbrella), but spatial ramifications as well. Because these different services exist in different spaces (note the upstairs/downstairs distinction), the compartmentalization gives the feeling of being in foreign space when attending a different service. The BAYT building is large enough that one might never go into a different part of the building unless absolutely necessary. A person without children would have no reason to go up to the third floor, where the children's programs are held. Similarly, someone who enters the building from the Clark Avenue entrance on the building's north side and attends the main service would conceivably never need to go downstairs to the first floor.

Unifying Religious Microspaces in Secular Time: The Kosher Grocery Store

Perhaps the biggest complaint about modern suburbia is the lack of shared public spaces. In contrast to the idyllic small town where everyone gathered in the general store, or the larger cities with grand public parks, suburban neighborhoods are chastised for lacking gathering spaces, especially spaces that support serendipitous meetings. Moreover, the dominance of the automobile means that people have fewer opportunities to interact, mostly because they are too busy driving everywhere to actually meet face to face.

Thornhill does not quite fit into this category of community placelessness, as there are a number of public (or semi-public) spaces that promote gathering. There are three community centers in Thornhill, with the Garnet Williams Community Center on Clark Avenue as the oldest. Constructed as part of the original Spring Farm development, the Garnet Williams Center was donated to the City of Vaughan by Joseph Tanenbaum to function as a gathering place. With a swimming pool, fitness center, ice rink, and space for community recreation programs, Garnet Williams does its job for the broader Thornhill community. Other gathering places include two public libraries and the Promenade Shopping Center, a major regional mall with over 175 stores.

While these spaces serve the entire Thornhill community, the various kosher establishments in Thornhill target the Jewish residents of Thornhill. The neighborhood's three pizza shops, five other kosher restaurants, two kosher bakeries, and two kosher candy stores all offer opportunities for intentional gatherings (such as making a lunch date) and unintentional meetings (running into your neighbor on Friday morning while buying *hallah* for Shabbat). The pizza shops are particularly important as gathering spaces for youth—especially on Saturday nights.

More than anything else, the true community center for the Orthodox Jews of Thornhill is Sobey's, the main grocery store in the Spring Farm plaza on Clark Avenue. In his original conception of his development, Joseph Tanenbaum wanted to create a central shopping space that would serve as a community magnet. Sitting at the heart of Spring Farm, this strip plaza certainly fulfills this role, with its three kosher restaurants, a Jewish gift and bookstore, bank, an ice cream shop, and other stores. The main attraction of the plaza, however, is Sobey's, part of a national grocery chain based in Atlantic Canada. Though not exclusively kosher, the Thornhill Sobey's has positioned itself so that it has become the largest kosher marketplace in Toronto, in Canada, and perhaps even all of North America, with an in-store kosher bakery, a kosher butcher, kosher fish department, kosher prepared-food section, and three aisles entirely devoted to kosher products. Furthermore, every product in the rest of the store that has a reliable kosher *hashgacha* (supervision) is marked on the shelf as kosher. One estimate claimed that as much as 65 percent of the food on the shelves is kosher.[18]

As the largest source of kosher products in the Toronto Jewish community, Sobey's offers a neutral, nonideological space for community gathering. Synagogue affiliation is irrelevant to the purchase of kosher bread or meat. Moreover, the variety of kosher products available at Sobey's, and the plethora of acceptable hashgachot means that Jews from a variety of religious backgrounds can almost always find acceptable products. As a meeting place, then, Sobey's—and the rest of the Spring Farm plaza—is the closest thing Thornhill has to a town square, a "public" space that supports chance interactions among loosely connected individuals. As far back as 1989, when the development was only a few years old, the plaza had already become "very much the 'main street' of the area, with shoppers and browsers greeting friends, shopping, discussing politics (governmental, religious, educational, and any other kind possible), admiring each other's children and dispensing advice."[19] Today, one can still almost always count on running into someone familiar at Sobey's, and, at the very least, one will always see Jews from other synagogues and other neighborhoods who otherwise would not be part of the same religious microspace.

Religious Unification in Sacred Time: The Neighborhood Park

On Shabbat and Jewish holidays, of course, Orthodox Jews refrain from shopping, and as such would not go to Sobey's or one of the kosher bakeries or restaurants. Nonetheless, there is a Sabbath equivalent to Sobey's in terms of a community-unifying space: the neighborhood park. The "Shabbat park" is likely a familiar term to observant Jewish readers, as a place for children to play and parents to interact on a Sabbath afternoon.

Thornhill is fortunate to have several neighborhood parks that serve this function. For the Spring Farm neighborhood, there is Heatherton Park on the east side of Atkinson as well as Campbell Park, just west of Netivot HaTorah. On the west side of Bathurst is Downham Green Park. The largest park in the area is York Hill District Park, one of the City of Vaughan's major park facilities. With two large playgrounds, basketball courts, baseball fields, and many benches and walkways, York Hill Park offers play facilities for the wide range of Sabbath-observant Jews. Those who permit their children to play in the playground on the Sabbath can do so, those who play basketball on the Sabbath can do so, those who play baseball can do so, and those who simply want to take leisurely walks can do so—and can do so in a generally non-threatening, non-confrontational manner among other Jews of different religious observance levels. Everyone is there (or so it seems) and, for the moment, the particular synagogue or day school affiliation is secondary to sharing a pleasant Shabbat afternoon.

Because the parks are on the routes to the synagogues, they serve as paths that almost all Sabbath-observant Jews will use to get to and from synagogue. Again, this creates a community-building experience, as families pass one another, extend Sabbath greetings, and continue to their particular congregation.

The "Hidden" Nature of Thornhill's Religious Microspaces: How Outsiders see Thornhill's Orthodox Jews

The web of religious microspaces within Thornhill's Orthodox Jewish community is largely hidden to the non-Jews of Thornhill, who have much larger concerns than the particular spatial arrangements of the Orthodox Jewish residents. We can best describe Thornhill's Orthodox Jewish community as living *in* Vaughan but not being *part of it*. There are three reasons for this, one historical, one political, and one geographical.

Through the 1970s, Thornhill had been a Protestant town with a keen sense of local tradition and history. The onslaught of new suburban development in the 1980s—Joseph Tanenbaum's Spring Farm being the largest example—overwhelmed the town's old-timers. The public discussions rarely made explicit mention of a "Jewish" invasion of this Protestant village, but we can find clues that betrayed a concern that the community's character was undergoing change. Consider a dispute in early 1980 over the designation of historic properties located within Spring Farm. Thornhill's leaders wanted to preserve these heritage homes, while Joseph Tanenbaum's company, Runnymede Developments, opposed the idea. "Developer fights like Khomeini," declared the *Thornhill Month*. Although the magazine got the ethnicity wrong, the reference was clear that an outsider—that is, a non-Thornhill Protestant—was coming in to challenge the status quo.[20]

Political circumstances have also hidden the Orthodox Jewish community's internal structures. Thornhill is not an independent municipality, but is actually the southeast section of the much larger City of Vaughan. Whereas Jews comprise about one-third of the population of Thornhill, and dominate the local political landscape, Jews are in fact only about 5 percent of the overall population of Vaughan, which is politically and demographically dominated by Italians. This is not to say that the city is antagonistic to the Jewish community. To the contrary, there are individual examples of cooperation: on the morning before Passover, the fire department provides the BAYT with an oversized metal garbage container for burning the *hametz*,[21] along with a team of fire fighters supervising the entire process. On *Yom Ha-atzma'ut* (Israeli Independence Day), the city closes Atkinson Avenue to allow the three day schools along that street to hold a joint parade.

However, other episodes remind the community of its minority status, such as the scheduling of a summer concert in York Hill Park during the period of the Nine Days when Orthodox Jews refrain from listening to live music.[22] This political imbalance means that, other than the normal activities of any municipality (garbage pick-up, recycling, etc.), Thornhill's Orthodox Jews have for the most part, little interaction with the broader political and economic decisions of Vaughan. Moreover, it means that the internal spatial dynamics of the Orthodox Jewish community remain hidden to the non-Jewish leadership.

Finally, the simple fact of Thornhill's physical location at the fringe of the main conduits of information and power in Vaughan has created a disconnection between the Orthodox Jewish community and the wider non-Jewish community of Vaughan. In Vaughan, information and business flows primarily in a west-to-east direction, emanating from the large Italian (much larger than the Jewish) population at Vaughan's western end in Woodbridge, and the Vaughan City Hall to the north, located in the town of Maple.

Vaughan's west-to-east orientation sharply contrasts with the north-south historical geography of Toronto's Jewish community along Bathurst Street. Even though one can live a thoroughly Jewish religious life within Thornhill—the schools, synagogues, and shopping are all within a few blocks of each other—the institutional, personal, and historical ties to the rest of the Jewish community further south remain firmly in place. In this way, Thornhill's Orthodox Jews care far more about developments within the Toronto Jewish community than they do about developments in the Italian community in Vaughan.

There is a degree of irony in the Thornhill Jewish community's southern orientation toward Toronto, because many Toronto Jews have long seen Thornhill as the great suburban frontier, as neighborhoods "up north" that were somehow different from the other areas of Jewish settlement. To some degree such perceptions were correct, in that the all-encompassing nature of Spring Farm and the BAYT project—with homes, stores, and a synagogue all put in place at one time—differed from the earlier neighborhoods where subdivisions, synagogues, and Jewish stores were developed separately and slowly by individual interests. Here, however, the whole package was put in place at once, a religious analog to the Levittowns of an earlier suburban era.[23] Adding to the distinctiveness is the fact that Thornhill has a different area code than Toronto. It is popular in Toronto to posit the urban "416ers" against the suburban "905ers," and it is no different within the Jewish community. In the late 1980s and into the 1990s, a popular rumor circulated that many of Toronto's *haredi* Orthodox schools looked down upon students from Thornhill. The reason? Families in Thornhill were likely to own televisions and other materialist possessions that were contrary to Orthodox Jewish values. In other words, 905ers were suspect simply by dint of their geography, whereas the 416ers were kosher.[24]

Conclusion

In a 2002 review of the state of American Jewry, Alan Mittleman perceptively described the experience of Orthodox Jews in suburbia:

> Traditional Jews live within walking distance of a synagogue. They reclaim suburban streets for walking and, on the way home, for socializing. By noontime on any given Saturday, the streets around a synagogue are full of women and men, baby strollers, groups of boys and girls. Cars have to slow down and navigate around groups of walkers, often to the annoyance of drivers. Because Orthodox Jews walk they must live within a reasonable radius from the synagogue, say one-and-a-half to two miles. Because lavish meals are also a part of the celebration of the Sabbath, they join together at one another's homes for the afternoon meal. Groups of people may be found walking to and from lunch; children go to visit their friends in one another's houses. Despite the idea that the Sabbath is a day of rest, Orthodox neighborhoods are alive with social activity on Saturdays.[25]

In addition to capturing the flavor of an Orthodox Jewish neighborhood, Mittleman notes that these shared communal experiences can teach us something "about the possibilities for renewing community in the era of 'bowling alone.'" In referencing Robert Putman's lament on the state (or, better, absence) of North American community structures, Mittleman challenges us to take a closer look at the experiences of traditionalist Jewish communities. But while Mittleman rightly points out that these "shared communal experiences" occur within shared spaces, those shared spaces are not necessarily experienced in the same way. As this essay has argued, the institutional variations within Orthodox Jewish communities play themselves out in spatial terms: what is "my neighborhood" is not necessarily the same space as "your neighborhood" even when we are living on the same block!

The challenge to students of Jewish communities, then, is twofold. First there is the imperative to look beyond the denominational labels within the Jewish community to see institutional variety, and to understand that variety has meaning. One might say that this is a "post-denominational" perspective, but it might be more accurately described as "intra-denominational." Second, and more relevant to this volume of essays, is the challenge to see the spatial dimensions to modern Jewish life. How and where we affiliate (and where we do not) shapes our geographical understanding of our environment. Places that we frequent and the other people who also frequent those places become familiar parts of our daily lives; conversely, where we do not go remains unfamiliar and even strange. By examining how our religious choices and experiences shape our geographic perceptions, we can begin to appreciate the dynamics of the modern Jewish experience.

Notes

1   Along with the English, this essay uses a number of common Hebrew and Yiddish terms in their transliterated form: *Shabbat* for Sabbath; *shul* for synagogue, and *minyan* for prayer service. I would like to thank a number of individuals who shared their insights into Thornhill and the Orthodox Jewish community: Rabbi Baruch Taub, Stephen Speisman, Paul Franks, Hindy Najman, Benjamin and Alison Pollock, Alan Shefman, and Adam Birrell. My wife Judy Snowbell Diamond graciously read early versions of this essay and provided an open ear to my ideas.

2   On Sephardic Jews in Toronto see Kelly Amanda Train, "Carving Out a Space of One's Own: The Sephardic Kehila Centre and the Toronto Jewish community," in *Claiming Space: Racialization in Candadian Cities*, ed. Cheryl Teelucksingh (Waterloo, Ontario: Wilfrid Laurier University Press 2006), 41–64.

3   A *hashkama minyan* is a Sabbath-morning service that meets earlier than a congregation's main service. I will discuss the phenomenon of the *hashkama minyan* later on in this essay.

4   Thornhill is also home to a large non-Orthodox Jewish population. Because there are no local synagogues, however, their spatial experiences are different from the Orthodox Jews, who have local synagogues (and who have the religious imperative to walk to synagogue on Shabbat). There is undoubtedly a geographical dynamic within the Conservative and Reform Jewish communities, but that will have to remain unexplored for the time being.

5   Etan Diamond, *And I Will Dwell in Their Midst: Orthodox Jews in Suburbia* (Chapel Hill: University of North Carolina Press, 2000).

6   Although this essay primarily focuses on the geographical dimension of the Orthodox Jewish experience in Thornhill, there is another side of the analysis, namely the suburban dimension. That issue is much larger that we have space for here, but it will suffice to say that the Orthodox Jewish experience in suburbia poses a fascinating challenge to the notion that suburbia is inherently community-less. After all, here are people who are fully immersed in the consumerist lifestyle of modern suburbia, but who intentionally choose to be part of a community structure with its attendant institutional affiliations and social networks. In other words, suburban communities can be found if one wants to find them.

7   *Canadian Jewish News*, September 3, 1981.

8   Note that the boundaries of Jewish Thornhill as used in this essay are slightly different from the generally accepted boundaries. For my purposes, the boundaries are as follows: CN rail tracks on the south; a western boundary through the neighborhoods west of New Westminster, until the hydroelectric lines and Highway 407 on the north. The hydro lines form the northern boundary, which then curves southeastward toward Yonge Street. The specific boundaries—which streets are in and which streets are out—are not so important to the broader argument and, in fact, the fuzziness of the boundaries only confirms my point that the Jewish religious spaces are as much based on perceptions as they are on any concrete definitions. As such, a particular family might live beyond these boundaries, and their religious microspace will be different as well. And the rest of the community will see this family as living "far" from the geographical core.

9   The city of North York had been an independent suburb that had become part of the Municipality of Metropolitan Toronto in the 1950s. It later was dissolved into Toronto in the late 1990s.

10 Leila Speisman, one of the earliest BAYT members, quoted in the *Thornhill Month* (September 1982).

11 Steven Speisman, interview, August 7, 2006.

12 *Canadian Jewish News*, April 7, 1988.

13 To say that suburbia is full of segregated spaces is in itself not a particularly surprising comment. One of the primary criticisms of the modern suburban environment is the sharp separation of land uses — residential neighborhoods here, commercial development there, industrial development nowhere. This is, of course, contrasted with the jumble of land uses in denser urban areas, the kind of complex streetscapes that urban proponents such as Jane Jacobs had long championed. Furthermore, sociological studies such as the landmark Crestwood Heights study of the 1950s provided detailed descriptions of how suburban space is divided up even at the household level. Even what has been written about religion and segregated suburban space has generally focused on the geographies across *different* religious groups — such as how the Catholic parish structure creates a different relationship to place than do the congregational models of Jews or Protestants. See Jane Jacobs, *The Death and Life of Great American Cities* (New York: Random House, 1961); John Seeley, R. Alexander Sim, and E. W. Loosley, *Crestwood Heights: A Study of the Culture of Suburban Life* (New York: Basic Books, 1956); Gerald Gamm, *Urban Exodus: Why the Jews Left Boston and the Catholics Stayed* (Cambridge: Harvard University Press, 1999). What I am discussing here is how even within a single religious group—even when of the same religious denomination—different institutional affiliations can lead to different spatial patterns and, consequently, different experiences and identities.

14 There are several apartment and condominium buildings in the area, and one street with townhouses. Although my discussion focuses primarily on the streets with single-family homes, there is somewhat of a social hierarchy among the apartment buildings as well. The Conservatory, a condominium building at the corner of Clark and Hilda, is considered the "nicest" building, both because of its proximity to the BAYT and to the grocery store, Sobey's, and because it contains a "Shabbat elevator" for residents. The Conservatory is contrasted with the various apartment buildings that surround the Promenade shopping centre, which are slightly further from the BAYT and lack the religious amenity of a Shabbat elevator.

15 The *hashkama minyan* phenomenon is linked, in part, to the larger right-wing shift in Orthodox Jewish communities since the 1970s. The comparable "authenticity" of the yeshiva world over the modern Orthodox world has resulted in an increase in behaviors that both demonstrate a more traditional religious outlook and the desire to distance oneself from things that "look liberal." Thus the built-in seats or pews in an Orthodox sanctuary are too similar to the kinds of sanctuaries in Conservative and Reform synagogues (bad), whereas tables and chairs are similar in style to yeshiva study hall (good). The popularity of studying in Israel among Orthodox Jews has also contributed to a perception that the Israeli-style synagogue is more authentic. In Israel, Shabbat services are typically held much earlier in the morning and have fewer "frills" (less congregational singing, no sermon). Upon their return to North America, many Orthodox Jews have sought to recreate the Israeli experience through the *hashkama minyan*. For a more detailed discussion on the right-ward trends in Orthodoxy, including the impact of studying in Israel, see Samuel Heilman, *Sliding to the Right: The Contest for the Future of American Jewish Orthodoxy* (Berkeley: University of California Press, 2006).

16 Yiddish for "worshippers." To *daven* (pronounced *daw-ven)* is to pray. *Davening* (daw-ven-ing) is the prayer service.

17 On the other hand, given the talkative nature of the parents in synagogue, perhaps the children *are* being unwittingly socialized into acceptable behaviors.

18 Kosher Today, "Kosher Today Weekly News Archives, 11/07/2005," http://www.koshertoday.com/archives/newsletter_2005/11_07_05.htm (accessed July 27, 2007).

19 *Canadian Jewish News*, June 22, 1989.

20 *Thornhill Month*, March 1980. In the end, Runnymede resolved the dispute by agreeing to move the historic buildings to a nearby park.

21 *Hametz* is the leavened food products that are prohibited during the week-long holiday of Passover. It is customary on the morning before Passover to burn one's *hametz*, so as to show that one has gotten rid of this prohibited food. (Whatever *hametz* is not burnt is symbolically sold to a non-Jew for the duration of Passover.)

22 The "Nine Days" is a period in the Hebrew month of Av, in which Jews recall the destruction of the Temple in Jerusalem in 70 CE. During the Nine Days, Orthodox Jews typically refrain from engaging in "happy" activities, such as listening to music. It is difficult to maintain this custom, however, when a loud concert is being held in a public park right behind your house.

23 The prototype of master planned suburban communities in the post-World War II period, Levittown was the brainchild of the land developer William Levitt. There were, actually, three Levittowns, the first rising from the potato fields of Long Island in the late 1940s, the second in suburban Philadelphia a decade later, and the third in central New Jersey shortly thereafter. Levittown's fame stemmed from three factors: the assembly-line speed at which the homes were built, the relative affordability of the neighborhood compared to urban areas or more upscale suburban enclaves, and the all-encompassing nature of the development, with schools, parks, shopping plazas, and even places of worship, erected alongside the houses. Each of these factors represented a change from earlier suburban developments and each became part of the standard repertoire of subsequent developments, Joseph Tanenbaum's Spring Farm included.

24 By the late 1990s, such language seemed to have dissipated, in large part because affluence levels in Thornhill had been superseded by Toronto neighborhoods populated by the *haredi* Orthodox; custom-built homes in the Orthodox Jewish-dominated Bathurst-Lawrence neighborhood in Toronto, for example, easily surpassed the typical subdivision home of Thornhill.

25 Alan Mittleman, "From Jewish Street to Public Square," *First Things* 125 (August/ September 2002): 29–37, http://www.firstthings.com/ftissues/ft0208/articles/mittleman.html.

7 Altering Alternatives

Mapping Jewish Subcultures in Budapest

Eszter Brigitta Gantner and Mátyás Kovács

"Judapest," blog header of Hungarian Jewish Internet portal of the same name, 2007

With the events of 1989, Central European countries and the former member states of the Soviet Union experienced major changes. In tandem with the democratization of these countries, a thaw began to occur in the rigid structures of the still existing Jewish communities.

This process is clearly visible in Budapest, the only city in Central Europe that has a large and well-established Jewish community. Within the former Soviet block, the continuous, noticeable presence of this Jewish community and of Jewish culture represents a unique phenomenon.

The political concept that defined Hungarian Jewry exclusively on the basis of religion and, as a result, homogenized its diversity, also dominated Hungarian Jewish institutions, and those institutions have been corroded by the processes of transition of the last two decades. It is hardly surprising that in response to these political changes previously unknown concepts of identity appeared within Hungarian Jewry after 1989.

Over the last several years, a number of scholars have examined this transformation.[1] An book edited by András Kovács emphasizes the plurality of Hungarian Jewish identities in the present and analyzes the possibilities of retrieval/consolidation of individual and social identity after 1989. This pluralism has determined self-definitions within the Jewish community, attitudes towards Jews and the Jewish community, as well as the development of local Jewish culture. Typical of Central European Jewish communities in general, this pluralism can be characterized as follows:

> After the 1989 transformation, the "re"-founded religious organizations and associations have tried to locate themselves on one of the poles of this ideological system, searching for their diachronic place in the history of the Hungarian Jewish interpretation of religion; however, their actual institutional framework and their financial support is determined by foreign patterns and institutions.[2]

This phenomenon clearly surfaced within the Jewish community of Budapest, which is—due to the remaining organizational structures—an arena of processes moving in divergent directions. The presence of a Jewish cultural space defined by "official" institutions is undeniable. However, the reading of Jewishness presented by the official Hungarian-Jewish associations, unsurprisingly conforms with the expectations of non-Jewish residents as well as the assimilated generations that adapted or were even brought up with the political program of being "culturally Jewish." This concept, adopted after the Hungarian Revolution of 1956, focused on secular Jewish culture and the successful moments of the Hungarian-Jewish symbiosis, while at the same time suppressing any religious or Zionist elements.[3] Whereas it is promoted by the established community institutions until this very day, it does not reflect the fact that newer generations have grown up who define their Jewishness in ways that differ from those of their parents, and who are familiar with Jewish life in Israel and possess a knowledge of religion and new forms of Jewish identity.

Accordingly, the forces of fragmentation or pluralization of Jewish identity in Budapest can be most clearly observed along generational lines. A new generation has grown up whose identity was not determined by a model inherited from the socialist system but rather has been informed by knowledge of Western models and Israel. This "new" Jewish identity is what simultaneously redefines the spaces used by the new generation. Based on this, we assume that these different uses of space interfere as the space is not only determined by the presence of official Jewish institutions—representing mostly the older generations, but also by the self-organized Jewish subculture of younger people who create their own interpretations of Jewish culture, and in so doing, make the plurality of Jewish life visible. Alternative Jewish subculture in Budapest is the product of our generation; of a generation with new experiences of Jewish identity.

In this essay, we examine the process of the shaping and configuring of "Jewish space" in Budapest on three different levels. The differentiation within the three levels is based on diverging identification lines:

- Historical spaces: there are two historical Jewish quarters, both located in Pest on the left (east) bank of the Danube—Inner-Erzsébetváros, or Elisabeth Quarter in the 7th district, which is the so-called Old Jewish Quarter,[4] and Újlipótváros, or New Leopold Quarter, which is part of the 13th district further up north and has served as the location for much of the Jewish middle class since its construction during the first decades of the twentieth century.
- Spaces of contemporary urban culture: official sites and alternative scenes.
- Virtual spaces: official Jewish magazines and the appearance of new online media that offer a forum to Jewish identity and culture, e.g., blogs and web magazines.

Historical Spaces

The fabric of the city is woven by physical places, cultural ideas and narrative spaces since cultural representations and ideas are also built in the buildings; at the same time, the built spaces generate experience and interpretations. Hence the immateriality of substantial elements and the ideas that become material appear simultaneously. This twofoldedness makes it necessary that we put beside the concept of the city in its physical sense, the concept of the symbolically created city that is experienced, imagined, walked and written down. The city conceived in this way is the pragmatic example of how a geographical place transforms into a cultural space.[5]

A conspicuous example of what is described in the above epigraph is the Jewish quarter of Budapest. It is not only a physical place, several buildings of which are part of the religious Jewish space, but also a space whose narrative projection and the ideas connected to it have determined attitudes held about the Jewish Quarter until the present day. These attitudes consist primarily of a set of clichés that are projected onto the quarter, blurring the historical frames that have determined the foundation and development of this neighborhood within the city.

We can talk about a Jewish quarter in Pest from the beginning of the nineteenth century. In the context of our work, we use the definition of the *Encyclopedia Judaica*,[6] according to which a Jewish Quarter is understood as a residential area that evolves spontaneously. It is not separated from its surroundings—its natural boundaries are determined by the built Jewish environment and its inhabitants. This was the case in Pest.

Jews were permitted to settle in Pest from 1786. The city council allowed them to rent permanent dwellings and to maintain shops as well. During the first decade of the nineteenth century the number of Jewish families tolerated in the city increased to fifty.[7] Jews settled east of the city wall, close to the so-called Újvásártér (New market square); a market square at the beginning of what is today's Király (King) Street. The old Jewish quarter[8] began to grow during this period, as did religious, social, and philosophical differences. These tensions became visible in the form of three buildings in this densely populated space of little streets: the Dohány Street Neologue[9] synagogue, which was completed in 1859; the Orthodox synagogue, built in 1911; and the Rumbach Street synagogue, designed by Otto Wagner and built in 1869. By the end of the nineteenth century, the Quarter was home to everyone from university professors and rabbis to Jewish workers, as well as members of the lower middle class such as shopkeepers, artisans, tailors, clerks, chandlers, agents, brokers, redcaps, and students.

In 1850, Jews in Terézváros (Theresia Quarter)[10] represented 26.9 percent of the population as a whole. In 1867, the number of the Jewish inhabitants of Terézváros had increased to twenty-six thousand. The Jewish population reached its peak in this district before the First World War, with more than fifty thousand Jews living in Terézváros and approximately seventy thousand in Erzsébetváros.[11] Of the total population of Budapest Jewry, 59.3 percent lived in this part of the city. However, between the two world wars, a new district emerged and caught up with the old quarter: Újlipótváros (New Leopold Quarter), built in the place of the former mill row, can be considered as the continuation of the downtown area Lipótváros (Leopold Quarter). This is the only area of Budapest that experienced a growth in the number of Jewish inhabitants after the 1920s.[12] It was primarily Jewish intellectuals, professionals, and private clerks who moved into the new district; thus, in contrast to the traditional Jewish quarter, this new neighborhood was home to those who had more recently achieved their current status. Even between the two world wars, the traditional Jewish quarter included a significant number of Orthodox Jews, and traditional professions such as merchants or artisans, who were also highly represented.

The darkest era of the Jewish community in Budapest arrived in November, 1944, after the unsuccessful attempt of the Horthy government to withdraw from the German alliance. Supported by the German invaders, the new Szálasi government erected ghettos in both areas, with the intention of deporting the Jews of Budapest. The larger ghetto was built in the Old Jewish quarter, crammed with tens of thousands of people living under inhuman conditions. At the same time, an "international" ghetto was established in certain streets of Újlipótváros; here stood most of the so-called protected buildings where Jews of foreign citizenship were assembled.[13] Újlipótváros was liberated on January 16, 1945, and two days later the Soviets also liberated the main ghetto in the old Jewish quarter.

Between 1945 and 1989

After 1945, in spite of huge losses, life in the Hungarian Jewish communities picked up once again. Unlike Jewish communities in the provinces, where most members had been deported and killed between 1944 and 1945, the community in Budapest had been spared major deportations, thanks to the impending defeat of the Wehrmacht and the German withdrawal from Hungary. As a result, the composition of Hungarian Jewry changed dramatically.

Since the majority of Hungarian Jews lived in the capital after the war— 96,537 as of 1949[14]—the Budapest Jewish community had a defining role in postwar Hungarian Jewish life. The social structure of the country's Jewish community changed partially because of the Holocaust, partially because of emigration. Whereas the religious, or at least traditionalist, provincial Jewish communities had been almost entirely destroyed, the assimilated middle-class had the highest survival rate, and was now trying to find its place in the "reborn" Hungary.

Religious life after the war was soon established anew, mainly by the reorganization of the neologue communities. The other types of community organizations, for example, associations and guilds, only functioned until 1950, when they were forcibly dissolved. Synagogue congregations were placed under considerable pressure by the state to form one unitary religious community—led by the National Representatives of Hungarian Jews. This centralization allowed for both more efficient and more stringent control of the community and tied in well with the concept of "cultural Jewishness" that was promoted by the official Jewish organizations after the 1956 revolution. Many Jews in socialist Hungary, out of their personal experiences and their fear of anti-Semitism, against which they hoped to be defended by the government, accepted this homogenizing, two-dimensional self-definition.[15] Although the political changes of 1989 shook the Jewish as well as the larger Hungarian society, remnants of this mindset can still be found in the policies of the official Jewish organizations almost twenty years after the transformation. Even if it seems probable that their monopolistic status will fade as independent new congregations are established, these official organizations are still the ones with which the Hungarian government and foreign Jewish and non-Jewish organizations negotiate. This is how a paradoxical situation arose, one in which a multifaceted Hungarian Jewry continues to be represented by a single association with religious and political functions as well: the MAZSIHISZ,[16] the Alliance of the Jewish Communities of Hungary, is the only one that is officially accepted. It is not our task to discuss the complicated questions of legitimacy in detail; we are only noting the dynamics that influence the aspects of space formation analyzed here. The above-mentioned practice of homogenization is also a factor in festival celebrations and other Jewish cultural events, including the Hanukah Ball and Holocaust commemorations, most of which, incidentally, take place in the so-called Jewish Quarter.

Thus, the shaping of the quarter may be conceived in two ways: one is the reading of the quarter, i.e., the representation of the quarter in publications, while the other is the use of the quarter as a space of legitimacy and a venue for cultural events.

Jewish Quarters: Old and New

The old and the new quarters, as historical spaces, together carry more than merely the built environment; hence, we have to place this 'walked' and 'written' district/space beyond the physical, built space, as described in the quote that began the section on historical spaces above.

Both districts, the Old Jewish quarter (today's Erzsébetváros) and Újlipótváros, the second area of settlement, represented a Jewish space for their contemporaries. Many literary works, articles, and memoirs[17] written by non-Jewish authors clearly identify these places as specifically Jewish quarters. In anti-Semitic literature prior to the First World War, and in the interwar period, these spaces were also inseparably interwoven with Jewish notions. The Jewish character of a place was determined by its users—mostly the inhabitants and the approved religious and communal buildings.

To this day, both have been retained in the collective memory as Jewish quarters. However, due to its historical development, the old Jewish quarter has slowly become a constructed "Jewish cultural space" for Jews and non-

Jews alike. This space functions as a screen onto which Jewish themes are projected by gentiles.[18]

At the same time, Jews have constructed their own reading: the secular Jewish reading perceives the Jewish quarter as part of its cultural heritage, a site at which several kinds of secular Jewish identities materialize. In the present phase of our research, it seems that two basic readings of this space within the secular Jewish environment are possible.[19] On the one hand, a *shtetl*, composed of romantic ideas of Jewish life at the turn of the twentieth century and of Holocaust elements, is being constructed here, a marketable reading for both Jews and gentiles. On the other hand, the phenomenon of an alternative Jewish subculture is also organized in the same built space. This phenomenon carries new kinds of self-expression and redefines the space as it is used by the subculture. So far the two readings we have mentioned have not conflicted with each other, although many of the subculture's activists have deliberately criticized the first reading, which is primarily associated with older age groups within the diverse Jewish community and is the image conveyed by official institutions.[20]

It is obvious that behind these generational rifts stand different historical experiences which have, together with possible self-reflections, influenced the interpretation of the spaces used. The Holocaust, followed by nearly four decades of state socialism led to the emergence of a concept of survival in Jewish organizations. Representing the community to, and fulfilling the role of negotiating partner with the government, these official bodies allowed no other forms of identity to succeed within their organizational frameworks.

Interestingly enough, the concept promoted by the official Jewish institutions leaves no room for an interpretation of Újlipótváros or for the fact that the generation that grew up with and after the political transformation of 1989 considers the quarter as a place of differing cultures and self-definitions. The use of Újlipótváros as a Jewish space is informal at the moment, operating on the level of personal relations rather than on any formally organized level. Although Újlipótváros has the same typical markers as the old quarter (Jewish inhabitants, Jewish infrastructure, a synagogue), the filling and forming of the space has not yet taken place here, neither officially nor in the way of its Jewish inhabitants' self-organization. And there appears to be no impetus for such organization, despite the fact that one-fifth of the Jewish population of Budapest lives here. The connection of the two quarters to each other can be grasped on two levels: traditionalist-assimilated and center-periphery relations. The answer to the question of why a Jewish subculture evolved in this space can be found first of all in its physical characteristics: central location and atmospheric buildings. Second, the Old Jewish quarter of Pest is known as a religious, traditionalist place. The elements of provocation, Jewish identity, and subculture—which incorporate young people as their own creators and consumers—are formulated by young Jews in comparison to, in opposition to, or as a reinterpretation of, the traditional way of life that requires closed and strict conduct.

Contemporary Urban Spaces: Official Institutions and the Alternative Scene

By 1945, Jewish life in Eastern and Central Europe had disappeared from the sphere of everyday interactions, and the resulting empty space did not become the subject of discourse: it was pushed out of the public arena by the silence and suppression imposed by the gradually evolving Communist state power and its anti-Zionist policies. In light of these transformations it is worth noting not only the way that Jewish life was forced into representation by a single organization and stigmatized because of that organization's religious nature, has been forming its plural social and free religious-institutional life, but also how the two Jewish quarters the Elisabeth and the Theresia Quarter as well as New Leopold Quarter, on the official level deprived of their Jewish character for half a century, could again be filled with Jewish content.

In several Central European countries, the built environment—the Jewish quarter—is clothed with nostalgic contents that push it back into the past. In the absence of Jewish communities that might have once had a voice in the discourse, the architectural outlook and the marketing of the Jewish quarters depends mostly on non-Jewish agents and associations; which has resulted often in clichéd and stereotypical reconstructions.

Therefore, by "constructed cultural space" we understand a space in which the Jewish elements are determined by non-Jewish agents. Furthermore, this space needs to be centered around former synagogues and other buildings with religious functions. The former function of these buildings (e.g., sites of ritual), the knowledge connected to them, and the fact that they once played a significant role in the life of the community renders legitimate the cultural spaces developed on these same sites, even if the original structures have disappeared. However, this concept takes into consideration neither the existence or respectively the nonexistence of the community, nor the reading by the community itself.[21] The Jewishness of the neighborhood proclaimed as the 'Jewish quarter' is defined by the buildings and the ideas connected to them. As we will discuss later, this Jewishness consists mainly in remembering a longed-for past and conveying a homogeneous image of Jewish society. Our interpretation draws on both the broad approach of Pinto, Bodemann, and Gruber,[22] as well as on Richárd Papp's concept of "symbolic space." The basis of Papp's concept "beside the narrower ritual community is created by a traditional-halachic value and norm system connected to the cultural space 'invisible to others' and to the practice of ritual life."[23] It is important to note here that apart from the studies of Richárd Papp and András A. Gergely, Hungarian academia hardly deals with urban anthropology, much less the special problem of "ethnoscapes" and "places of identities" in the context of Hungarian Jewry. This is why we can barely depend on or refer to Hungarian research literature or discourse.

However, the cultural space that has evolved in Budapest only partly corresponds with the above mentioned concepts. On the one hand, there is Jewish cultural space as an existing community has shaped it according to its own identities: the sum of its manifestations constructs the cultural space. Jewish cultural space in Budapest thus differs from other Central European cases in that its creators are official Jewish institutions. At the same time, however paradoxical it may seem, Hungarian Jewish organizations since the last third of the nineteenth century, being the institutionalization of a long-standing assimilationist strategy convey a homogeneous image of Jewry that obstructs the unfolding of any kind of meaningful plurality. Therefore, monopolization and the two-dimensional nature of Jewish cultural space in Budapest can be attributed to non-Jewish agents as well as the one-sided conceptions within the establishment of the Jewish community. Accordingly, what makes Budapest different from other cities in Eastern and Central Europe is not the lack of constructed Jewish space but the plethora of alternative readings that appear at the level of subcultures.

In Budapest, the culture constructed by official organizations, as well as that of the subculture constructs, appear in the same physical space—in the old Jewish quarter. Whereas the former has its historical causes, since the representational-political structures of emancipated Jewry and the ritual institutions suitable for the growing population evolved in the district throughout the nineteenth century, the emergence of subculture here was a matter of choice, based on the decision of its creators. One of the owners of the club "Szóda"—regarded by people as a Jewish place—chose the old Jewish quarter as its location for personal reasons, namely because he and his co-owner grew up there.[24]

The Venues of Jewish Subculture

The places of the subculture include Szóda (Soda), the ad hoc stages of the Budapesti Zsidó Színkör (Jewish Theatre of Budapest), and Sirály (Gull),— the old Jewish quarter. These sites also represent different forms of cultural activities: while Szóda is a meeting place, Sirály organizes different concerts (e.g., alternative Jewish music), roundtable discussions on contemporary Jewish problems (e.g., homosexuality and Judaism), and screens contemporary Israeli films. Sirály understands itself as an open cultural workshop where

Jewish programs are a part of a more general, alternative cultural approach. With the exception of Szóda, all others have functioned until now as NGOs.

Yet, the phenomenon of these organizations (re)turning the Quarter into a "Jewish place" was not determined exclusively by their location in the Quarter. Interactions occur here because the Quarter has retained its Jewishness for the generations who grew up in the 1980s and 1990s, while at the same time, with the regular presence and unconcealed Jewish identity of those who organize and participate in the activities of the alternative scene, the Jewish character of these meeting places is reinforced, providing the Quarter and what goes on there with a new interpretation compared to the traditional one determined by religion.[25] Signs that these alternative entities have every intention of legitimating themselves are obvious, and in that regard there is no difference between them and the organizations belonging to the Jewish institutional structure. The rift between the subculture and its space on the one hand and the Jewish institutional structures and their space on the other, is apparently defined by audience: the shapers of the subculture unequivocally create/offer content that presents and represents their own generation.

> We proceeded from the idea that for individuals the meaning of Jewish origin and the affiliation to Jewry is determined by the historical and social context. Since this context has gone through dramatic changes over the last century and a half, we supposed that the identity strategies have been greatly influenced by generational affiliations.[26]

Taking the categorization of András Kovács as a basis, we classified the creators of the subculture as the generation born after 1966; by and large, they positively identify as Jews, chosing certain aspects of Jewish identity while rejecting others.[27] The programs are also shaped according to the needs and wishes of this generation, modeled primarily on contemporary progressive designs of Western Europe, the U.S., and Israel[28] in contrast to the three official main topoi: Holocaust, Yiddishkeit, and a nostalgia for the good old days.[29] A perfect example is the above-mentioned Sirály.

The programs offered by the subcultural scene do not seem to fit into the image of the institutionalized official Jewish organizations. At the same time, several subcultural initiatives, Marom, for example, have grown out of these institutions. Led by Ádám Schönberger, this student organization was established on the initiative of young people organizing themselves in the synagogue in Bethlen Square. This congregation, together with its rabbi, made it possible for the group to found a rock band. While the institutionalized Jewish political organizations do not respond to the existence of the subculture, certain formations within the subculture do respond with a deliberate political attitude to the deficiencies, inflexibilities, and anachronisms of the rigid, traditional community structures. The director of the Budapesti Zsidó Színkör, Róbert Vajda, told us in an interview that in a performance called "Tartuffe"[30] they intentionally responded to the activity and self-concept of MAZSIHISZ[31]—the association considered as representative of Hungarian Jewry on a governmental level. This activity "is quite detrimental, (...) Mazsihisz only knows of Holocaust and anti-Semitism, (...) it is not in their interest that young people should express their opinion and show that now there are other alternatives."[32]

Despite the static image outlined above, some kind of interaction does seem to be taking place, especially in the field of culture. However, these interactions occur within such a limited arena and affect so few people that they lack the power to really alter the character of the space maintained by the official structures and the self-representations within that space. A good example, again, is Sirály, several of whose programs have been adopted by Bálint Ház, which, in turn, belongs to the official structure.[33]

Virtual Spaces: New Forms and New Mediums of Jewish Culture and Identity

The brand new form of Hungarian-Jewish self-representation is happening online.[34] (Izidor Kohn and Manó Schönberger)

This was the headline of an article by Kohn and Schönberger, which provided the first detailed analysis and critique of Hungarian Jewish Internet activity, published in June 2005 in *Szombat* (Sabbath), a Jewish political-cultural weekly. The authors gave a brief overview of the local Jewish online scene through a comparative introduction of four websites—Judapest and Sófár Média blogs on the one hand, and WiW and Jewish Meeting Point community sites on the other. Since then—apart from some interviews in the general media[35]—not a single academic study has tried to review the Hungarian Jewish web, and except for some very recent initiatives, no Hungarian-Jewish website Goliath has emerged to challenge these earlier ones.

As opposed to Ruth Ellen Gruber's thesis, which understands the Jewish virtuality[36] of Central and Eastern Europe as an absence and fake refill with substitute activities (ghetto tours, reuse of former Jewish spaces, etc.),[37] Budapest is in a unique situation. Unlike Cracow, the Scheunenviertel in Berlin, or Prague, virtual spaces in Budapest are not pseudo-structures in real, but vanished places[38] since Jews here act as productive members of the virtual scene. Jewish life in Budapest is not showcased only by Jewish Quarter tours, festivals, and museums, but also by a rich and proliferating online culture. This kind of representation, too, mirrors the above mentioned diversity of Hungarian Jewish identity.

In the following we will give a comparative overview of the contents and orientations of these websites, basing our analysis on the four listed below:

- Judapest.org—pulsing net ghetto (www.judapest.org)
- Sófár Média (www.sofar.hu)
- Pilpul alterJew webmag (www.pilpul.net)
- +1: Matula Magazin (www.matula.hu)

Judapest.org (www.judapest.org)

Every active member of the Jewish subcultural scene in Budapest with whom we conducted interviews in connection with this study, had one thing in common when it came to identifying the places and spaces of their (sub-)culture: Bruno Bitter and his blog *Judapest.*

> When I'm online, my name's Shadai, I'm a Jewish blogger. I've been writing my post for one and a half years, with some of my friends at the judapest.org blog. Most of my posts are about Jewish pop-culture, postmodern identity-politics, alternative and radical Jewish initiatives, Israeli public and street art. Offline my name is Bruno Bitter, I work as a marketing- and market researcher at a firm specialising in qualitative measures. I live in Budapest as a true patriot of Újlipótváros, about which I have another community blog under www.ujlipotvaros.hu."[39]

Although since Bitter's interview for Szombat, the Újlipótváros "super-local site"[40] has been closed down, and its author has moved a little bit closer to the Old Jewish Quarter in the 7th district, the *ars poetica* of Judapest remains the same. In the blog's FAQ, this is how they explain their philosophy:

> The word 'judapest' was invented by Karl Lueger, Mayor of Vienna at the turn of the 19th century: Karl used it with an anti-Semitic (—and by the way anti-Hungarian) sting. The local humour magazines of the epoch quickly took over the topos; 'judapesting' has in fact until today been an expression of anti-Semitic discourse. Nevertheless, we declare a semantical war. We will take over the expression and fill it up with positive contents. Judapest: creative energy, reflection, a positive way of seeing things, inspiration. Jewish identity and culture for the 21st century.[41]

This self-definition reflects the Budapest Jewish subcultural avant-garde with a high degree of accuracy, and confirms the previously mentioned findings of András Kovács concerning the generation of Hungarian Jews born after 1966.[42] In the semantical fight, the Judapest blog aims to deconstruct crumbled identities,[43] as well as link the identity-pluralism through the views of the editorial board, emerging as a source of creativity and content-pluralism on the Internet:

> The youngest 'judapest-blogger' is 22 years old, the eldest 31; about half of our team is made up of girls. We have for example an anarcho-capitalist business punk, a Lubavitcher Japanese film expert, a doctor transformed into a writer, a book-keeper, a beginner master of the art of living and a city-biker psychologist. [Concerning religion,] the judapest blogger team is again far from homogenous. We have a Shomer Sabbath Chabadnik and an anticlerical

libertarian as well. After reading my posts, a modern orthodox friend of mine called me a 'l'art pour l'art Jew'—this sounds funny, but actually, it's true.[44]

The blog consciously tries to ignore daily Hungarian and Israeli politics, the question of the Holocaust in terms of politics, the problem of Hungarian extreme right-wing politics, and those mainstream topics that constitute the profound elements of the discourse about Jewry in the "big media," e.g., the daily political press and mainstream Jewish publications. Instead, they feature Hungarian, Israeli, and Western Jewish pop culture, gastronomy, theater and film reviews, and contemporary reflections on religion. A post about the Jewish Summer Festival clearly outlines Judapest's attitude towards the Jewish mainstream:

> Let me begin with a tiny outburst: the Jewish Summer Festival has for years been based on retrograde, perfectly tasteless productions and skansen-like Jewish kitsch. It is wholly infiltrated by some outworn and deceptive attitude: this at least flawlessly represents Hungarian Jewish public life. Until now, we haven't talked about this, but let's make it clear once and for all: KLEZMER SUCKS. 99% polka, for German and Danish tourists, as well as for indolent, mid-aged Hungarian Jews, who of course have no bloody idea about Jewish culture and good taste as such.[45]

Sófár Média (www.sofar.hu)

Sófár Média, though started as a blog-like news site, grew to become a complex enterprise. It contains a Jewish event calendar, a Jewish community radio program, photo albums and videos, a Jewish lexicon, a virtual folder for articles about the former Jewish Quarter, a Jewish dating service, community blogs and online forums—and in addition it has a major profile as news provider. Sófár Média not only differs from other Hungarian Jewish sites in its complexity and comprehensiveness: the operators of the website became political actors when the head of the Sófár initiative announced a flashmob for a "Jew-friendly Jewish Quarter" in 2006.[46]

Concurrent with the "culturally Jewish" concept, these actions outline Sófár's narrative of the former Jewish Quarter, containing nostalgic elements ("with us, the Jews of Pest," "Király utca was Budapest's first cobbled public road, where nearly a century ago our ancestors were entertained in dozens of cabarets and theatres") and Yiddishkeit-reminiscences ("Explore the Secrets of the Pest Jewish Triangle", etc.). Even the flashmob-ad represents this highly nostalgic attitude ("A century ago Jews and Hungarians made up a bustling, multi-coloured and multi-cultural crowd in the Jewish Quarter." The event closed with the collective singing of "The rooster is crowing," the most famous Hungarian-Jewish folk song). All in all, Sofár topics underlie the Jewish self-definition of an older generation that determines the cultural mainstream and a secular reading of Jewish issues.

Pilpul: www.pilpul.net

Pilpul is perhaps the first Hungarian-Jewish site since the launching of Judapest to specialize (also) in subculture. However, in comparison to Bruno Bitter's blog, this webmagazine is not at all independent. Its two main sponsors are *The American Jewish Joint Distribution Committee* and *The L.A. Pincus Fund for Jewish Education in the Diaspora, Israel*. Moreover, the website and its hard copy fanzine are the official media of the Marom conservative Jewish youth movement. It is chaired by Adam Schönberger, who on the other hand is one of the chief organizers of Sirály "the newest nightspot in Budapest in the heart of what used to be this city's Jewish ghetto." [47]

The first among Pilpul's main columns is called *dossier*, in which one is able to browse articles in thematic groupings, e.g., cinema, Judaism, or book reviews. Then there are the *serious* and the *easy*, presenting articles that appear in the dossier, but collected separately into these two categories. These again are a Marom initiative, which—although defining itself as a conservative movement—also belongs to the progressive Jewish scene, at least on a local Hungarian level. Marom also organized a religious round-table called *Pluralism in Jewish Tradition*, in which they managed to bring together representatives of all contemporary Hungarian-Jewish movements[48] to discuss such topics as assimilation, the role of women in Judaism, mixed marriage, and redemption.

The Pilpul editorial board represents the new generation familiar with Western and Israeli patterns, the type of identity that has heavily influenced the Budapest Jewish subculture. Compared to the cultural approach of Judapest, Pilpul is the flagship for contemporary reflections on tradition and wider perspectives on religion. As a former synagogue youth group under Schönberger's leadership, they were the first to organize Jewish theater performances and to set up a klez-hip-hop band called Hagesher. Today they are determined to contribute to Sirály programs. As of this writing, their most recent action was a "matzo-snack-dinner" mix-Haggada prepared for Passover 2007, to be freely downloaded from their website.[49]

Matula Magazin: www.matula.hu

In comparison to the first three websites, Matula Magazin is a subcultural, but not at all Jewish page, at least not in terms of Judaism, which is why we marked it with a "+1" in the introduction to this section. A fanzine written by twenty-somethings, Matula is one of the most controversial independently-run subcultural sites. The style of the site could be described as a mix of punk and anarchism, and the authors deal mostly with religious, cultural, and lifestyle issues. Matula's reputation for controversy is well-deserved, boosted by articles such as the one that announced Pope John Paul II's death as he was going through his first serious illness, or the portrayal of Jesus, as he is first entering the world, sporting pink sunglasses. Matula became an important agent in the mainstream discourse about Hungarian Jewry when the political dailies discovered their service IZSDB, *The Internet Zsidó [Jewish] Database:*

> The Internet Zsidó Database is a machine by which anyone can find out if a living Hungarian public figure is Jewish. A Jew is who we think is Jewish. A public figure is who we think he/she is.[50]

The database was meant as a kind of "alternative" and satirical reaction to one of the most troubling issues in Hungarian political discourse in the last seventeen years, namely, the "hidden" anti-Semitism of the right-wing and conservative parties. However, the database led to a political scandal,[51] in which the ombudsman for data protection, speakers of the March of the Living 2005 Budapest, as well as the extreme right-wing Jobbik party voiced their opinion. Linking politics and Judaism made another brilliant satirical appearance via Matula: *Dejudaising Újlipótváros*[52] was a fictive report from German television ARD, broadcast from Budapest in 2005 and parallelling the Israeli evacuation of the Gaza Strip. The basic idea of the imaginary program—released in a written format on the webpage—was to explore the possibility of the same action taking place in Budapest.

Matula Magazin's satiric article on Újlipótváros and the "repatriating" of its residents, as well as the contributions in Pilpul and Judapest would be unpublishable in the mainstream media. They are the self-expressions of a new generation, which instead of using the institutionalized channels of freedom of speech and Jewish intracommunity communication, reach their audiences via the Internet. The low costs of launching and maintaining such sites, the potentially enormous outreach, and the pluralism of content and reactions democratize the discourses *in* and *about* the community. All this shakes the mainstream to its foundations, changing not only the implications of a subcultural position, but also the sources of self-definition of the shapers and recipients of this cultural sphere, their cultural and other self-expressions, and the interpretation of (their own) Jewishness as well.

Conclusion

Since the 2006 publication of the first results of our research,[53] we have observed that the institutions and representatives of the Budapest Jewish subculture have become firmly embedded in the former Jewish quarter, giving undeniable impulses to its cultural palette. During this same period, with the establishment of the new and hip venue Sirály, the emergence of

Jewish subculture has been striking. Its struggle for survival clearly outlines generational, institutional, and (cultural-)political fractures, which on the one hand characterize the uses of Inner-Erzsébetváros, and the official Jewish organizations and alternative cultural associations on the other. In physical as well as virtual space, new demands of Jewish self-representation and institutionalized forms of Judaism have arisen. Meanwhile, Sirály is becoming a symbol of the encounter of youth organizations with virtual experiments, and its struggle for legitimation.

This study is a snapshot, since space in a physical, symbolical, religious, and virtual sense, as well as the actors of the subculture, constantly change—the latter much more dynamically. Nevertheless, this snapshot allows conclusions to be drawn not just about the present and the past: Judapest, the Jewish Theatre of Budapest, Hagesher, Sirály, the appearance of increasing numbers of cultural initiatives—these all imply that there is an ever-growing number of youth rushing through the streets of the former Jewish quarter, open to Jewish subculture(s) and sharing similar values and beliefs.

Sirály, the best example of the institutionalizing and maintaining of these processes, can in the long run function as a milestone in the creation of a critical (subcultural) mass, its transparency serving as a reference for the legitimacy of the alternative approaches within Jewish (mainstream) culture.

In our explorations in the fields of Jewish subculture, the *mainstream versus alternative* comparison has to some extent obviously oversimplified flexible and sensitive self-representations, although, as seen in the comparison of Inner-Erzsébetváros to Újlipótváros, dichotomic contemplation may serve as an adequate basis for the description of these phenomena and the systematic mapping of the given relations. Analyzing religious subcultures and the preferences of smaller Budapest-based communities—egalitarian, Modern Orthodox, a congregation led by a female rabbi, etc.—did not fit the narrow frames of this study. This chapter, *Altering Alternatives,* has focused primarily on secular Jewish cultural space. However, in our experience these dichotomic systems can also be applied to the religious sphere. The latter raises a further issue, namely, the role and legitimacy of the representative Hungarian-Jewish political-cultural organizations in an overwhelmingly assimilated environment. As one of the interviewees put it, "Why don't we call the representative Alliance of the Jewish Communities of Hungary a sort of subculture instead of using the term for the alternative cultural events mobilizing hundreds and thousands?"

The emergence and spread of the World Wide Web—both the forever increasing speed and greater access utilized by a highly qualified social segment of professionals, and the experience of temporarily living abroad served as the primary basis for our research—has not only provided a new surface for the unfolding of the Jewish subculture. It has also changed the role and orientation of the creators and recipients from the ground up, in the Budapest alternative scene as well as among consumers of traditional and new media. The internet, especially the so-called web 2.0, the read-and-write web, ignoring the channels of intracommunity communication and opening new forums for expression, has lead to a stronger democratization within the Jewish community while at the same time strengthening the well-known plurality of Hungarian Jewishness. Nowadays, views, works, and productions have come to light that ten years earlier would never have had a chance to find either an audience or patrons from conventional backgrounds. The virtual meeting points, amplified by the clubs and cafés established in the former Jewish quarter, signify real alternatives to the institutionalized cultural outlets such as the Jewish Summer Festival or programs of the traditional "Bálint" Jewish Community Centre. The virtual origins of relations and opinion-exchange along with the Inner-Erzsébetváros bars create intimate communities open to everyone, and which in the long run will be able to give shelter to the constantly renewed forms of alternative self-expression.

Notes

1   Zvi Gittelman, "Reconstructing Jewish Communities and Jewish Identities, in Post-Communist East Central Europe," in *Jewish Studies at the CEU: I. Yearbook (Public Lectures 1996–1999),* ed. András Kovács (Budapest: CEU, 2001); András Kovács, ed., *Zsidók a mai Magyarországon* [Jews in Hungary Today] (Budapest: Múlt és Jövő, 2002). Note: All translations from Hungarian into English in this article are by the authors.

2   Kata Zsófia Vincze, "A zsidó valláshoz való 'visszatérők' Budapesten: A hagyománytól való elszakadás és a *báál tsuvá* jelenség kérdései" [Ba'alei teshuvah in Budapest: The break with the tradition and the questions of the ba'alei tshuvah phenomenon] (PhD diss., Eötvös Loránd University of Sciences, Budapest, 2006), 43.

3   András Kovács, "Magyar zsidópolitika a második világháború végétől a kommunista rendszer bukásáig" [Hungarian policy and the Jewish community from the end of WWII until the collapse of the Communist regime], *Múlt és Jövő* 3 (2003); see also László Csorba, "Izraelita felekezeti élet Magyarországon a vészkorszaktól a nyolcvanas évekig" [Jewish community life in Hungary after the Shoah until the 1980s] in *Hét évtized a hazai zsidóság életében II* [Seven Decades in the Life of the Hungarian Jewry], ed. Horváth et al. (Budapest: MTA Filozófiai Intézet, 1990) 61–191.

4   Its inner parts, as the outer shells of the historical downtown, represent the Old Jewish Quarter.

5   "Tér, kép és tér-kép: A modern város" [Spaces, images, and space-images: the modern city], Scientific Conference, Pécs, Hungary, May 7–8, 2004. http://www.hermes.btk.pte.hu/letoltes/URBS.RTF (accessed July 25, 2007).

6   "From the beginning of the 16th century the name given in Italy to the Jewish quarter which was separated and closed off by law. From the other parts of the town by wall and gates the word ghetto has also been used to designate Jewish quarters which were officially set aside in other countries and erroneously this name has also been regularly applied to quarters,

neighborhoods and areas throughout the Diaspora which became places of residence for numerous Jews." *Encyclopedia Judaica*, Vol. 7 (Jerusalem: Keter, 1971), 542–43.

7   Géza Komoróczy et al., *Jewish Budapest* (Budapest: CEU Press, 1999), 215.

8   The Elisabeth as well as Theresa Quarter are separate, official local district entities in Budapest, namely the 7th and the 6[th.] They are divided by King Street. (Even house numbers in the 6th district side, odd in the 7[th.]) The old Jewish quarter is situated in the inner parts of both districts, rather though in the 7th, because all the three synagogues which mark the area symbolically, are located in the 7th district, Elisabeth Quarter. Thus we refer to this very quarter (parts of two districts in administrative sense) as *one*, for King Street isn't a physical border. The other so-called Jewish quarter we refer to means the inner parts of the 13th district, called New Leopold Quarter.

9   Neologue: the name of the Hungarian Reform Jewry after the Nationalite Israelite Congress 1868/69. Today neologue is equal to conservative.

10 Terézváros, literally meaning Theresatown, is the 6th district of today's Budapest. Király (King) Street, right at the borders of the 6th and the 7th district, was the main street of the Old Jewish Quarter.

11 János Ladányi, "A zsidó népesség térbeni elhelyezkedésének változásai" [The changes of the mobility of the Jewish Population], in *Zsidók a mai Magyarországon*, ed. Kovács, 79.

12 Ladányi, "A zsidó népesség térbeni elhelyezkedésének változásai," 81.

13 Komoróczy et al., *Jewish Budapest*, 462.

14 Tamás Stark, "Zsidóság a vészkorszakban és a felszabadulás után" [Jews during and after the Shoah] (Budapest: MTA Történettudományi Intézet, 1995), 95.

15 For what happened to those who did not follow this example, see Csorba about Sándor Scheiber. Csorba, "Izraelita felekezeti élet Magyarországon."

16 Magyarországi Zsidó Hitközségek Szövetsége: Alliance of the Jewish Communities of Hungary.

17 For example Tamás Kóbor, *Ki a gettóból* [Out from the Ghetto] (Budapest: Singer und Wolfner, 1911); Margit Kaffka, *Állomások* [Stations] (Budapest: Franklin kiadás, 1917).

18 Brigitta Eszter Gantner and Mátyás Kovács, "A kitalált zsidó: A konstruált zsidó kulturális tér Közép-Európában; Egy új értelmezési lehetőség" [The constructed Jew: A pragmatic approach for defining a collective image of Jews in Central Europe], *Café Babel* 53 (2006), 77–88.

19 The religious environment—the identities that materialized in the religious environment and the reflections of these on the quarter is another phase of the research.

20 The empirical manifestation of this is the monopolistic Jewish summer festival organized by the festival office of the Jewish Community. (See its analysis: Brigitta Eszter Gantner and Mátyás Kovács, "Zsidniland: A zsidó kulturális tér Közép-Európa városaiban; Egy új értelmezési lehetőség" [Zsidniland: The cultural Jewish space in Central Europe] *Antropolis* 3, no.1 (2006).

21 If the building of the synagogue lacks the Torah scroll, it loses its ritual function for religious Jews. In everyday language, the synagogue is often referred to as a church, as the "House of God" in Christian interpretation, instead of the authentic "place of assembly" definition. This is demonstrated by the case of the synagogue in Óbuda, a district of Budapest: although it is being used as a television warehouse, in vernacular it is referred to as a synagogue.

22 While Diana Pinto emphasizes the rediscovery and the characteristics of interpreting Jewish culture in European countries, Michael Bodeman takes the activity of non-Jewish agents as a basis for his definition. Incorporating this into her own theory, Ruth Ellen Gruber expands the notion of Jewish cultural space ad infinitum, which makes an adequate description of

the phenomenon even more difficult. Diana Pinto, "The Third pillar? Toward a European Jewish Identity," in *Jewish Studies at the CEU: I. Yearbook (Public Lectures 1996–1999)*, ed. András Kovács (Budapest: CEU Press, 1999), 177–99; Ruth Ellen Gruber, "A Virtual Jewish World," in *Jewish Studies at the CEU: II. Yearbook (Public Lectures 1999–2001)*, ed. András Kovács (Budapest: CEU Press, 2001), 64–70; Y. Michal Bodemann, *Gedächtnistheater: Die jüdische Gemeinschaft und ihre deutsche Erfindung* (Hamburg: Rotbuch Verlag, 1996).

23 Richárd Papp, Van-e zsidó reneszánsz? Kulturális antropológiai válaszlehetőségek egy budapesti zsidó közösség életének tükrében [Is there a Jewish Renaissance? Reflections of cultural anthropology by surveying the life of a Jewish community in Budapest] (Budapest: Múlt és Jövő, 2004), 164.

24 Interview with Dávid Kautezky, co-owner of the club and café "Szóda" (Soda); Budapest, February 28, 2007.

25 One of the examples for this is the Tartuffe-reinterpretation of the Budapesti Zsidó Színkör presented in Sark Café. In the Molière play the actors and actresses—using the framework of the classical piece—reflect on the genius loci and on their own Jewishness as well.

26 András Kovács, "Zsidó csoportok és identitásstratégiák a mai Magyarországon" [Jewish groups and identity strategies in Hungary], in *Zsidók a mai Magyarországon*, 16.

27 Ibid., 32.

28 Interview with Ádám Schönberger, Art Director of Sirály and leader of the Jewish Youth Organisation MAROM; Budapest, March 2, 2007.

29 Gantner and Kovács, "Zsidniland."

30 Interview with Róbert Vajda, director of the Budapest Jewish Theater, March 2, 2007.

31 Alliance of the Jewish Communities of Hungary.

32 Interview with Róbert Vajda, Director of the Jewish Theater, March 4, 2007.

33 Interview with Ádám Schönberger.

34 Izidor Kohn and Manó Schönberger, "Zsidók a hálón" [Jews on the net], *Szombat*, June 2005.

35 Exactly one year after the Kohn-Schönberger article, an interview with Brunó Shadai Bitter, the founding blogger of Judapest was released. Márton Csáki, "Zsidó blogger vagyok: Interjú Bitter Brunóval" [I'm a Jewish blogger: An interview with Brunó Bitter], *Szombat*, June 2006, http://www.szombat.org/2006/0606zsidoblogger.htm (acessed July 26, 2007).

36 Ruth Ellen Gruber on virtual Jews as quoted before.

37 Ruth Ellen Gruber, *Virtually Jewish: Reinventing Jewish Culture in Europe* (Berkeley: University of California Press, 2002).

38 For a critique on Gruber's theory and a new interpretation of Jewish space see Gantner and Kovács, "A kitalált zsidó."

39 Csáki, "Zsidó blogger vagyok."

40 Headline at "Újlipócia," http://www.ujlipotvaros.hu (accessed September 16, 2007).

41 Shadai, "Ez mi ez? (FAQ)."

42 Kovács, *Zsidók a mai Magyarországon*, 32.

43 In the quoted *Judapest* interview, Shadai describes his identity as follows: "I can't really do anything with these religious labels, but I guess most of today's youth just feels the same: our identity is too much fragmented for this. From this point of view, the whole judapest. org project is a sort of experience for deconstructing these fragmented identities. This sounds serious but actually it's great fun, as we refer to Ron Jeremy, Ali G or Matisyahu, and things seem to get their right place—if even only for a minute." Márton, "Zsidó blogger vagyok."

44 Ibid.

45   Shadai, "Zsidó Nyári Fesztivál" [Jewish Summer Festival], *Judapest*, August 23, 2006, http://www.judapest.org/?p=702.

46   "Ütött az óra!" [The time came!], *Sófár Média*, August 17, 2006, http://www.sofar.hu/hu/flashmob.

47   Nathaniel Popper, "Budapest Ghetto Gets Facelift," *The Forward Online*, October 27, 2006, http://www.forward.com/articles/budapest-ghetto-gets-facelift (accessed September 16, 2007).

48   Rabbi Kata Kelemen (progressive-reform), Dr. Gábor Balázs (modern Orthodox), rabbi Slomó Köves (Chabad Lubavich), Tamás Lózsi (Orthodox) and rabbi Tamás Verő (neologue).

49   "Széder: pászkavacsi" [Seder: Passover meal], *Pilpul*, March 30, 2007), http://www.pilpul.net/konnyu.shtml?x=35903 (accessed September 16, 2007).

50   Internet Zsidó Database, Matula Magazin, http://matula.hu/izsdb (accessed July 25, 2007); see also "Nyomoz a rendőrség az IZSDB 'zsidóadatbázis' miatt" [The police investigates in the matter of the Jewish database], *Index*, March 21, 2006, http://index.hu/politika/belfold/0321izsdb (accessed July 25, 2007).

51   In the contemporary Hungarian political discourse the question of "Jewish origin" of certain politicians is an important issue.

52   Matula News Network, "Zsidótlanítják Újlipótvárost" [Újlipótváros without Jews] *Matula Magazin* 10, no. 35 (2006) http://matula.hu/index.php?section=article&rel=35&id=424.

53   Gantner and Kovács, "Zsidniland."

Part III

Cityscapes and Landscapes

8 Poland

A Materialized Settlement and a Metaphysical Landscape in Legends of Origin of Polish Jews

Haya Bar-Itzhak

This figure has intentionally been removed for copyright reasons.
To view this image, please refer to the printed version of this book

Yiddish Map of Poland and Lithuania from a CYSHO Atlas, 1922

Introduction

This article focuses on legends of Polish Jews as a cultural production that constitutes an important and fascinating object of scholarly study and provides a window onto the creative genius of the community, its collective memory, and the way in which it created and consolidated its identity during the course of its history.[1] These are place legends that center on Poland and reveal how its Jewish inhabitants perceived it. From this angle, Poland-as-place is a cultural category and a space of Jewish consciousness.

The specific legends that I deal with are legends of origin. The historical myth recounted by the storytelling society as part of real and authentic life. Hence one can understand the intense emotional opposition aroused when an attempt is made to reject or refute these myths, either from the outside or from the inside.

The legends of origin of Diaspora Jewry, including those of Poland, constitute a genre that combines and emphasizes the dimensions of time and space. These legends tell of the initial settlement in a new territory, with all that this involves; they are about "making a place." Because this is the first settlement, however, the legends deal with time as primeval time and describe this epoch as it was perceived by the narrating society in the various periods when the legends circulated. The birth of settlement in a strange place involves various and sometimes conflicting qualities. On the one hand, it is accompanied by anxiety in the face of the unknown and unfamiliar; the community's relative success or failure in coping with this anxiety will have a major impact on its chances for survival. On the other hand, it is also a period of renewal and hope, of an eruption of creative force and activity aimed at getting to know, understand, dominate, and assimilate the new reality. Legends of origin are recounted not only by the generation that actually lives through the encounter with a new place, although we may assume that this is the period when they begin to take shape. These legends were told throughout the centuries of the Jews' residence in Poland, metamorphosing from time to time into new shapes. In this sense they manifest the collective memory of the primeval time. Because there is a difference between time experienced and time remembered, every period lends a different quality to the primeval time as the age when everything began.

The age of initial settlement occupies a central place in the mind of the community as a time that stands out from the normal course of the centuries and is perceived not quantitatively but qualitatively. It is a period that determines the nature of the days that follow.[2] It is the age when the very identity of the society is molded and defined. It then follows that legends of origin are important in every generation, throughout the life of the community. The primeval era is reshaped in order to mold the present and future while deriving legitimacy from the distant and hallowed past. A society is always molding its first days in a way that can be used to justify how it lives in the present—or alternatively, if it wishes to change this life in a way that can justify change or even revolution. Hence, tracing the legends of origin of Polish Jewry, as they crystallized and were told in various periods, allows us to expose the narrative of the Jewish community in Poland and the changing cultural awareness of the narrating society.

Finding and Documenting the Sources

The first task was to find and document the stories. As will be seen later, a significant proportion of the legends of origin of Polish Jewry are no longer being related orally; therefore, it was necessary to refer to written sources. This posed the first difficulty. As was rightly argued by Israel Zinberg, even though every cultural historian is aware of the major role played by Jews as intermediaries in the field of European folklore and as artists who wove their hopes into legends and wonder tales from the heritage of folklore they received from previous generations, "almost all of this has remained an oral tradition, and only by chance have a few elements of this folklore material been preserved."[3] When students of folk culture wish to uncover these sources, they must rely on written material that was transcribed and preserved "by accident."

The first source is tales found in chronicles and historical studies, starting with Zemach David, by the sixteenth-century astronomer and historian David Gans.

Another source is the stories preserved at YIVO, including those gathered by the famous An-Ski expedition of 1912–1914. Some of them have been published in Yiddish periodicals and anthologies.

A third source is stories that were published in German and Polish, that is, outside the narrating culture itself. Some of these narratives were published in periodicals such as Am Urquell: Monatsschrift für Volkskunde, Der Orient, Wisła, Lud, Izraelita and, Jutrzenka; others in anthologies assembled by non-Jews, such as Klemens Junosza's *The Miracle in the Cemetery*.[4]

The transcription and publication of Jewish material in non-Jewish languages was the result of the heightened interest in ethnographical studies in Germany and Poland before the First World War and during the inter-war period, spurred by Romanticism and especially the rise of nationalism in Europe. The drive for emancipation also led to the publication of legends of origin in the Gentile vernacular, allowing their use in that struggle.

A project to record the folk tales of Polish Jews in Israel began in the 1950s. Its files, in Hebrew and Yiddish, are located in the Israel Folktale Archives (IFA), at the University of Haifa.[5] This material, along with stories in Hebrew,

Yiddish, and Polish that I recorded during my field work with Polish-born Jews, comprise the fourth source. These reflect the stories as they are told today by Jews of Polish origin.

Another important source is the memorial volumes of the Polish Jewish communities destroyed in the Holocaust, both the collective works and the memoirs of individuals. These books are a treasure-trove of narrative material; some of it contains older stories that have been reprinted, while others are told in the post-Holocaust context and perspective.

Historical Background

The first appearance of Jews in Poland and their adventures during their early years of settlement in the country are concealed in the undocumented shadows of history. The written record indicates that starting in the thirteenth century, and particularly in the fourteenth century, Poland was a destination for Jewish refugees from Germany and from the areas along the southeastern borders of Kievan Rus.[6] Going back further, late-twelfth century documents refer to Jewish communities in Poland.[7] Some historians, however, believed that Jews reached Poland as early as the tenth and eleventh centuries—and perhaps even the ninth century.[8] The tendency to represent Jews as Polish autochthons was prominent mainly in Jewish intellectual circles, in support of their sociopolitical thesis that the Jews, as original inhabitants of Poland, merited full civil rights.[9] This teaches us something about the nature of historiography; or, as Weinryb puts it, the historians of Eastern European Jewry created their own myths.[10]

Nineteenth century historiography was influenced by the struggle for liberalism and progress; for minority groups like the Jews, this was focused in their campaign for emancipation. The writing of history became part of a political contest. Historians who wanted to present the Jews as Polish autochthons, but lacked written sources to ground their claim, sometimes built on the legends of origin of Polish Jewry. These scholars transcribed the legends and contributed to their preservation—as did their opponents, who cited the same legends while seeking to refute their historical reliability.

Even if Jews did not settle in Poland until the thirteenth and fourteenth centuries, all agree that here they enjoyed rights and privileges denied them elsewhere in Europe. In this context mention must be made of the privileges Jews were granted by Bolesław of Kalisz in the thirteenth century and expanded in the next century by Casimir the Great of the House of Piast. These privileges brought Jewish prosperity, accompanied by a spiritual flowering reflected in the theological production of the rabbis and sages who arose in Poland.

By "Polish Jews" I mean the historical community that first coalesced in the kingdom of Poland, without reference to the vicissitudes of political geography. This kingdom, large and powerful until the mid-eighteenth century, ceased to exist as an independent political entity in 1795, when it was partitioned among Russia, Prussia and Austria. The independent Polish republic was not reconstituted until after World War I. Communist Poland after World War II, was assigned different boundaries. Hence, although it is difficult to speak of Polish Jewry in geopolitical terms, we may accept Rosman's contention about the Jews of post-partition Poland that Jews did not undergo any radical changes, rather they remained steadfast to the way of life that had characterized Polish Jewry: "The descendants of the Jews of Poland preserved the unique features that were immediately evident to them, to other Jews and to the non-Jewish world."[11]

Polish Jewry was annihilated during the Second World War. Although there were still Jews in post-Holocaust Poland, one can no longer speak of Jewish communities. Hence, according to the legendary chronicles, we are dealing with a millennium of Jewish life and creativity in Poland.

Drawing a Mental Map of Jewish Poland: The Judaization of Place Names

An analysis of the legends shows that Poland was viewed not just as a materialized settlement, but also as a metaphysical landscape. In other words, the geography of Poland became the geography of the Jewish imagination. Humanist geography recognized that the meaning of a place is constructed by a person's beliefs and feelings. How people relate to a place is what turns the space from what Relph designates "no place" into a "place"; that is, from a space devoid of meaning to one pregnant with meaning. Viewed from this perspective, the concept of "place" transcends the geographical space. The real place is embedded in human consciousness by means of memories, thoughts, and affections and becomes an "inner place."[12] A "place" is a space to which human beings give meaning and thereby endow it with its essence. This is the process that Poland goes through in Jewish folk legends and which convert it from a foreign geography to a Jewish place.

Settling in an unfamiliar place, asserts Mircea Eliade, is "equivalent to a new act of Creation." The alien regions are assimilated to the primordial chaos. Settling a new territory turns the chaos into cosmos by means of a ritual that makes it real and valid.[13] The legend of origin about the migration and settlement is an expression of the ritual act as well as part of it.

The legends of Polish Jewry recreate Poland according to a Jewish archetype whose most conspicuous expression, in the traditions and legends of the Jews of Poland, involves the Judaization of place-names by means of homiletical explanations (midrashim) of the alien-sounding names.

The homiletical exposition of names, including place-names, goes back to the Bible, which attaches great importance to explaining names and to exploiting the phonetic and semantic components of a name.[14] The preoccupation with expounding names and the creation of name-midrashim continued in the post-biblical period. The Apocrypha preserves and expands name-midrashim. The liturgical poetry of the first millennium frequently incorporates name-midrashim. There is an abundance of midrashim on biblical and post-biblical names in the rabbinic literature of the Talmudic era and thereafter in the traditional commentaries over the generations.[15] Consequently, we are dealing with an ancient tradition incorporated into a literary work.

In the case of the midrashic exposition of Polish place names, we are dealing with a Slavic toponym explained on the basis of Jewish languages. This Slavic place name is understood not as a random and arbitrary set of phonemes, but as a concatenation that conveys a meaning in a Jewish language—Hebrew and/or Yiddish. The name-midrash unveils this meaning, which allows an identification to be made with the place by Judaizing it. The name-midrash or explanation is incorporated into a literary work and creates a legend of origin. Let us consider a few examples:

> What is the source of the name Shebershin? Tradition has it that the town of Shebershin was first of the nine communities where Jews settled after the expulsion from Spain. So they called it Shebershin, meaning 'settle first' (Hebrew: shev rishon).[16]

The Judaization of the place-name begins with its pronunciation, as Mahler notes in his article on the Jewish names of places in Old Poland.[17] The Jews pronounced the name Shebershin, whereas the Poles pronounced it Sczebrzeszyń.

The name is expounded in Hebrew. The sounds that make up the Slavic name, as it was pronounced by the Jews, make possible a midrash that can be incorporated into the tale of the Jews' flight wanderings from Spain and their arrival in Poland. Sczebrzeszyń is turned into a Jewish city by means of an internal Jewish code—the Jew's own language—and is consecrated by the power of the Holy Tongue.

Stories like this keep cropping up in various forms and various eras, motivated by the desire to express significant events in the life of the community. In this last example, only Hebrew was called on to explain the Polish name. The next example adds Yiddish to the mix.[18]

> What is the source of the name Ostre? Hundreds of years ago, when Talmudic scholars, geonim, and rabbis like the great Maharshal Luria, the holy Maharsha, and others lived there, the city earned the name Ostre—that is, Os Toreh [Ashkenazi Hebrew: 'mark of

Torah']. Today, by contrast, when the Jews no longer devote themselves to learning and the houses of study stand empty, the city is indeed Ois Toreh [Yiddish: 'without Torah'].[19]

Here, too, the explanation is based on the Jewish pronunciation, Ostre, rather than the Polish Ostróg. The midrash exploits the homonymity of the Yiddish ois 'out' and the Ashkenazi Hebrew os 'mark', 'sign'. It is no coincidence that Hebrew is used to express the holiness of the city in bygone days, whereas Yiddish, the profane vernacular, denotes its deterioration in the later era. The story employs the double midrash on the name of the town to convey a facet of internal Jewish history—Torah study and its place in the Jewish world. In this way, geography is Judaized and placed at the disposal of Jewish history.

Name-midrashim and explanations are intended to provide a theological imprimatur for Jewish residence in these places. Conferring such approval on residents of the Polish diaspora is a central theme of the legends of origin, meant exclusively for Jewish consumption. The same applies to Ostre when its name is linked to Torah study, or to Kolomyja, explained as kol m-ya (Hebrew: "voice of God").[20]

Po lin! Lodge Here! Legends of Origin

This brings us to the legend of origin that deals with the first arrival of Jews in Poland, the Po-lin ("Poland, lodge here") legend that expounds the Hebrew name of the country.[21] The oldest written source in which I have found a midrash on a Hebrew name for Poland is the "Elegy on the Massacres in Polonia,"[22] an elegy on the pogroms of 1648–49 by the seventeenth century Jacob b. Moshe Halevy, first printed in Venice in 1670/1. The name-midrash appears in the third stanza of the elegy:

> The glory of the earth has now become
> the shame of all towns, a disgrace among cities.
> A place of Torah learning—here God lodges (Hebrew: po lan Yah).[23]
> is now full of sadness and great mourning,
> instead of the poetry and song that used to be
> in them. Just as by the waters of Babylon,
> the joy of my heart has been exiled, and all my prosperity.
> Behold, is there any anguish like my anguish?

Here Poland is described as "the glory of the earth (…). A place of Torah learning—here God lodges." The devastation wreaked by the pogroms of 1648–1649 creates a contrasting analogy to the former greatness and the glory.

Stories that incorporate the name-midrash or are based on it were first transcribed at a much later date. Lewin cites the legend from an oral tradition of the late nineteenth or early twentieth century.[24] He merely summarizes the story, which refers to the exiles from Spain: when the Jews were expelled from Spain and their wanderings had brought them to the east, they said, po lin-"Poland, lodge here"; this is the source of the name Polin. The legend attempts to explain the name of the country, associates it with the exiles from Spain, and incorporates it into the history of the Jews and the role that Poland played for them, as the place where they found a safe haven and refuge.

Marek cites another version,[25] which refers to refugees from Germany. According to his summary of the legend, a Jewish survivor of the massacres in Germany arrived in Poland, where he heard a heavenly voice: po lin— "Lodge here". This version adds the element of divine intervention in the form of the heavenly voice that declares that the survivor should stay in Poland. This turns the story into a sacred legend and gives a theological imprimatur to the act of settling in Poland.

A folk version that must have been transmitted orally was printed by Gershom Bader.[26]

> If you want to know how it suddenly occurred to these Jews in Germany to seek refuge in Poland, legend has it that after the Jews had decreed a fast and beseeched God to save them from the murderers, a slip of paper fell from heaven. On it was written: 'Go to Poland, for there you will find rest. (…) The Jews set out for Poland. When they reached it, the birds in the forest chirped to greet them: "Po lin! Po lin!" The travelers translated this into Hebrew, as if the birds were saying: "Here you should lodge. (…)." Afterwards, when they looked closely at the trees, it seemed to them that a leaf from the Gemara was hanging on every branch.[27] At once they understood that here a new place had been revealed to them, where they could settle and continue to develop the Jewish spirit and the age-old Jewish learning.

In Yiddish literature, the legend is recounted by I. L. Peretz in his Reise-Bilder, first published in 1891.

> The stillness of a summer night. At the edge of the sky, the nearby forest grows dark. On its trees our ancestors engraved the names of the Talmudic tractates they finished studying on the road. Not far from there they once encamped in the evening, and the Exilarch said: Po lin! And to this day, the country is called Polin; but the Gentiles can't explain why![28]

Echoes of this legend also resound in the poem by Noach Pniel, "Po-lin":[29]

> Po lin! chirped the birds from your trees.
> Indeed, my ancestors lodged in you for a thousand years
> And thrust my roots deep into a land that is my step-mother
> I hung Gemaras in the top of every tree in your forest.

In all these legends, Poland is recreated as a Jewish land. It is the Jews who name it to commemorate the role it plays in their history. The crux of the matter, though, is that their arrival and residence in Poland receives a divine stamp of approval, whether in the form of a heavenly voice that guides the refugee or of a slip of paper that falls from Heaven and guides them to sanctuary in Poland.

The legend seeks to give a sense of continuity and permanence to the Jews' residence in Poland; the ability to preserve the Jewish spirit and engage in Jewish scholarship in this new place are the justification for settling there.

In general however, in all versions of the legends the name-midrash conveys the temporary nature of the sojourn in Poland, and I will return to it later.

Three Spaces: The Legends of Wanderers and the Land of Israel

The Po-lin legends are the legends of wanderers, dealing with a community that has left its former home with its re-establishment in a new place. It follows naturally that space should have a special significance in these legends. In fact, three distinct spaces are invoked. The first space—the country that has been left behind—hardly appears. The need to abandon it suffices to characterize it as a perilous place. The narrators do not bother to describe this space since its mere mention awakens the terror latent in the collective memory.

The second space is that of the passage, the places through which the refugees wandered while searching for a new haven. Providence does not forsake the wanderers and directs them to Poland, the sanctuary where they can find repose.

The descriptions of Poland, the landscapes of their refuge, their cynosure, the land of their desires—the third space—are typified by allusions to and associations with sacred Jewish concepts. The birds in the Polish forests chirp in Hebrew; the boughs of the trees bear leaves from the Gemara. This association appears in S.Y. Agnon's retelling, too: "When they reached the land, they discovered a forest of trees, and a tractate of the Gemara was carved into every tree."[30]

The Jewish legend borrows its spatial apparatus from the world of wonder tales. In wonder tales, the boughs of the marvelous trees that grow in the wondrous space bear leaves of gold or fruit of precious stones, whereas in the Jewish legend the sparkling jewels of the material world are replaced by the priceless gems of the spiritual world.

Legends describing idyllic landscapes were common in various Jewish communities. Let us consider one example set in the area of Sczebrzeszyń.

> The path leading to the village of Kawenczynek, in the Shebershin district, goes through a steep valley between high mountains, almost as deep as a pit. In the valley there is a cave, which the locals call the Żydowska Szkoła or 'Jewish Shul.' The story is told that when the Jews made their way from Spain to Poland they camped for a while in this valley; during their stop they studied and prayed. You could see traces of Hebrew words carved into some of the

trees in the area, which were to be read: 'Here we finished studying the tractate Shabbat.' There is also a brook there, which is still known as the Brook of the Prophet Samuel.

In the same area there is a cave which, so tradition has it, leads to the Land of Israel. People say that roughly a hundred years ago there lived in Shebereshin a righteous man, who was known as the 'White Rebbe' on account of the white garments he always wore. He is still renowned throughout the district for the wonders he worked. He very much wanted to bring the Messiah. Once he went out to the forest and said he was going to the Land of Israel. All the young ragamuffins went with him, as did a few pious women. He came to the cave and stopped outside. First he sent a kid [baby goat] into the cave—and waited three days. When he saw that after three days the kid had not returned, he said that this must certainly be the path to the Land of Israel. Then he went into the cave and never came out again.[31]

In this story, a forest in Poland becomes a landscape of the Jewish study. The cave of Kawenczynek becomes an ancient Jewish house of study, while the trees in the forest are consecrated by Hebrew inscriptions indicating which Talmudic tractates the Jews studied en route. The legends show how landscape and human beings create each other and influence each other. Human beings project their inner world on the topography and landscape of the place, giving them a new symbolic meaning for both the individual and the community. It was this endowment with meaning that gave the Jews a sense of deep rooted connection to Poland.

This application of sacred Jewish iconography to Polish landscapes could be troubling. The legends ought not to detract from the myth of redemption, in which the Land of Israel plays the central role. In other words, such landscapes should be reserved for the Land of Israel, the cynosure of the future redemption. As I have shown elsewhere,[32] in Diaspora folk narratives the Land of Israel is treated as an extension of supernatural space that penetrates into human space.[33] The folk legends of Polish Jewry construct spatial devices that link Poland to the Land of Israel. In this story the device involves well-known motifs of Polish Jewish folk legends—the cave and the kid.

In other stories, the spatial connection between Poland and the Land of Israel is via a subterranean passage that leads from synagogues in the Diaspora to the Land of Israel, or through the medium of stones from the Holy Temple that are incorporated into the walls of the local synagogue.[34] In all these cases the Land of Israel remains the sanctified space for which the Jews yearn; it radiates a measure of sanctity on the landscapes and synagogues of their present abode in Poland.

As mentioned, the stamp of approval on residence in Poland, as an act of divine choice, ought not to detract from the myth of future redemption or come at the expense of the Land of Israel, the site of that future redemption. Space and time join forces in the legends to express the problem and to solve it. In the folk legend, Poland is recreated according to the Jewish archetype: the sacred Jewish world inheres in its essence, name, and landscapes. The archetype of settlement—the exodus from Egypt, the wandering in the wilderness, during which God accompanies and guides His people, and the entry into the Land of Canaan—is repeated in the settlement of Poland; but an awareness of the myth of redemption informs the depiction of space and time. The Land of Israel remains the Holy Land, the navel of the world, the lost paradise, and thus the cynosure and object of desire. The name-midrash on Poland associates it with rescue from persecution and with repose—but also with night, darkness, sleep, and transience; whereas the Land of Israel is associated with eternity. The spatial devices that link the two countries—especially the subterranean passage—constitute a sign that the Land of Israel remains the region of absolute reality and holiness; through this link it emanates a portion of its sanctity onto Poland. The road to the holy precinct is always arduous and fraught with perils[35]; in our story it assumes a garb that recalls the subterranean labyrinth, found in so many myths, which symbolizes the difficulties of every quest. This arduous path represents both the aspiration towards and the difficulties of the transition from the profane to the sacred, from the temporary and illusory to the real and eternal, from death to life, from man to godhead.

In Yiddish literature, it was Sholem Asch, in Kiddush Hashem, who stated this problem by linking two legends—Po-lin and that of the Polish synagogues and houses of study that will be magically transported to the Land of Israel in the days of the Messiah.

Reb Mendel is talking with Reb Jonah:

"Do you think, Reb Jonah, that there will ever be a settlement here with Jewish towns and synagogues?"

"Of course! How, then? The place is specially intended for Jews. When the Gentiles had greatly oppressed the exiled Jews and the Divine Presence saw that there was no limit and no end to the oppression and that the handful of Jews might, God forbid, go under, the Presence came before the Lord of the Universe to lay a grievance before Him, and said to Him as follows: 'How long is this going to last? When You sent the dove out of the ark at the time of the flood, You gave it an olive branch so that is might have support for its feet on the water, and yet it was unable to bear the water of the flood and return to the ark; whereas my children You have sent out of the ark into a flood, and have provided nothing for a support where they may rest their feet in their exile.' Whereupon God took a piece of Eretz Yisroel, which He had hidden away in the heavens at the time the Temple was destroyed, and sent it down upon the earth and said: 'Be My resting-place for My children in their exile.' That is why it is called Poland (Polin) from the Hebrew po lin, which means: 'Here shalt thou lodge' in the Exile. That is why Satan has no power over us here, and the Torah is spread broadcast over the whole country. There are synagogues and schools and yeshivas, God be thanked."

"And what will happen in the great future when the Messiah comes? What are we going to do with the synagogues and the settlements which we have built up in Poland? asked Mendel as he suddenly thought of Zloczow."

"How can you ask? In the great future, when the Messiah comes, God will certainly transport Poland with all its settlements, synagogues and yeshivas to Eretz Yisroel. How else could it be?"[36]

As Asch recounts the legend, Poland is a piece of the Land of Israel, which the Holy One, blessed be He, set aside for His children as a place of repose in their exile. This explains why the Jews flourished in this land and, in particular, why it became a center of Torah study. The tension between Poland and the Land of Israel as the site of the future redemption is expressed in Mendel's question: "And what will happen in the great future when the Messiah comes?" The miraculous transport of Poland, with all its towns and houses of study, to the Land of Israel, invoked alongside the Po-lin legend, reflects a problem that troubled the Jews of Poland and their legendary solution to it.

Legends of origin endeavor to provide an internal Jewish theological imprimatur to the Jews' settlement and residence in Poland. The rhetorical and poetical devices employed in these legends promote the adoption of the new land by associating it with the sacred Jewish concepts. In this way, the geography of Poland becomes the geography of the Jewish imagination, and its landscapes—the landscapes of its most heart-felt desires.

Recounting Legends of Origin After the Holocaust

The identity of a place evolves over many years, in a long and protracted process that influences how human beings perceive the place. As I have pointed out elsewhere,[37] although legends of origin are set in the distant past, the legend itself is a dynamic creation subject to continual change and expresses the problems that troubled the society at a given time. The "springtime" of the community in which the plot unfolds, is portrayed in a manner that reaffirms the community's way of life in the generation that recounts the story. The legend expresses the present in which it crystallized and was told. Legends of origin that deal with the acceptance of Jews in Poland, for example, construct different models of relations between Jews and Gentiles. The eighteenth century legend constructs a model of segregation. In the nineteenth century, by contrast those who wanted emancipation and Polonization told stories that described the earliest relations in a mode of cooperation based on compassion and equality.[38]

What happens to these legends after the Holocaust, in the wake of the community's physical and spiritual annihilation? I tried to answer this question by analyzing legends of origin of Jewish synagogues that are still told by Jews of Polish origin in Israel.

Before the Holocaust, these legends fulfilled a number of functions: (1) They legitimized the synagogue and the community it represented to the outside Gentile world—the synagogue was erected in conformity with the law by rulers or with their consent, following donations by Jews and sometimes as an expression of gratitude to them. (2) Internally, for the Jews themselves, they provided spiritual and theological warrant for their residence in Poland—subterranean passages to Jerusalem, the construction of a synagogue with the assistance of biblical figures like King David and the prophet Elijah, the use of stones from the Holy Temple.[39] (3) They reinforced the link between individuals and their place of residence and community and provided a sense of belonging and continuity.

Given that their erstwhile functions have lost all significance, should we assume that these legends would disappear? Anyone close to the world of the survivors of those communities can answer this question with a vigorous "no." The survivors' sense of commitment to their dead and their community produces a sense of obligation to tell their stories and that of the community. Now only the narrative can confirm and sustain the community's existence. Henceforth, the community is a community in memory and of memory, and its continued existence lies in the transmission of the memory to future generations through the telling of the story. This commitment produced the memorial books for communities wiped out in the Holocaust, and these books frequently note that they stand in place of the grave and tombstone denied to the dead.

Legends of origin about Jewish synagogues are still being told, but they have undergone a transformation. Today the origins of Polish synagogues can be recounted only through the lens of the Holocaust. The geography of the narrative is now exclusively a geography of memory; the same place that is sanctified in memory is often considered to be cursed in real life. No longer fulfilling their erstwhile functions, the narratives have become a means for expressing and working through what post-Holocaust Polish Jews see as the true story of the community—the story of its destruction.

Let us consider two legends of this type.

The ancient synagogue in our town was built more than 900 years ago. They built it over a period of several years but were unable to finish it. Suddenly a Jew appeared from far away. No one knew who he was or where he had come from. He pledged to the community leaders that he would complete the synagogue.

When construction was complete the man abruptly disappeared. The next day the congregation found all the money the community council had paid him for his work in a corner of the synagogue.

People said it was none other than King David, may his merit defend us and all Israel, who built this splendid synagogue, for it was impossible that normal flesh and blood, a gevaynlikher mentsh, could build such a glorious holy place.

I myself cannot believe that I ever merited to see with my own eyes this remarkable and magnificent synagogue, which had all the hues and colors of the sun and the moon and the rainbow.

And when I remember and call to mind the Great Synagogue, the ancient synagogue in our town, which was destroyed by the Germans, may their name be blotted out, then my eyes shed tears because the enemy has overcome; my sighs are many and my heart is sick.[40]

The first part of this narrative is a legend of origins. The synagogue is said to be 900 years old, which gives a stamp of legitimacy to the community's existence in Poland. Its beauty and splendor dignify the town and its congregation, which had the merit of having such a synagogue. The attribution of the completion of the synagogue to King David gives a spiritual and theological seal of approval to the community's presence in Poland, for King David, the greatest king and hero of Israel, is also the ancestor of the Messiah.

This part, which was the focus of the legend before the Holocaust, becomes merely the introduction and excuse for what the post-Holocaust narrator wants to tell about himself and his community—destruction, eulogy, and lament. The following story also exemplifies the new paradigm:

The splendid ancient synagogue in the town of Shebershin was built a thousand years ago, in all its glory and splendor, by the Jewish king of Poland, Saul Wahl. Ever since, a new light shone on the Jewish community of Shebershin.

The Jewish king of Poland reigned for only a brief time, but he took advantage of the royal diadem to benefit his people. He issued a decree that synagogues be built for the Jews, and, mutatis mutanda, churches for the Christians.

This great synagogue distinguished the holy congregation of Shebershin. Its ornaments and murals were splendid to behold. The ceremonial objects, too, beautified the Lord's sanctuary. Anyone who visited this synagogue could not stop feasting his eyes on its great beauty and splendor. He was struck with amazement at the colors and hues revealed to his eyes when he entered it. There were all kinds of figures that had been drawn and carved by great artists. The synagogue glinted in the sunlight with all the colors of the rainbow. The ceiling was painted sky-blue and added a charming note.

But now the candles in the synagogue have been extinguished. The Germans burned it, with its worshippers and admirers and hundreds of Torah scrolls. The Jews were gathered in the blazing synagogue, wearing their kittels and tallises, praying and chanting and praising their God. They hurled themselves into the flames with the Torah scrolls to sanctify God's name.

All the candles in the Great Synagogue of Shebershin have gone out. Only one light burns: the memorial candle, the yohrzait-likht, hissing and casting a shadow on the wall. In my heart, too, a small flame burns; whenever I am not working, and during sleepless nights, it recalls to me the past and the destruction.[41]

Thus legends of origin turn into legends of destruction. The paradigm is as follows:

- First comes the narrative of origins, rooted in the distant past and distinguished by its splendor and glory. This part is generally recounted in the third person and the past tense.
- This is followed by the threnody for the destruction. Here the narrator switches from the past to the present and from the third person to a more intimate first-person narrative, concluding with a dirge and lament, always in the elevated style of traditional Jewish dirges.[42]

Sometimes the order is reversed: the narrator begins in the present, describing the destruction, and only then flashes back to the glorious past. But the rhetoric described above is still there.

Whatever the sequence, the legends of origin have become a means to tell of the destruction of the community and to mourn for it. It is no coincidence that the narrating society selected legends of origin as a way to deal with the trauma of destruction. The antithetical analogy created by these stories make it possible to probe the trauma without using explicit language. The mechanism of analogy leads to the conclusion while heightening the disparity: the full intensity of the calamity is emphasized by the contrasting background of the original grandeur and glory.

The narratives generate a chain of opposites that mediate between the fundamental poles of life and death.

The communal life of Polish Jewry lies forever in the past. There, in the past, are found the origins, the construction of the synagogue, the happiness and glory. The present in which the synagogue legends are recounted is the era after the death and destruction of the community, and an expression of the pain and lament. So much for the verbal level. In the folk text, however, the context is part of the text, and the context generates an inversion that exceeds the power of words. On the one hand, the context makes the destruction tangible, since the narrators are survivors and the narrative is told repeatedly at Memorial Day gatherings. On the other hand, the very fact that there are survivors who can recount the story to their fellow survivors, along with the families they established after the Holocaust, is a sign of life. The narrator is alive, in a place felt to be sheltered against the dreads of the past; in particular, the narrator is a man or woman who is remembering. Both the destruction and the glory are preserved in memory, and memory is survival. The narrative becomes a ritual that makes it possible to restore and renew the bygone origins and to link the historical time of the present with the mythical time of the beginnings. The human capacity to overcome the destructive force of time lies in the link created between the new start and memory. The narrative is the concrete expression of memory, which, by constructing a myth of creation and destruction, stresses life and resurrection. In Eliade's terms,[43] the terrible chaos experienced by the narrators becomes a cosmos and expresses the desire to give life order and meaning.

In legends, place is a space to which both the individual and the community attribute some kind of meaning according to their memories, feelings, and thoughts. The essence of Poland as a place changes in post-Holocaust legends from a land of refuge typified by allusions to sacred Jewish concepts in the pre-Holocaust legends of origin into a perilous place. The context of the storytelling event turns Israel into the place of the eternal refuge, both material and spiritual.

To borrow from Pierre Nora, legends become a "site of memory," preserving in the collective memory of Polish Jews both the beginnings and the final destruction. In this way, the legends of the origins of Polish Jews, having become legends of destruction, construct the myth of the Jewish people, which includes its origins on Polish soil, its catastrophe, and its rebirth in the Land of Israel.

Notes

1   This article is based on my studies that were published in *Jerusalem Studies in Jewish Folklore* and my book *Jewish Poland: Legend of Origin; Ethnopoetics and Legendary Chronicles* (Detroit: Wayne State University Press, 2001).

2   Yonina Gerber-Talmon, "Ha-zman ba-mitos ha-primitivi" [Time in the primitive myth], *Yiun* 2, no. 4 (1952): 201–214.

3   Yisrael Zinberg, *Toldot sifrut Israel* [The history of the Israelite literature] (Tel Aviv: Sifriat Poalim & Sheberk, 1958–1971), 4:89.

4   Klemens Junosza, *Cud na kirkucie* [The miracle in the cemetery] (Warszawa: Skład Główny w Ksiegarni E. Wende i Spółka, 1905).

5   The Israel Folktale Archives named in honor of Dov Noy (IFA) was established in 1955 by Dov Noy and is located at the University of Haifa. IFA houses 23,000 folk narratives recorded from narrators from different ethnic groups in Israel. It is the largest collection of Jewish folk narratives in the world.

6   Shmuel Arthur Cygielman, *Yehudei polin ve-lita ad shnat 1648 mevo'ot u-mekorot mevo'arim* [Jews in Poland and Lita until 1648 introduction and interpreted sources] (Jerusalem: Merkaz Zalman Shazar, 1991), 26.

7   Elazar Feldman, "Di eltste yedies vegen Yidn in poylishe shtet in XIV–XVI y.h." [The oldest information regarding Jews in Polish cities in XIV–XVI c.], *Bleter far Geshihte, Yunger Historiker* 3 (1934); Bernard D. Weinryb, *The Jews of Poland: A Social and Economic History of the Jewish Community in Poland from 1100 to 1800* (Philadelphia: The Jewish Publication Society of America, 1973), vii.

8   Hermann Sternberg, *Geschichte der Juden in Polen unter den Piasten und Jagellonen* [History of the Jews in Poland under the Piast and the Jagiellon Dynasties] (Leipzig: Duncker & Humblot, 1878), 6–15; Joachim Lelewel, *Polska wiekow srednich* [Poland in the Middle Ages] (Poznań: Nakł. J.K. Żupańskiego, 1851–1856), 18; Alexander Kraushar, *Historya Żydów w Polsce* [History of the Jews in Poland] (Warszawa: Drukarnia Gazety Polskiej, 1865), 40,

57, 65; Meir Bałaban, *Beit israel be-polin: Mi-yamim rishonim ve-ad limot ha-hurban* [Jewish House in Poland: From the first days until destruction], ed. Y. Halperin (Jerusalem: World Zionist Organization, 1948), 1–5; Ignacy Schipper, *Di virtshaftsgeshikhte fun di yidn in Poyln beysn Mitlalter* [Economic history of the Jews of Poland in the Middle Ages] (Warsaw: H. Bzshoza, 1926), 15.

9   Cygielman, *Yehudei polin ve-lita ad shnat 1648*, 27–28.

10 Bernard D. Weinryb, "The Beginnings of East-European Jewry in Legend and Historiogaphy," *Studies and Essays in Honor of Abraham A. Newman* (Leiden: E. J. Brill, 1962), 10–12.

11 Moshe Rosman, "Ha-yishuv ha-yehudi be-polin: Yesodot geografiyyim, demografiyyim u'mishpatiyyim" [The Jews of Poland: geographic, demographic, and legal bases], in *Polin: Perakim be-toldot yehudei mizrah-eiropah ve-tarbutam* [Poland: chapters in the history and culture of East-European Jewry], 1–2 (Tel Aviv: The Open University, 1991), 19.

12 Edward Relph, "Geographical Experience and Being-In-The World: The Phenomenological Origins of Geography," in David Seamons and Robert Mugerauer, eds., *Dwelling, Place and Environment: Toward a Phenomenology of Person and World* (New York: Columbia University Press, 1985), 15–31.

13 Mircea Eliade, *Myth and Reality* (New York: Harper and Row, 1968), 9–10.

14 Moshe Garsiel, *Midrashei shemot ba-mikra* [Names Midrashim in the Bible] (Ramat-Gan: Revivim, 1987).

15 Ibid., 20–21.

16 YIVO, Inv. C 47,115 (Shebershin Lubliner); published by J. L. Cohen, *Yiddisher folklor* [Jewish folklore] (Vilna: Yidisher Visenshaftlekher Institut [YIVO], 1938).

17 Rephael Mahler, "Shemot yehudi'im shel mekomot be-polin ha-yeshana" [Jewish place names in old Poland], *Reshumot N.S.* 5 (1953), 146–161. According to Mahler, it is important to publish lexicons of the Jewish geography of countries that had large concentrations of Jews, to facilitate the study of both language and history (Ibid., 146). On the Yiddish versions of Polish place-names, see Edward Stankiewicz, "Yiddish Place Names in Poland," in U. Weinreich, ed., *The Field of Yiddish, II* (The Hague: Mouton & Co., 1965), 158–181; on Yiddish place-names in other countries, see, for example, Florence Guggenheim-Grunberg, "Place Names in Swiss Yiddish: Examples of the Assimilatory Power of a Western Yiddish Dialect," in U. Weinreich, ed., *The Field of Yiddish, II* (The Hague: Mouton, 1965), 147–157.

18 It is interesting to note that research among Polish immigrants in Israel indicates that the same strategy was used after their aliya. Consider the case of the Polish Jews who settled in Upper Nazareth in the late 1950s. At the time, the town was called Kiryat Natzeret, but the olim referred to it as Kiryat Na-tzores, or "City on Troubles"—combining Hebrew kiryat ('city'), Polish 'na' = 'on', and Yiddish tzores ('trouble'). See Haya Bar-Itzhak, Israeli *Folk Narratives: Settlement, Immigration, Ethnicity* (Detroit: Wayne State University Press, 2005), 57–69.

19 YIVO, Inv. C 57,452 (Ostre, Vohlyner), published by Cohen, *Yiddisher folklor*. Compare Shmuel A. Horodezky, *Yahadut ha-sekhel ve-yahadut ha-regesh* [Judaism of reason and Judaism of emotion] (Tel Aviv: Tverski, 1947), 452.

20 I heard this midrash from Dov Noy, to whom I am grateful. I have not found it incorporated into a story.

21 The midrash is based on the Jewish pronunciation Polin instead of the Polish Polska.

22 Poland has a number of different names in medieval and early modern Jewish literature: Polin, Poloniya, Polania, Polonio, Polenden. Compare Shimeon Huberband, *Kiddush Hashem: Writings from the Time of Holocaust*, eds. N. Blumental and I. Kernish (Tel Aviv: Zakhor, 1969), 325.

23   Written with asterisk and an explanation "Polonia".

24   Lewin cites the story as told by Professor Berliner from Berlin, who heard it in his youth in Owersitcko: Louis Lewin, "Deutsche Einwanderungen in polnische Ghetti" [German immigrations to Polish ghettos], II, *Jahrbuch der Jüdisch-Literarischen Gesellschaft V* (1907): 75–154; here 147.

25   Berl Marek, *Di geschichte fun Yidn in Poyln biz sof fun XV y.h.* [The History of Jews in Poland until the fifteenth century] (Warsaw: Farlag Yiddish Buch, 1957), 190.

26   Gershom Bader, *Draysig doyres Yidn in Poylen* [Thirty generations of Jews in Poland] (New York, [O.f.g] 1927), 2–3.

27   In Yiddish "blatt" means both leave and page.

28   I. L. Peretz, *Ale Verk* [Complete works], II (New York: Tsiko Bikher, 1947), 170.

29   Noah Pniel, *Siah le-et erev: Shirim* [Poems] (Haifa: Pinat Sefer, 1980), 53–54; first published in Hapoel Hatza'ir 1968, 29.

30   Sh. Y. Agnon, *Kol sippurav shel Shmuel Yoseph Agnon* [All stories by Shmuel Yoseph Agnon] (Schocken: Jerusalem, 1967).

31   YIVO, Inv. C 47, 115 (Shebreshin, Lubliner), in Cohen, *Yiddisher folklor.* An English version of the story can be found in Beatrice Silverman-Weinreich, *Yiddish Folktales,* trans. Leonard Wolf (New York: Pantheon Books, 1988), 347. It is regrettable that in this translation the name Kawenczynek was omitted.

32   Haya Bar-Itzhak, "Ha-merhav be-agadat k'doshim amamit yehudit" [Space in Jewish saints legends], *Tura II* (1992): 121–133.

33   Thus, for example, the Saints Legend (Shevah) of the Yemenite Jews speaks of a subterranean passage leading to Jerusalem from the tomb of R. Shalom Shabazi in Ta'ez (Ibid.).

34   On this, see Haya Bar-Itzhak, *Poland: Legends of Origin; Ethnopoetics and Legendary Chronicles* (Detroit: Wayne State University Press, 2001). See also Haim Schwarzbaum, "Beit ha-knesset be-agadot am" [The Synagogue in folk narratives], in *Roots and Landscape: Studies in Folklore,* ed. Yassif Eli (Beer-Sheva: Ben-Gurion University, 1993), 137–149. For specific stories from other Diaspora communities, see Nahum Sluszec, *Ha'i pliah* [The island Pliah] (Tel Aviv: Dvir, 1958); Zeev Vilnai, *Agadot eretz-israel* [Legends from the Land of Israel] (Jerusalem: Kiryat Sefer, 1959), 521; Yehuda Bergman, *Folklore yehudi* [Jewish folklore] (Jerusalem: Mas, 1952), 122; Wolf Pascheles, *Sippurim: eine Sammlung jüdischer Volkssagen, Erzählungen, Mythen, Chroniken, Denkwürdigkeiten und Biographien berühmter Juden aller Jahrhunderte, insbesondere des Mittelalters unter Mitwirkung rühmlichst bekannter Schriftsteller* (Prag: J. B. Brandeis, 1850), II:14; David Solomon Sassoon, *The History of the Jews of Baghdad* (Letchworth, England: S. D. Sassoon, 1949), 165.

35   Mircea Eliade, *The Myth of the Eternal Return or Cosmos and History,* Bollingen Series XLVI (Princeton NJ: Princeton University Press, 1991), 18.

36   Sholem Asch, *Kidesh Hashem un andere ertsehlungen fun Shalom Ash* [Kiddush Ha'shem and other stories] (New York: Forverts, 1919), 63–64.

37   Bar-Itzhak, *Poland: Legends of Origin,* 45–88.

38   Ibid.

39   On the use of stones from the Holy Temple to build synagogues, and the link between synagogues and the prophet Elijah or with sacred objects brought from Jerusalem, see Schwarzbaum, "Beit ha-knesset be-agadot am", 137–149. For specific tales of this sort from other communities, see: Sluszec, *Ha'i pliah,* 97; Vilnai, *Agadot eretz-israel,* 521; Bergman, *Folklore yehudi,* 122; Pascheles, *Sippurim,* 1850, 2:14; Sassoon, *The History of the Jews of Baghdad,* 165.

40   The legends have been transcribed and preserved in IFA (Israeli Folktale Archive), IFA 5219.

41 IFA 5217.

42 On the lament as a Jewish response to disaster, see, for example: David Roskies, *El mul p'nei ha-ra'ah: Teguvot la-puranut ba-tarbut ha-yehudit ha-hadashah* [Confronting evil: Responses to disaster in modern Jewish culture] (Tel Aviv: Hakibbutz Hameuhad, 1993), 9–22; Alan Mintz, "The Rhetoric of Lamentations and the Representation of Catastrophe," *Prooftexts* 2 (1982): 1–7; David G. Roskies, ed., *The Literature of Destruction, Jewish Responses to Catastrophe* (Philadelphia: The Jewish Publication Society, 1988), 5–6.

43 Eliade, *Myth and Reality*, 15.

9 A View of the Sea

Jews and the Maritime Tradition

Gilbert Herbert

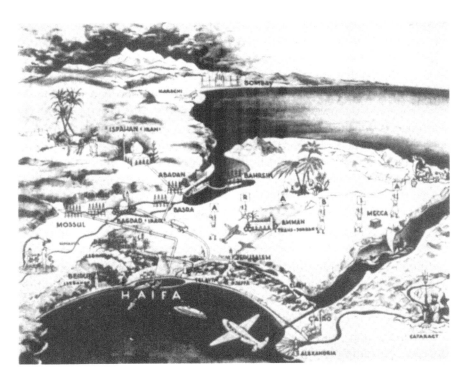

Haifa at the crossroads, artist unknown. Dye-line print, 1930s

And God said unto Noah: '(...) Make thee an ark of gopher wood; with rooms thou shalt make the ark, and shalt pitch it within and without with pitch. And this is how thou shalt make it: the length of the ark three hundred cubits, the breadth of it one hundred and fifty cubits, and the height of it thirty cubits. A light shalt thou make to the ark, and to a cubit thou shalt finish it upward; and the door of the ark shalt thou set in the side thereof; with lower, second and third stories shalt thou make it.' (Genesis 6:13–16)

Introduction

The thrust of this article is to examine Jewish connections to the sea, Jewish perceptions of the sea, and the role that the oceans have played in the survival of the Jewish people. The place of the sea in the evolving territorial consciousness of the Jews reflected the ever-changing circumstances of Jewish history. On the face of it, the orientation of the Jews has always been towards the land. Long before modern Israel achieved sovereignty as an independent state, Eretz Israel — the Land of Israel — was an enduring element in the collective memory of the Jewish people, and the focus of its hopes and yearnings. Israel, the Holy Land, a land flowing with milk and honey, formed not only a physical land-based territoriality but an indefinable spiritual primal landscape, ever-present in Jewish consciousness. Against this centrality of the land, how do we assess the Jewish view of the sea? This question has two facets, one dealing with the past, the centuries of dispersion, the other with the realities of the present. For a people who for most of their history were deprived of a land base of their own, the question is what connection Jews as a people had to the sea, or whether we can talk of a Jewish maritime tradition in any meaningful sense. However, when we consider the unique geopolitical configuration of the State of Israel, whose sovereign territory, to a large extent cut off from its natural *hinterland*, comprises a narrow strip along an extended shore-line on the Mediterranean — that is a country with an extremely high coast-to-land ratio — the question of perceptions of the sea must be confronted not as a Jewish generality but in more specific Israeli terms. Can Israel today be regarded as a maritime nation, a nation materially and emotionally linked to the sea, such as Great Britain, Norway, or Japan have been in recent history, Spain, Portugal, Holland in the Age of Discovery, or the Vikings, the Polynesians or the Phoenicians in more remote times?

The Anatomy of a Maritime Nation

A maritime nation emerges when the motivating forces are insistent, and where the necessary enabling conditions prevail. In addition to overvaulting ambitions of domination, imperial power, and the lust for wealth, which over the centuries drove many states and their rulers to cross the oceans, the more rational motivations of sea power have traditionally derived from the imperatives of survival, security, and prosperity. Naval power protected commercial traffic at sea and defended the coastal cities at home from attack. Merchant fleets were needed for international intercourse, the export trade upon which a nation often depended, the importation of food and vital raw materials. It also facilitated the movement of peoples, essential for the exchange of knowledge and the cross-fertilization of ideas upon which the growth of civilization depended. Of course there were basic geopolitical considerations. Sovereignty was the essential prerequisite. Secure access to and political control of deep-water harbors was a prime consideration. Equally important was control of the *hinterland*, and especially of the land routes—the caravan trails, the highways, eventually the canals and railroads—which converged on the ports.

Power projected at sea demanded adequate financial, material, technological and human resources at home. Although today, in the age of globalization, international centers of shipbuilding—in the Far East, for instance—have contributed to the collapse of home-based shipyards, traditionally a maritime state depended upon its own material resources. The Armada would not have been defeated without England's access to its expansive forests of oak. Along with materials went the need for an organized industrial infrastructure, traditional, generation-deep skills enhanced by a trained work force capable of handling emerging technologies, and efficient management, all powered by investment capital. A maritime nation required a trained and dedicated body of seamen, prepared to face the dangers and hardships of the deep waters, and the long periods of separation from their families.

A maritime nation, in the final analysis, demanded the most profound kind of commitment. The ships of such a nation were manned by men who had the ocean in their blood, the sea fever so poetically evoked by John Masefield in his "Salt Water Ballads."[1] The men of a maritime nation hunted Moby Dick with Captain Ahab, were fishermen on tossing trawlers in the North Sea, the lifeboat volunteers braving the storm, the smugglers landing their contraband on dark lonely shores, a nation of people who sailed their small craft for the sheer joy of the lap of the waves and the rasp of crisp sea air on their faces. A nation with a living maritime tradition not only sailed with Nelson at Trafalgar, but mobilized the armada of small ships for the miracle at Dunkirk. The core ingredient of a maritime nation was psychological—an abiding, emotional link with the sea.

Jews, the Sea, and History

During the long centuries of the Diaspora, when the Jewish people were deprived of nationhood, to what degree was there such a thing as a Jewish maritime tradition? This is not a contradiction in terms, for one can envisage important connections with the sea, independent of the infrastructure subsumed in the concept of a maritime nation. During the millennia of dispersion Jews always had a link of some sort to the sea, as individuals and as special interest groups, but we shall have to see whether this was an integral part of Jewish culture.

The people of Israel, despite their many migrations, were not renowned in ancient days as seafarers. At the time of the Kingdoms of David and Solomon (1000–925 BC), it was Phoenicia rather than Israel that was the primary maritime nation dominating the Mediterranean Sea routes. From the small strip of coastline to which the Jews had access (between present-day Netanya and Haifa) some trade was reported with Egypt. While Israel was barred by the Phoenicians and the Philistines from the Mediterranean Sea, to the south it had access to the Red Sea. As it is told in the Bible, "King Solomon made a navy of ships at Ezion-geber, which is beside Eloth [Eilat], on the shore of the Red Sea, in the land of Edom." These ships were manned by the servants of Hiram, King of Tyre, "shipmen that had knowledge of the sea, together with the servants of Solomon." In addition to ships of Tarshish, these fleets brought to Solomon gold from Ophir, and exotic cargoes of silver, ivory, apes, and peacocks from Arabia, East Africa, and India.[2]

In the post-biblical period, through the long subsequent history of exile and return, subjugation and expulsion, as well as a succession of foreign conquests by Greece and Rome, Islam and Crusader, the Jews were not masters in their own land. Paradoxically, exile and dispersion generated increased Jewish maritime activity, related to the growing—although sporadic—Jewish involvement in international commerce, not only on land, but also at sea. The ports of origin were in Europe and the Americas, rather than the Holy Land,[3] and the Jewish connection was primarily commercial, concerned more with financing or owning—even occasionally building—the vessels, rather than manning them on the high seas. This is not the occasion to give a comprehensive history of the maritime trade, but two examples will suffice to illustrate its scope.

In the ninth century, multilingual Jewish merchants, the Radhanites,[4] functioned as neutral go-betweens in the trade wars between the Islamic nations of North Africa and Christian Europe. For an extended period of time, long before Marco Polo, they crisscrossed Europe and Asia, trading in the widest variety of exotic goods, from spices, perfumes, fabrics, and jewelry to—sad to say—eunuchs and slaves. A typical journey might start from France, usually the Rhône Valley, or Spain, sail to Egypt, then proceed by camel train across the Sinai Peninsula, and finally involve a long sea journey from the Red Sea to India or China. Other routes took these traders along the North African coast, across the Mediterranean to Constantinople, or down the Volga to the Caspian Sea.[5]

Less romantic perhaps, but commercially more significant were the Marranos (or *conversos*)[6] who left the Iberian Peninsula, mainly after the 1492 edict of expulsion. These forced converts to Christianity, compelled to conceal their Jewish origin or continue to practice their religion in secret, served an important function in the development of trade between Europe, America, and the Levant, transporting the gold and silver mined in South America to Spain and Portugal, where these precious metals were traded for goods from the East that eventually found their way to Antwerp, at one time the financial center of Europe. *Conversos* in the seventeenth century had family connections in Hamburg, England, Austria, the West Indies, Barbados, and Suriname. *Converso* families settled in Bordeaux, and "because they had several firms that could outfit ships for oceangoing commerce," they were active in trade with the French colonies across the Atlantic.[7] According to one source, "by 1753 the greater part of the British trade with the Spanish West Indies was in the hands of the Jews, especially the trade of Jamaica with the Spanish main."[8] The most important port was Newport in Rhode Island, where many Jews, encouraged by the founder of the colony, Roger Williams, settled and engaged in commerce. "The ships of the Sephardic merchants Aaron Lopez and Jacob Rivera exported colonial products to Jamaica, Barbados, and London. Lopez's (…) fleet totaled thirty ships."[9]

Whether one considers the forced migrations of Jews, the consolidation of new centers of Jewish life in Europe—Amsterdam and Hamburg became important nodes of Sephardi Jewry—or the role of Jews in the growth of international commerce, the common factor is the sea as a vital channel of communication. Jews were shipowners, shipbuilders, and entrepreneurs of maritime trade. Perhaps the most significant Jewish connection with the sea was intellectual rather than hands-on—in the fields of map making and navigation—which played an important role in the age of the great explorations: the age of Amerigo Vespucci (who owned the famous map of the known world, the *mappa mundi* made in 1431 by Gabriel Vallsecha, son of the renowned Jewish cartographer Hayyim ibn Rich of Majorca);[10] Vasco da Gama (whose ships were outfitted with Abraham Zacuto's astronomical tables);[11] and Christopher Columbus. Whether Columbus himself was a *converso* is a matter of ongoing debate—although the evidence is strong—but it is undeniable that his voyage was vigorously advocated by Jewish and *converso* courtiers and funded by Jewish *conversos* Luis de Santangel and Don Isaac Abrabanel, and that there were at least six known *conversos* on his crew, including his translator, the apothecary, and the physician.[12] However, without the Jewish contribution to navigation these historic missions might never have been accomplished. Critical navigational instruments were the astrolabe, "an instrument introduced into the Arab-speaking world by a remarkable Jewish genius, Mashala of Mosul," and the quadrant, an advanced model of which, designed by Rabbi Jacob ben Machir, became known as the *Quadrant Judaicus*.[13] These navigational aids, and the underlying astronomical knowledge, were Jewish underpinnings of the voyages of discovery. However, there was a more fundamental Judaic contribution, and that was the understanding that the world was a globe rotating on its axis like a ball, an insight contrary to accepted wisdom, which goes back to the Talmudic days of the fourth century.[14]

If nautical instruments and astronomical tables demanded knowledge of the heavens, then map making, the other essential ingredient, demanded a knowledge of the coasts and the oceans beyond. Jews, the perennial wanderers, were well-placed to collect geographic data through their personal knowledge and their network of connections, and became prominent in the field of cartography. One family renowned for its contribution to cartography was that headed by Abraham Crescas, who despite his Jewish origin was appointed "Master of Maps and Compasses" by Juan of Aragon. We have also mentioned the Vallsechas of Majorca in connection with the *mappa mundi*. Majorca, an island, "strategically placed midway between Africa and Europe, became a beehive of Judaic map making. The Majorcans were unrivaled seafarers, knowledgeable about coastal configuration. The Jews among them became cartographers *par excellence*."[15]

Does this historic connection to the sea include a body of Jewish mariners; those who go down to the sea in ships? There are reports of Jewish sailors, but just how many Jews were active seamen actually manning the vessels is difficult to ascertain. There were certainly Jewish navigators, as is attested to by the existence of an astrolabe actually inscribed in Hebrew. We do not know of many Jewish master mariners—Captain Ribeiro of Hamburg is one recorded case[16]—but one more general case must be mentioned. "At the time that many Sephardim were on the move everywhere in Europe, many New Christians departed incognito from Spain and Portugal for the New World. Since they were officially barred from settlement in Spain's colonies, utmost secrecy was essential. Frequently they signed on as ship captains or crew, which they were permitted to do, then illegally transported other co-religionists or jumped ship in South America itself. It was generally acknowledged that Jewish captains knew of all of the secret coves where illegal passengers and goods could be landed in the New World."[17] If this suggests that there were some Jewish mariners who combined Jewish solidarity with smuggling, then there is a related breed of Jewish seamen whose similar activities outside of the law have been recorded. The exploits of Jewish pirates have been chronicled up to the nineteenth century, both by Haim Finkel,[18] and in an as yet untitled book by Ed Kritzler.[19] In an interview with Kritzler, reported in the on-line *Jewish Journal*, several interesting points arose. Most of the pirates were Ladino-speaking Sephardic Jews, *conversos* or crypto-Jews, and their motivation for becoming buccaneers provides an intriguing commentary from an unexpected angle on the period following the tragedy of the expulsion. According to Kritzler, "some Jews, like Samuel Pallache, took up piracy in part to help make a better life for expelled Spanish Jews (…) others were motivated by revenge for the Inquisition. One such pirate was Moses Cohen Henriques, who helped plan one of history's largest heists against Spain."[20] Henriques set up his own pirate kingdom off the coast of Brazil. Another such 'pirate colony' was that of the renowned Sephardic Jewish pirate Jean Lafitte, whose *converso* grandmother and mother fled Spain for France in 1765 after his maternal grandfather was put to death by the Inquisition for "Judaizing." Lafitte, known as the Corsair, established a pirate kingdom in the swamps of New Orleans, and led more than 1,000 men during the War of 1812.

In the late nineteenth century Jews began to play a more socially constructive part in the shipping industry. In 1870, W. Kuntsmann of Stettin founded the largest shipping firm on the east coast of Prussia, while Albert Ballin, scion of a Danish-Jewish family, was the director of the Hamburg-Amerika Line from 1899 to 1918. Ballin developed the enterprise into Germany's biggest shipping line, at times, in fact, it was the largest line in the world, serving the market created by the emigration from Europe to the United States. Ballin was first to introduce the so-called "middle passage" on all long-distance passenger ships, a revolutionary plan which made it cheaper for emigrants to travel. Through Hamburg-Amerika's fast and affordable service innumerable Jews reached the New World. It is a wry thought that this patriotic German-Jew's participation in the maritime industry contributed—not as a cause, it is true, but as a facilitator—to the dramatic change in the demographic balance of world Jewry that took place in the years prior to the Great War, a change that eventually was to keep significant numbers of the Jews of Europe beyond Hitler's fatal reach.

Having said all this, we must still retain a sense of proportion. Despite the significant role played by individual Jews in the development of international commerce by sea, the Jewish communities of the Dispersion—unless compelled by harsh circumstances to migrate or escape persecution—were not yet a seagoing people. The majority of Jews in the New World, even if they settled in a port city like New York, figuratively speaking, had their backs to the sea. In North Africa, Arabia, Central and Eastern Europe, and the vast expanses of Imperial Russia, they settled in the villages and cities, where most found their livelihood in agriculture, as craftsmen, and as petty traders. Jews were to be found in the sweatshops of the Lower East Side, in the alleys of the Jewish Quarter (*mellah*) of Fez,[21] the Jewish area (*Dar Al-Yahud*) of Baghdad, the crowded housing of Whitechapel, the land-locked industrial cities of Warsaw and Lodz, and the prototypical small community (*shtetl*) of which thousands were spread across the Pale of Settlement in Russia. Despite their dependency upon the sea for escape and survival, and notwithstanding the significant contribution made by many Jews to the science and techniques of navigation and to the development of international maritime commerce, as a people the face of the Jews had not yet turned towards the sea, and they were not to be found in significant numbers as seamen astride the heaving decks of ocean-going vessels. If and when they undertook long sea voyages, they did so out of necessity.

Let me illustrate this phenomenon with a typical example from my own family history. My maternal grandfather, Solomon Miller, had considerable experience traveling by sea. As a young man he made his passage from Riga to London, where he married in 1887. The following year, the Millers crossed the Irish Sea to Belfast, where Solomon worked in his brother's mackintosh factory. Apart from the birth of my mother, this was apparently not a positive interlude, and a year later the family was back in London. In March 1901 the Millers sailed to Cape Town in search of a more congenial climate and a more prosperous future, but again this venture did not succeed (grandpa's poor health deteriorated in Cape Town's damp winter climate), and in July 1903 they returned to England on the *Walmer Castle*, a 6,463-ton vessel. Sometime between 1907 and 1909 they returned once more to South Africa, this time moving to Johannesburg, where Solomon died in 1918. In just over twenty years my grandparents had made the 1,000-mile journey from Latvia to England, had twice crossed the Irish Sea, and had four times taken the 6,000-mile London-Cape route (each journey a three-week ordeal), traveling under miserable conditions in relatively small vessels in steerage class. None of these journeys were undertaken for pleasure, or as an adventure, or out of love of the sea. They were undertaken as pure necessity, at great cost to health, and on very limited budgets. In my memoirs I wrote an entire chapter entitled "Transports of Delight" in which I recounted in lyric terms my joy in traversing the oceans, a joy my grandparents' generation was unlikely to share.[22] In this account, I believe I was expressing the view of an unrepentant romantic and the sensibilities of an architect responding both to the natural and designed environment, rather than as a Jew. A more characteristic Jewish view at the turn of the century could hardly be termed romantic. My grandparents' perception of the sea, as for countless thousands of other poor immigrants, was purely utilitarian, as a means to an end. Moreover, their extensive travels did not give them a world view. For my grandparents, home (*der Heim*) was still Dvinsk, for my mother (who was born in Belfast, educated in London, who married and raised a family in Johannesburg, and who died in Safed, Israel), home was eternally England.

A Maritime State in the Making

While European Jews were streaming across the Atlantic to the "goldene medina" in the West, Eretz Israel, the "promised land" in the East—always the spiritual center of Jewish life—began to assume a physical, geopolitical, significance of direct relevance to our story.[23]

After the liberation of Palestine from Turkish rule in 1918, the country came under British administration. The strategic importance of Haifa as a focus of regional and international communications was immediately recognized. By 1920 Haifa had become the center of a new regional railway network connecting to Kantara in Egypt, Damascus in Syria, Amman across the Jordan, and possibly Saudi Arabia, if the Hedjaz Railway to Medina (sabotaged by Lawrence of Arabia) could be restored. Haifa was the logical point of location for a major land-sea interchange. In a communication to Whitehall in July 1919, General Allenby, the conqueror of Palestine, wrote: "(…) in the event of H.M.'s Government accepting a mandate for Palestine the question of port construction at Haifa will become of such importance that it is considered that a preliminary study should be made now."[24] Surveys were undertaken and proposals made, including that of the renowned engineer Frederick Palmer, in 1923. Complex political considerations both in Palestine and in Whitehall, and lack of funding delayed the project until 1927, when the green light was given for the construction of a harbor according to Palmer's proposal.[25] By this time plans were well advanced for Haifa to become the terminal for a pipeline from the new oilfields of Kirkuk and Mosul. From the British point of view the deep-water port at Haifa, inaugurated in 1934, was a significant asset, both in terms of strategic value (as a naval base and major oil port in the Eastern Mediterranean) and of its commercial interests. However, it was also to prove of importance in subsequent maritime history, first, of the Jews of Mandated Palestine, and later, of the State of Israel. The expansion of Jewish settlement in Palestine had consequences for the country's external lines of communication. "The increase in Jewish immigration heightened awareness of the need for maritime knowledge and experience, since the sea was the natural means of communication with the rest of the world, as well as an important source of income."[26] By the 1930s it was clear to both the Imperial Government in London and the Jewish *yishuv* that Palestine stood at the crossroads of the world, linking east and west, and that from a maritime point of view—and at the time that was the dominant form of international contact—Haifa stood at the focal point.

Various attempts were made by Jewish interests to establish shipping services between Palestine and its neighbors, first in 1921, and again in 1928, with limited success. It was only with the establishment of the Orient Shipping Line in 1933 that services ran regularly from Haifa to Egypt,

Lebanon, Syria, Cyprus, and Greece. In 1934, the Atid freight service was established, and from 1934 to 1939 Palestine Maritime Lloyd operated a regular service between Haifa and Constanta with the 2,500-ton *Har Zion* and *Har Carmel*. This growing activity created a need for skilled manpower at sea in a nation without a true maritime tradition, for "the Jews of Eastern and Central Europe had had little opportunity to learn anything about maritime operations."[27] There was a psychological as well as a practical problem to be overcome, for if the nation was to be made sea-minded a change in mental attitude had to occur. "Going to sea," it was explained, "involves absence from home for long stretches of time and the sacrifice of family life, to which Jews are known for being traditionally devoted."[28] The need for a change in orientation from land to sea became apparent to the Zionist movement.

In Germany, a unique venture was initiated, the establishment in 1935 of a merchant marine training center (in Hebrew, *Hachshara*) in Hamburg for young Jews contemplating settling in Palestine.[29] The idea was conceived and carried out by Naftali Unger, a young emissary from the Histadrut Trade Union Federation of Palestine, and Lucy Borchardt, owner of the Fairplay Tugboat Shipping Company of Hamburg.[30] The plan was to use the ships of the Fairplay Company as training for young Jews in order to obtain the immigration certificates of the Palestine Office, for which a professional or technical qualification was a prerequisite. If achieving immigration certificates was the official goal of the venture, its long-range aim was to build a Jewish merchant fleet in Palestine. Even after Naftali Unger returned to Palestine, Lucy Borchardt continued to promote the building of a Jewish merchant fleet until the enterprise was disbanded by the Nazis in the summer of 1938. The British mandatory power, which accepted merchant marine training as satisfying the formal prerequisites for an immigration certificate, nevertheless maintained that qualified Jewish seamen would have no work opportunities in Palestine. Fortunately for the Zionist enterprise, they were mistaken. Not only were the first foundations of a Jewish merchant fleet being laid in Haifa, as we have seen—Lucy Borchardt's son Jens, who had emigrated to Palestine in 1934, was involved in this venture—but preparation for a Jewish pool of trained mariners was also underway.

In Palestine, various organizations encouraged maritime training, including two kibbutzim, Sdot Yam, near Caesarea, and Glil Yam, near Tel Aviv, whose focus was preparing a Jewish corps of sailors. Eventually the Palestine Maritime League was established in 1937. In 1938, it was this League, together with the Jewish Agency, that founded a nautical school affiliated with the Technion, the Israel Institute of Technology in Haifa, with a total of forty students enrolled in its four departments: navigation, marine engineering, radio, and boatbuilding. A 300-ton sailing ship, the *Cap Pilar*, was acquired by the school as a training vessel, followed some years later by the 106-ton yacht, the *Valdora*.[31] With the outbreak of the war in 1939, some 1,200 Jewish volunteers served in the British Royal Navy.[32] The experience gained during the war was to provide an invaluable source of trained manpower; many of the returning servicemen formed the nucleus of the growing merchant marine, destined to play a critical role in the unfolding drama of Jewish nationhood.

The period from the 1939 White Paper, which limited Jewish immigration to Palestine, until the establishment of the State in 1948 was a particularly difficult one in the relationship between the Jews and the British Government, a time when a conflict in aims and a battle of wills was being played out between them, a conflict whose arena was the sea itself. The episode to which we refer is a very special chapter in Israel's maritime history during the mandate period: the tragic and heroic saga known as the "Illegal Immigration" (*Aliya Bet*).[33] While there are historic precedents for Jewish clandestine operations at sea,[34] the climax of such efforts occurred in the period of British rule. There is irony here, for Article 6 of the Mandate under which Palestine was being administered had charged the government with the duty to "facilitate Jewish immigration under suitable circumstances." "[S]uitable circumstances," defined by Whitehall in terms of political expediency, became tragically irrelevant when immigration became a matter of life and death for the beleaguered Jews of Europe. "Illegal Immigration" was the *yishuv's* response to this challenge.

It is not my intention here to enter into the political or moral debate on the campaign, but rather to consider its impact on the maritime history of Israel. This was not a case of isolated incidents, but a venture which saw the conveyance by sea to Palestine of some 115,000 immigrants, mostly in the decade before the establishment of the State in 1948. It was a campaign opposed by the British, who eventually used the full power of the Royal Navy to interdict the flow of refugees; it was a campaign with its bitter disappointments and its tragic losses of life. The *Struma*, one of several vessels refused entry by the authorities, sank in the Sea of Marmora with the loss of all but one of the 769 souls on board, and over 50,000 immigrants who had succeeded in reaching the shores (or at least the coastal waters) of the Promised Land were expelled and interned in Cyprus, a bitter experience for the traumatized survivors of the concentration camps. Two of the vessels of this, the "shadow fleet," were the 4,750-ton *Pan Crescent* and *Pan York*, originally built in the USA in 1901. They were purchased on behalf of Aliya Bet and renamed the *Atzma'ut* (Independence) and *Komemiut* (Sovereignty). In December 1947, the *Atzma'ut* sailed from Bulgaria with no less than 15,000 refugees on board. Intercepted by the Royal Navy and sent to Cyprus, the crew and passengers decided to remain on board, and the ship was maintained and kept in running order, until Independence was proclaimed in May of the next year, when the Israeli flag was hoisted, and the appropriately named *Atzma'ut* sailed to Haifa. The experience gained in acquiring and managing ships, and in training—in the most difficult of circumstances—a generation of mariners, was invaluable for the future development of an Israeli merchant fleet.

It was at the conclusion of the war that the most significant step was taken in developing the merchant fleet with the establishment in 1945 of the Zim Palestine Navigation Company, Ltd.[35] This was on the initiative of the Jewish Agency, the Histadrut (General Federation of Labour), and the Palestine Maritime League. Zim's entry into the shipping trade, although historically significant, was a modest one. In 1947, in partnership with the London-based firm of Harris & Dixon, it purchased an old passenger ship[36] (refitted and renamed the *Kedmah*), a 3,500-ton vessel that plied between Haifa and Marseilles. The primary aim of the company, to "make the prospective Jewish state independent of foreign shipping,"[37] was ambitious. However, neither the political nor the economic environment was particularly congenial to this venture. As a Zim publication later pointed out, "the Mediterranean trade was fiercely competitive, and the British Administration in Palestine saw no reason to support Jewish shipping."[38]

This was a paradoxical time in the maritime history of the Jewish people. On the one hand, for the first time since Solomonic days, it saw the beginnings of a fleet of ships, albeit modest, owned and largely manned by Jews, and serving the national cause in both its economic and political aspects. It was also a period in which the training of Jewish seamen took on new dimensions of importance and urgency. Finally, it was a time when Jewish conceptions of the sea changed significantly. Rather than being a means of escaping persecution, of finding refuge, a foothold on strange shores; rather than using the sea pragmatically as a source of individual economic well-being; rather than generating the astronomical knowledge, charting the oceans and devising the navigational instruments for others to use; instead, the Jewish people began to understand the sea as an instrument for the ingathering of the exiles, as a channel for the sustenance and strengthening of the Jewish National Home, the State in the making. Finally, a career at sea, a "life on the ocean wave" as the old song has it, became a viable alternative life style for many young Jews living in Palestine. This was the beginning of what could in the right circumstances become a maritime tradition. But the circumstances were not yet right. Jews in the Holy Land were not yet masters of their own destiny. The harbors could be closed to them at will by the British Administration, and they had no assured means of breaching the blockade of the Royal Navy in the Mediterranean. Further development depended upon achieving national independence.

The State of Israel

After the War of Independence and the establishment of the State of Israel in 1948, the country found itself in a unique situation. Ringed by hostile states officially at war with it—Lebanon to the north, Syria to the northeast, Jordan to the east, Egypt to the southwest—on all sides its land borders were sealed, and Israel was artificially cut off from its traditional *hinterland* and vital resources. Rail links were discontinued, the oil pipe line shut down. It had become a virtual island, whose connection with the outside world depended upon as yet undeveloped air links and an embryo fleet of ships. It had in Haifa a major port on the Mediterranean, itself an unfriendly sea, with the whole of the North African coast from Egypt to Morocco in hostile hands. Access was denied to the Suez Canal, and for nearly a decade the Red Sea was a no-go area. The conditions were far from favorable, but the imperative to further develop maritime resources was undeniable.

The critical factor in this development was a declaration in the Bundestag on 17 September 1951 that the Federal Republic of Germany, acknowledging that "unspeakable criminal acts were perpetrated against the Jewish people during the National Socialist regime of terror," was determined to make good, "the material damage caused by these acts." The agreement signed in Luxembourg on 10 September 1952 as a consequence of this declaration, under which the Bonn Government, as reported the following day in the *Jerusalem Post*, "agreed to pay 3,450 marks as reparations for material damage suffered by Jews at the hands of the Nazis," was the instrument that enabled the first substantial expansion of the Zim Company's fleet.

Between 1953 and 1955, with purchase credits received under the reparations scheme, Zim ordered eighteen new ships from German shipyards.[39] Four passenger vessels were constructed in the yards of the Deutsche Werft AG of Hamburg between 1955 and 1957. These comprised two pairs of sister ships, passenger-cargo vessels the *Israel* and the *Zion*, and passenger liners in the full sense of the term, the *Theodor Herzl* and the *Jerusalem*. At the beginning of the 1960s, when Zim's dependence on German reparations—and consequently German shipyards—came to an end, the action moved from Germany to France, with whom at that stage Israel had a warm cooperative relationship. The *Moledet* was constructed in the shipyards of Ateliers et Chantiers de Bretagne (A.C.B.). However, for their next (and, as it was to turn out, last) venture into the building of passenger liners, Zim turned to the large Chantiers de l'Atlantique shipyards in neighboring St. Nazaire, where the Loire opens out to the sea. It was here that the famed French liners the *Normandie* and the *Ile de France* had been built. This shipyard was entrusted with the Zim Company's most ambitious project, the commissioning of a large trans-Atlantic passenger liner, the *Shalom,* with a gross tonnage of 24,500 tons,[40] as the flagship of the fleet.

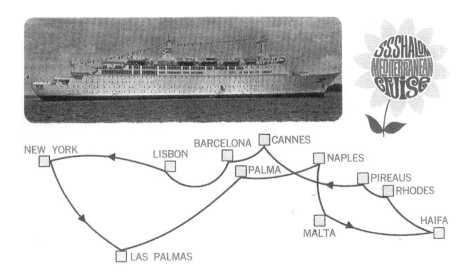

Journeys of the ss. Shalom, flagship of the Zim Line, 1960s

These passenger vessels were unique, both in their character and the symbolic purpose that brought them into existence. Their interiors were designed by some of Israel's finest architects,[41] and they were decorated by artists of world renown, including Ben Shahn of the USA, Bezalel Schatz, Yitzhak Danziger, and Yaacov Agam of Israel, and Rufino Tumayo of Mexico. In their planning and interior design, their materials and furnishings, they represented modern architecture at its cutting edge. The modern in art and architecture was seen as a symbol of Israel as a progressive modern state. Both in the character of the interior architecture and in the choice of furnishings—icons of the International Style—the architects projected an image of the country that was outward-looking, cosmopolitan, seeking its place amongst the advanced nations of the world. A writer in *Travel World* of November 1955, commenting on the *ss. Israel,* noted:

> The inner architecture of the *Israel* reflected the spirit of Israel, its landscape, its pioneering efforts, its traditions and struggles (…) This is not a ship that could belong to any country. It is literally part of the Israel that the Jews eventually hope for and are struggling to create. It is Jewish territory.

It is interesting to note that national identity was a concern not only of the young state of Israel, but of ancient seagoing nations as well. The architect Hugh Casson, in a scathing critique of British ship design, made a parallel claim: "A ship is a travelling piece of Britain and the British way of life."[42] For a few glorious years, until the popularity of air travel made the trans-Atlantic passenger trade unprofitable, Israel in this way became a member of the international family of maritime nations. It was to maintain and expand that role in the coming years in the field of commercial traffic.

Sailing from Haifa, Zim ships maintained services to Europe and across the Atlantic. After the Sinai campaign of 1956 and the opening of the shipping lanes from the southern port of Eilat, extensive seaborne trade was also maintained with the Persian Gulf, East Africa, Southeast Asia, and the Far East. After a hiatus of some three millennia, King Solomon's ancient sea routes were once more being traversed by Jewish mariners. Ships carrying freight were obviously the backbone of the fleet, and by 1963, of the forty-six vessels in service, 40 were cargo ships, bulk carriers, and tankers. Recognizing the national significance of a strong merchant navy, the Government of Israel acquired a one-third interest in the company in 1959, and by 1965 this share had risen to eighty percent. In 2004, reflecting a changed economic climate, the company was privatised. Today, operating about one hundred vessels, supertankers, and container ships—the 13th largest fleet in the world—Zim ships flying Israeli colors, and servicing the three deep-water ports of Haifa, Ashdod, and Eilat, bring in the essential goods and materials, and support the export trade, upon which economic survival depends. To enhance security, the small but highly sophisticated Israeli Navy guards the country's Mediterranean and Red Sea approaches. There are shipbuilding facilities—Israel has built not only some merchant ships, but has designed and constructed some of the world's most advanced missile boats—and a growing flotilla of leisure craft fill the coastal marinas. From all these points of view, Israel may now be counted amongst the maritime nations of the world.

Conclusion

I have argued that one way or another there has always been a Jewish connection to the sea, a connection that in its varied nature and intensity reflected the vicissitudes of Jewish history. Whether there is a Jewish view of the sea is a more difficult question to answer. The modern Jew today, much like his mediaeval counterpart, or those participating in the mass migrations of the late nineteenth century, first regards the sea with a pragmatic eye: the sea as facilitator, as a means of escape from persecution, a conveyor of goods and people, a gateway to security, and an assurance of prosperity. It is a hard-headed view, the product of millennia of endemic uncertainty, eternal wandering, and new perils awaiting. In our hedonistic age, Jews having access to the sun-drenched beaches and surf of California, Sydney, Durban or Phoket, Jews on the cruise liners in the Caribbean or the Adriatic, Jews with small boats in the marinas of the world, view the sea as a purely recreational resource. But this view is no different than that of their gentile neighbors, and in my understanding is not specifically Jewish.

What of Israel? Here the view of the sea is much more complex. Jews in Israel certainly share a worldview of the sea as a source of pleasure, whether active, passive, or social. But for Israelis the sea has added dimensions. In a land restricted in area, narrow (less than ten miles between Netanya and Tulkarem), hemmed in, and predominantly arid, the sea has a psychological value beyond measure. It is an unspoiled natural resource, akin to the wilderness[43], a breaching of claustrophobic boundaries, a widening of the horizon. It is no accident that the majority of Israelis have settled on the coastal plain, nor that proximity to the sea, whether physical or visual, has considerable real-estate value. In addition, the sea, ever since the reclamation of the Haifa foreshore in the 1930s, has also been regarded as a potential source of additional land, with artificial islands featuring in many visionary architectural projects. One such project is an off-shore international airport, currently under consideration, recently advocated by then-Vice-Premier Shimon Peres in a conference on "The Sea as an Economic Resource," as compensation for the abandonment of the West Bank, a political policy of which he has been a long-time proponent.[44] In addition, as I have described, Israel today has acquired many of the characteristics of a maritime nation, with all its impact on the security and economic development of the country, a factor that should be assessed in terms of Israeli national interest rather than as an expression of a more universal Jewish perception.

Was there a view of the sea that went beyond the concept of the sea as thoroughfare, a divide to be bridged? Was there a Jewish concept of the sea as a unifier, as a bringing together? We get a glimpse of this in the age of discovery, when a world view opens up. We see the sea as a universal highway as the voyages of the Radhanites brought the cultures of East and West together; as the great merchant adventurers pooled the abundant riches of the globe; as the vessels of the Jewish State gathered in the scattered remnants of the tribes of Israel, and brought their Jewish brethren in their thousands to re-establish their historical ties with the Holy Land. With the entire North African littoral still inherently hostile, it will take a political earthquake before Israel, turning its eye outward to the oceans as a matrix of unification, can regard itself not as an outpost of Europe in the Middle East, but as a full member of the Mediterranean family of nations.[45] Ecological and cultural imperatives may accelerate this process.

Notes

1 *The Collected Poems of John Masefield* (London: William Heinemann, 1932).

2 Kings I, 9:26–28, 10:11. See also Martin Gilbert, *Atlas of Jewish History* (Jerusalem: Steimatzky, 1976), 5. For a more complex analysis of this early period see Harry Bourne, "From Red to Med," in an essay on the website on Phoenician History: http://phoenicia.org. Bourne refers to the work of the anthropologist Rafael Patai, who was responsible for two important

studies: "Jewish Seafaring in Ancient Times," *The Jewish Quarterly Review* 32, no. 1 (July 1941): 1–16, and *Children of Noah: Jewish Seafaring through the Ages* (Princeton University Press, 1998).

3   Gilbert, *Atlas of Jewish History*, 23.

4   It is unclear whether this term applies to a specific group of merchants or to Jewish traders in general.

5   *Encyclopaedia Judaica*, s.v. "Radaniya" (by Douglas Morton Dunlop), CD-ROM edition, Version 1.0, 1997.

6   The term 'Marrano' (pig) for the Jews of Spain and Portugal who adopted Christianity under compulsion is a derogatory one. Other more acceptable terms are 'Conversos', 'Crypto-Jews', or 'Anusim'. See Jane S. Gerber, *The Jews of Spain: A History of the Sephardic Experience* (New York: The Free Press, 1992).

7   Gerber, *Jews of Spain*, 193.

8   *Jewish Encyclopaedia*, s.v. "Commerce" (by Richard Gottheil, citing M. J. Kohler, Publication of the American Jewish History Society, X:62), online edition, http://www.jewishencyclopedia. com.

9   Gerber, *Jews of Spain*, 207–08.

10 *Encyclopaedia Judaica*, s.v. "Map Makers" (by Herrmann M. Z. Meyer), CD-ROM edition.

11 We are indebted to Samuel Kurinsky, "Jews and Navigation," *Hebrew History Foundation Fact Paper* 9, http://www.hebrewhistory.info/factpapers/fp009_navigation.htm (accessed July 28, 2007) for much of our information about the Jewish contribution to the science and practice of navigation. Kurinsky in turn acknowledges his indebtedness to Cecil Roth, "The Great Voyages of Discovery," chap. 4 in *The Jewish Contribution to Civilization* (New York: Harper and Brothers 1940).

12 See Gerber, *Jews of Spain*, xvi-xxi, and Howard M. Sachar's article on Columbus in *Farewell Espana: The World of the Sephardim Remembered* (New York: Alfred A. Knopf, 1995), reprinted on the website http://www.myjewishlearning.com/history_community/Medieval/TheStory6321666/ Christendom/InquisitionI.htm (accessed July 28, 2007).

13 Kurinsky, "Jews and Navigation."

14 In "Jews and Navigation" Kurinsky cites the fourth-century Jerusalem Talmud (Aboda Zara, 42c) and the Zohar (Leviticus 1.3).

15 Kurinsky, "Jews and Navigation."

16 According to Emil Hirsch (et al.), in "Hamburg and Amsterdam several Jewish seamen followed their calling. Captain Ribeiro died in the latter place in 1623; and the family Ferro in Hamburg have an anchor for their crest." *Jewish Encyclopaedia*, s.v. "Navigation" (by Emil Hirsch et al.) online edition.

17 Gerber, *Jews of Spain*, 179–180.

18 I had the privilege many years ago of reading the English draft of the late Haim Finkel's intriguing narrative, which was later published as a Hebrew-language book, *Shodedei yam yehudi'im* [Jewish Pirates] (Tel Aviv: Devir Publishers, 1984). In the Summer 2002 exhibition catalog of the National Maritime Museum, Haifa, entitled *Pirates: The Skull and Crossbones*, eds. Ruth Gertwanger and Avshalom Zemer, there are several references to the Jewish connection to piracy, both as victims and as perpetrators, from biblical times to the nineteenth century (see 22ff.).

19 This book is due to be published by Doubleday (New York) in 2007.

20 Adam Wills, "Ahoy, mateys! Thar be Jewish Pirates," *Jewish Journal*, 15 September 2006, http://www.jewishjournal.com/home/searchview.php?id=16490 (accessed July 30, 2007).

21 On the development of the *mellah*, see Susan Miller's article in this volume.

22 Gilbert Herbert, *On My Way: Memoirs of an Anglo-Saxon Litvak* (Haifa, 1995).

23   In part this chapter depends upon Gilbert Herbert, *Symbols of a New Land: Architects and the Design of the Passenger Ships of Zim* (Haifa: Architectural Heritage Research Centre, 2006).

24   Gilbert Herbert and Silvina Sosnovsky, *Bauhaus-on-the-Carmel and the Crossroads of Empire* (Jerusalem: Yad Itzhak Ben-Zvi, 1993), 38. For an account of the geopolitical significance of Haifa see chap. 2, "Crossroads: Imperial Priorities and Regional Perspectives," 34ff.

25   Herbert and Sosnovsky, *Bauhaus-on-the-Carmel*, 39–40, for an account of the history of Haifa Port.

26   *Encyclopaedia Judaica* IX, s.v. "Shipping," 835–36.

27   *Zim Israel Navigation Company Ltd.*, brochure, ca. 1963, 4.

28   Ibid.

29   The story is told in two articles, "Merchant Marine Hachshara in Hamburg (1935–1938)," and "Lucy Borchardt. The Only Jewish Female Shipowner in the World," taken from the original German text: Hans Wilhelm Eckardt et al., *Bewahren und Berichten: Festschrift für Hans-Dieter Loose zum 60. Geburtstag* (Hamburg: Verl. Verein für Hamburgische Geschichte, 1997), accessed from http://www1.uni-hamburg.de/rz3a035/borchardt.html, July 30, 2007.

30   The Fairplay Company was formed by Richard Borchardt, a Hamburg Jew. When war broke out he enlisted in the German Navy, and his wife Lucy became his deputy. On his death in 1930 she took over sole management of the company.

31   Carl Alpert, *Technion* (Haifa, 1982), I:192; picture of yacht, 214.

32   The captain of the *Atid* for example, a British supply ship landing armaments on the North African shore for General Montgomery's army, was a Jew from Palestine, and its all-Jewish crew consisted of recent immigrants from Germany.

33   The Hebrew word for immigration is 'Aliya', literally ascent, for in coming to Israel a Jew "goes up to Zion."

34   History repeats itself. We have earlier recounted how Sephardi sea captains engaged in the illicit immigration of Spanish Jews into South America, at the time of a previous persecution of the Jews, the Inquisition. Even in Turkish times Jews had attempted to evade the restrictions placed upon immigration to the Holy Land.

35   My references to the history of the Zim Israel Navigation Company come from the following sources: *Encyclopaedia Judaica*, XVI, s.v. "Zim," 1023; the brochure: *Zim Israel Navigation Company Ltd.*; Chaim Bar-Tikva, *Zim Israel Navigation Company Ltd.: The Fifty-Years Success Story of a Shipping Company, 1945–1995* (n.p.), also summarized on the official website of the company, http://www.zim.co.il; and *All Ways Zim: 50th Anniversary Album 1945–1995*, eds. Gideon Selinger and Nehama Duek (Tel Aviv: Kinneret Publishing House, 1995).

36   The motor vessel *Kedmah*, described as "a 'gallant ship' of long history" (Zim website).

37   *Encyclopaedia Judaica*, s.v. "Zim," CD-ROM edition.

38   *Zim Israel Navigation Company Ltd.*, 4

39   For the history of the design and construction of these ships see Herbert, *Symbols of a New Land*.

40   There is some confusion about the gross tonnage. The *Lloyd's Register of Ships* gives the tonnage as 17,844 tons; the Zim pamphlet illustrating the deck plans gives the tonnage as 24,500 tons; and the Zim Website gives the tonnage as 25,320 tons. The author Robert Wraight, "The Seagoing Gallery," *Tatler*, April 15, 1964, 147, writes: "Conceived as a 23,000 tonner and registered as a 24,500 tonner she finally settled for a Junoesque 25,338 tons."

41   They were designed by the collaborating teams of Al Mansfeld and Munio Weinraub of Haifa, and Dora and Yeheskiel Gad (and after Yeheskiel's death, Arieh Noy) of Tel Aviv.

42   Hugh Casson, "Ship Decoration," in *Shipbuilding and Shipping Record*, International Design and Equipment Number (1956) (typescript in possession of Dora Gad).

43   On the role of the desert as a symbolic landscape in early Israeli culture, see Yael Zerubavel's article in this volume.

44   At the conference, sponsored by the Ruppin Academic Center in April 2007, Peres said that instead of investing in the territories "we must invest in the sea, and stretch our western border in that direction by building artificial islands."

45   There is a public and academic debate in Israel that deals with Israel as a Mediterranean country—a concept sometimes termed 'Mediterraneanism' ('Yam Tikhoniut')—with all its associated cultural and political implications. Interested readers may refer to two works by Alexandra Nocke, each with its own extensive bibliography. These are: "Yam Tikhoniut: The Place of the Mediterranean in Modern Israeli Identity" (PhD diss., University of Potsdam 2006; forthcoming: Leiden: Brill, 2008), and "Israel and the Emergence of Mediterranean Identity: Expressions of Locality in Music and Literature," *Israel Studies* 1/11 (Spring 2006): 143–173.

10 Desert and Settlement

Space Metaphors and Symbolic Landscapes in the Yishuv and Early Israeli Culture

Yael Zerubavel

Eretz (Land) by the artist Pesach Ir-Shai, postcard from the 1930s from David Tartakover's collection

The desert begins where my street ends... (Amos Oz, *All Our Hopes*)[1]

A postcard with a painting by Pesach Ir-Shai from the 1930s[2] features a layered landscape: An outline of a camel's head and long neck constitutes the front layer, revealing behind it two radically different landscapes. On one side it features a round hill with a miniature mosque and a palm tree on its top and three small tents and a tiny cactus plant on its slope; on the other side a modern, square building dominates the space and its angular positioning accentuates its straight, symmetrical lines. The bright Hebrew letters in front of the modern building mark it as *Eretz* [literally, "land"], using a common Zionist nickname for *Eretz Israel* [the Land of Israel]. The two landscapes co-exist within the larger picture space and are partially overlapping, yet the camel and the hill are transparent and the modern building is seen through their layers. The painting juxtaposes two symbolic landscapes and emphasizes the difference between them: The Oriental landscape remains detached from a specific geographical locus or historical time and is barren, and its transparency suggests a lack of substance or stability. By contrast, the urban landscape is associated with modern architecture and the Zionist term "Eretz" that defines the land as a Jewish national space. The building at its center is large and solid and projects a sense of permanence. The smooth and round hill with the domed mosque on the right evokes an image of a woman's breast and highlights the effeminate character of the Oriental landscape whereas the angular geometrical forms of both the modern building and the Hebrew script under it inscribe power and masculinity onto the Zionist space.

This visual representation articulates a highly dichotomized conception of space that was typical of the Hebrew culture of the pre-state period. Zionist discourse portrayed the country as a desolate land and a desert, and emphasized the significance of the settlement of the homeland as a transformative process that was the key to Jewish national redemption. Hebrew culture of the *yishuv* period emphasized the importance of the settlement as the core of the Zionist project of national revival and defined the desert by its opposition to it. The Settlement, the *Yishuv*, was constructed as *The Place*, the territory that was marked, named, and historicized, and hence became the national space.[3] Jewish society of Palestine chose to define itself in terms of the settlement process. The Hebrew word for settlement, *yishuv*, was used as its collective self-reference and Israeli historiography refers to the period from the 1880s to the establishment of the State of Israel in 1948 as "the *Yishuv* period."[4] The use of the term *yishuv* implies not only a place that is populated but also a civilized place marked by it social and cultural life.[5]

By contrast, the desert constituted a mythical territory that lay outside of Jewish space and was marked by its inherent opposition to the settlement.[6] "Desert" therefore does not relate to a specific geographical region or a space defined as such by certain scientific criteria related to amount of rain fall; nor does it follow the biblical use of the term relating to grazing ground.[7] Rather, desert emerged as a *symbolic landscape*[8] that was defined as a symbolic void. Such a void did not imply nothingness but rather marks "a place in which there are none of those things which we expected to find there."[9] Desert was therefore a residual category that was defined by its difference from the settlement. As the Hebrew poet, Yehuda Amichai, brilliantly articulated, desert was the embodiment of "the other side."[10] Hence it could be applied to a wide range of geographical regions.

The two concepts—the settlement and the desert—thus served as oppositional yet interdependent space metaphors and both were integral parts of the cultural construction of space in the emergent modern Hebrew culture. Indeed, I would argue that it would be impossible to fully understand the intricate meaning of the former without appreciating the multiple meanings of the latter. This essay thus sets out to explore the complex, and at times contradictory, meanings of the settlement and the desert as dynamic symbolic landscapes prior to the establishment of the State of Israel and the early formative years of statehood. I first examine the interplay of history and space in the construction of the desert as a symbolic landscape. This discussion provides the context for the following section focusing on the emergence of settlement as a key concept of the new Jewish society of Palestine. This analysis of the oppositional meanings of the desert and the settlement as space metaphors is then juxtaposed with their construction within the discourse of romantic nature. By following this interpretive route in the Israeli cultural landscape during the formative years of Israeli society, this essay attempts to present the inherent tensions that underlie the various meanings of desert and settlement as key concepts that articulated and shaped Israelis' understanding of space within the exceedingly volatile context of the contemporary Middle East.

Space and History: Homeland, Desert, and Galut

Hebrew culture placed the Jewish settlement at the center of its geographical map and related to the territory surrounding it as a symbolic desert. Within the context of the settlement process, these space metaphors assumed historical connotations. The long period of Jewish exile from the homeland, *galut*, was negatively portrayed as a symbolic void or vacuum in Jewish history, a state of homelessness and lack that was contrasted with Jewish national life during Antiquity and in the modern period.[11] That state of lack was, in turn, inscribed onto the physical landscape of the homeland that Jews found upon their return. During centuries of life in exile Jews carried in their minds the famous biblical description of "a land of milk and honey." The landscape they encountered was far from that idyllic, mythical description. The country appeared to them dry and desolate, and its state of lack was emphasized in comparison to the European landscape that many of the Zionist pioneers had left behind.

Hebrew settlement discourse used the concepts "desert" (*midbar*) and "desolate land" (*shemama*) interchangeably to describe landscapes that appeared as desolate, wild, or occupied by Arabs. Travel narratives from the end of the nineteenth century and the first decade of the twentieth century demonstrate the frequent application of the term "desert" to a diversity of landscapes to describe their state of "lack." A well-known writer and educator, Ze'ev Yavetz, noted that the spreading sand dunes threatened to turn the shore into a *"desolate desert,"*[12] and Yehudit Harari, a young teacher traveling in the northern part of the country referred to the landscape of the lower Galilee on the way to Haifa as a wasteland and described it as "desolate and full of swamps, a *desert of white sand."*[13] Similarly, Nechama Pohachevsky, another settler and writer of the First Aliya, described the Galilean landscape seen along the road from the Sea of Galilee to Rosh Pina as "filled with rocks and a *desolate desert,"*[14] and other travelers felt anguish at seeing the barren mountains of Judea or other parts of the country that show that "our country has turned into a desert."[15] *The Frontiersmen of Israel*, a settlement novel published in 1933 by the educator and writer of the Third Aliya, Eliezer Smoli, follows the settlement process of a single family in a forest in the Galilee. Visitors from a nearby Jewish settlement, who come to see their comrades' farm, express their concerns and doubts about the prospect of survival in that *"desolate desert."*[16] This use of these terms was not meant as a literal description of the landscape in which the settler family established its farm, which was depicted as heavily forested and lush. Rather, it articulates the friends' view of the space as standing outside of the accepted territory of the Zionist settlement. From this perspective, the area was culturally defined as a symbolic desert.

The frequent allusions to "desolate desert" [*midbar shemama*] as noted above and in other texts[17] served as a rhetorical device to highlight the symbolic lack of the land, its wild nature and uncivilized character vis-à-vis the ancient landscape of the country or the new Zionist *Yishuv*. Narratives of the early settlement period often compared the devastation the settlers found with an earlier, fertile state of the land before Jews had been exiled from their homeland. Thus, a First Aliya settler, Hayim Hissin, noted in his diary in 1882: "The Land of Israel was once a fertile land, but this was in olden times. Since then, and as a result of the neglect of the land, the winds continued to carry incessantly clouds of sand that have piled up and have created such a layer that the Land of Israel, once one of the most fertile lands in the world, has turned into a bereaved country that has only barren hills and stretches of sand."[18] Yizhak Ben-Zvi, a leading figure of the Second Aliya who became Israel's second President, articulated the same view that the cultivated land became a desert as a result of the Jews' forced exile and the subsequent inhabitants' limited means and negligence: "When the land was conquered by desert tribes, camel drivers, and shepherds, the desert once more pushed back the cultivated land. Settlement shrank and the wasteland expanded; for the Bedouin made his livelihood mainly in wild growth and not in cultivated plants."[19] A children's textbook published in the 1930s presents this view by describing a Jewish guard's shocked response at the sight of the enormous desolation of the desert land. The guard attributes this condition to the Arab presence and compares it to the biblical descriptions of the fertility of this land during Antiquity and the archeological ruins that provide evidence of earlier human activities there.[20] A similar contrast appears in a textbook that was published three decades later: "In the ancient past, the valley was the storehouse of crops of the entire country, yet when the Jewish people were exiled from the country and the land became deserted, the fertile valley was transformed into the site of poisonous swamps."[21]

The emphasis on the growing deterioration following the Jews' exile from the land shaped a clear outline of a "decline narrative"[22] and further highlighted the importance of the Zionists' return to the land. The Zionist settlement discourse implied that the damage caused by men could also be repaired by human efforts. The mission of "making the desert bloom" was thus designed to repair the state of lack that had been brought on by exile and recover an earlier state of national bliss. Zionist collective memory portrayed Antiquity as the Golden Age of the Jewish people and that past became a source of inspiration and pride. Exile, on the other hand, was regarded as a prolonged period of historical regression associated with disintegration and national shame. The depiction of the country in a state of devastation thus produced a further incentive to promote the Zionist settlement, representing it as a moral imperative and a debt to the national past as well as a means to advance the cause of future national redemption.

The political and educational discourse of the pre-state and early state periods emphasized these themes. As A.D. Gordon, a spiritual leader of the Second Aliya, wrote: "Our country, which had been a land of milk and honey, and at any rate carries the potential for high culture, has remained desolate, poorer than other civilized countries and empty—this is sort of confirmation of our right to the land, a sort of hint that the country has been awaiting us."[23] Israel's first Prime Minister, David Ben-Gurion, called upon his countrymen "to remove the shame" of the history of the desert and contribute to the settlement effort.[24] And a textbook of the 1950s reprinted a Zionist pioneer's story describing how he took note of the emptiness of the landscape surrounding a new settlement: "Everything is so desolate, barren, mournful and still. There is no house, no tree, no stake—only desolate rocks and barren hills (…) And we need to revive all this?"[25] The momentary doubt served to emphasize the solution: constructing new settlements and working the land was the way to promote the redemption of both the people and the land. This double model was also articulated in the words of a highly popular Hebrew pioneering song: "We came to the Land to build it and be rebuilt in it."

The decline narrative from Antiquity to Exile was helped by the use of anthropomorphism in describing the state of the country. The land was depicted as suffering from shame, stricken by acute sickness, fallen into deep slumber, or engulfed by death and mourning following the Jews' departure. Ben-Gurion and Ben-Zvi concluded their 1918 book on the history of the Land of Israel with a similar trope: "The country awaits the people, its people, to come back and renew and reconstruct its old home, cure its wounds with its sons' love (…)."[26] In an educational play, children receive a similar message: "The land is weeping and it cries out to you to come and redeem it from its desolation. Only then would it stop crying. Would you do this for its sake?"[27]

Zionist "propaganda" photographs and films of the *Yishuv* era juxtapose images of desert landscapes that project bleakness, emptiness, and death with a vivid display of new settlements, settlers working the land, and children playing outdoors, highlighting the theme of "renewal."[28] The Jewish National Fund provided the writer of a settlement narrative an explicit directive to emphasize the desolation and neglect prior to the appearance of the Zionist pioneers, the quick change brought by the Jewish settlement, and the universalistic value of the work of draining the swamps and curing the land.[29]

The attribution of agency to the land in support of the settlement cause was particularly popular in children's literature. A story in a textbook for the fourth grade describes how four springs anguished over the Jews' exile from the land and vowed to turn their water into a curse during the Jews' absence. The springs voluntarily transformed their water into malaria-spreading swamps that brought death to the land. Only when they realized that the Zionist pioneers had returned to the land were the springs ready to remove the curse and become the source of life, thereby taking an active part in the Zionist revival of the land.[30] Another children's story tells of two lumps of soil that witnessed Yosef Trumpeldor, a Zionist military hero whose arm had been amputated, plowing the land by himself in spite of his disability. Impressed by his devotion, the two lumps decide to voluntarily jump into the turning blade in an act of self-sacrifice, emulating Trumpeldor's own heroism and readiness for sacrifice.[31] In this literature, land and nature do not constitute passive objects of Zionist cultivation but rather emerge as willing agents of change that act out of their recognition of the importance of the settlement mission.

The portrayal of the landscape as a "desolate desert" was clearly based on a selective view of the reality at the time. These descriptions focused on unproductive lands, barren areas, and the malaria-spreading marshlands while ignoring Arab villages and towns and other settlements built by European settlers, as well as the existence of cultivated fields, plantations, and orchards around the various settlements. Other sources, however, indicate that the settlers clearly saw the inhabited and cultivated parts of the land and developed relationships with members of other communities around whom they lived. In 1885, Zalman David Levontin, a leading figure among First Aliya settlers, learned about the great potential of the land from the fields, the olive groves, and the citrus orchards he saw when traveling from Jaffa to Gaza.[32] Other Zionist residents and visitors similarly observed the abundance of olive, palm, citrus, pomegranate, almond, and fig trees that could be seen in various parts of the land.[33] David Ben-Gurion and Ben-Zvi took note of the extensive areas covered by olive trees around Lod and Ramla and described Gaza as surrounded by large gardens, vineyards, and orchards on all sides. Nonetheless, they qualified this achievement by citing the local legends that "the olives of Gaza were planted by Alexander the Great and the assumption is that since the Arab conquest not a single olive tree has been planted."[34] At times, these observations disclosed the Zionist settlers' pain of watching parts of the homeland owned and cultivated by non-Jews. Thus, Yehudit Harari expressed her shock at hearing German settlers refer to "their beautiful Carmel" given the Jewish past associated with that area; and Nechama Pohachevsky described how her excitement at seeing the beauty of the Galilee was marred by sorrow that "this wonderous valley does not belong to us and that foreigners dominate it."[35]

Settlement as the Civilized Space

Arab and Jewish settlements existed next to each other in physical space, but in the cultural construction of space the Zionist settlers put the Jewish settlement at the center and saw it as surrounded by a desert. The famous Jewish writer and polemicist, Ahad Ha'am, dryly captured this view during his 1891 visit to Palestine. A Jew who is visiting the *Yishuv*, he noted,

> (…) moves from one *moshava* [Hebrew for colony] to *moshava*, and these might be separated by a journey of many hours, and fields and villages of Gentiles fill all that space in-between, but he sees this space between them as if it were a desert devoid of human beings. And after the 'desert' a settlement [*yishuv*] appears, and he can breathe again the national Hebrew air that revives his soul.[36]

Although the early Zionist settlers clearly saw and acknowledged the reality of different settlements and cultivated lands in Palestine, they relegated them to the periphery of their cognitive map as part of the desert domain.

As Ahad Ha'am's observation reveals, the fundamental view that constructed the desert and the settlement as opposing symbolic landscapes led to the perception of the latter as an oasis or an island surrounded by hostile space. This view was projected on the landscape at the level of a single settlement and on the Settlement, the *Yishuv*, as a whole. The Jewish settlement was thus seen as separate and distinct from its environment and hence in need of defending itself from it. The analogy between the island and the oasis also implied an analogy between the desert and the sea: both were large forces of nature that surrounded the *Yishuv*, the embodiment of civilization. This perception, as we shall see below, heightened the sense of a struggle for survival.[37] In spite of these shared characteristics, there were some critical differences between the desert and the sea as symbolic spaces bordering on the Jewish Settlement. The sea separated the *yishuv* from Europe but was also the gateway to it; it was the lifeline to the West and represented the potential for future ties and for additional Jews who would join the *Yishuv* society.[38] The desert, on the other hand, symbolized the East and was the connecting tissue to it. As such it evoked a much deeper ambivalence in the Zionist settlers, whose cultural orientation was first and foremost to the West.

The construction of the desert and the settlement as oppositional symbolic landspaces were clearly influenced by predominantly European views of the Orient, which European Zionist immigrants brought with them to the Middle East.[39] The Zionist settlers regarded themselves as the carriers of Western civilization into an area that many saw as primitive and backward. The First Aliya settler and well-known writer, Moshe Smilansky, referred to a "semi-wild region" inhabited by "savage" Bedouins;[40] and Yosef Weitz, a key figure in leading the Jewish National Fund's forestation and settlement efforts, used similar terms to refer to the challenges confronting the settlers during the early years of Zionist immigration. Accordingly, the First Aliya settlers had "to stand up in the struggle against nature and the savages around them, with no protection from the outside."[41] Yosef Klausner, a literary critic and historian, issued a warning that "we, Jews, who lived for two thousand years among cultured people; we cannot, and should not, deteriorate to the cultural level of semi-savage peoples."[42] Hemda Ben-Yehuda, a First Aliya writer and wife of the famous promoter of Hebrew revival, Eliezer Ben-Yehuda, described her first tour in 1892, shortly after her arrival to Palestine: "I was shocked by the Arab village I saw: houses made of mud, without windows, housing both men and animals. Piles of garbage everywhere and half naked children (…). Old blind women and dirty girls sit in front of the houses working, grinding wheat as was done a thousand years ago."[43]

Similarly, a Zionist settler described how he pointed out to his friend the difference between the Zionist settlements and the Arab villages while traveling:

> 'Look', I told my friend, 'how great is the distinction between our fields and theirs, our villages and theirs. Our fields are cleared of stones, fertilized, plowed in depth; and theirs are covered with many stones and their soil is used up. Our village is green with trees, surrounded by a strong fence, and built of mortar; their houses are gray—without windows and colorless, only dusted olive trees are seen in the village.'[44]

Another settler referred to a visible transformation from a landscape dominated by caravans of camels to one that was filled with houses, trees, and people building and working the land.[45] His description thus evokes the duality of the desert/settlement landscape that Pesach Ir-Shai's work displays.

The juxtaposition of the desert and the settlement landscapes also assumes social and moral dimensions. Zionist settlers characterized the Arab settlement by disorder, filth, and neglect. The Arab village represented a total disregard for the social distinctions that informed the European notion of order: inside and outside, hygiene and dirt, human beings and animals. Within the dichotomized framework of culture versus nature, the Arabs were seen as part of nature. The Zionist settlement, in contrast, introduced order into the chaotic local landscape, thereby transforming the desert into a civilized space.

Eliezer Smoli's youth novel mentioned above, which became part of the Hebrew canon for the youth, articulates these ideas through its main protagonist, the Zionist settler:

> This is the kind of life we'll make for ourselves. (…) We'll turn these *barren valleys* into *gardens of Eden*. They'll be covered with corn and barley, oats and hay, and we'll plant vineyards and orchards on the hillsides in place of these thorns and thistles. We shall have to drain the marshes to get rid of the mosquitoes and then turn them into vegetable patches.[46]

At another point, however, the settler tells his family: "If the soil in these hills is good enough to grow such fine forest trees, it will be good for fruit trees too. We'll plant figs and vines here in the winter, and turn this barren countryside into a Garden of Eden."[47] This depiction of the Garden of Eden is particularly interesting in light of an earlier description of that same area as covered by a lush forest, including indigenous fruit-bearing trees such as figs. Although the family is well aware of these trees and has eaten their fruits, the father's vision emphasizes the planting of new trees and crops as both innovate and transformative. The settlement of the land is thus achieved through the process of cultivation that includes the acts of building and planting. Similarly, a novel by the prominent Israeli writer S. Yizhar, published in 1945, describes an idyllic vision of a future settlement in the desert: "We'll have here plenty of trees, and grass, and houses, with red roofs, and you know, green fences, and the sprinkler will spin, and birds (…)."[48]

One of the most prominent features of the Jewish settlement was the introduction of straight lines in the landscape, which gave it a modern character. The imposition of the grid system on a natural landscape represents a Western view associated with rationality and technological progress and contrasts with the narrow, curved streets of the old towns and villages.[49] The new Hebrew city, Tel Aviv, was praised for its contrast with the old Arab town of Jaffa. It was hailed for its broad boulevards and straight streets, bright light, cleanliness and order. A geography book published in 1918 noted its positive qualities within the larger Middle Eastern landscape: "Tel Aviv is the culmination of the Hebrew settlement. It is *a European oasis within an Asian desert.* As if in a dream, one is suddenly transferred into one of the beautiful and quiet corners in one of the shore towns in South-Western Europe: Rows of straight, paved streets. Trees and flower gardens grow along them. Everything is new and shining (…)."[50] Similar references to the city's modern European style were in abundance.[51] The development of a new industrial site in the desert was similarly described as involving the creation of a network of straight lines in the open landscape.[52]

The early settlers also took pride in mastering the art of plowing in straight lines, and the new forests of the Jewish National Fund featured trees planted in straight lines.[53] Indeed, the view of lined trees became a sign that one was approaching a Hebrew settlement.[54] The Jewish National Fund posters of the late *Yishuv* and early state periods advertise the agenda of making the desert bloom by featuring straight lines of plowed fields, trees, and houses. In some posters, workers' erect bodies and work tools add a powerful vertical line, reminiscent of the artistic style of Soviet social realism. Pesach Ir-Shai's juxtaposition of the oriental and the Zionist landscapes is evident in other representations of that period. Wide and open spaces are presented in contrast to densely built settlements; tents, wild cacti, and herds of sheep are juxtaposed with modern buildings, cultivated trees and flowers, cars and machines; Bedouins passing through the space, walking or riding idly are juxtaposed with Zionist pioneers rigorously engaged in working their land.[55]

Zionist pioneering narratives, as I discuss elsewhere in greater length,[56] depict the settlement process as an uphill battle, a long and difficult struggle against the desert landscape. These narratives typically use protomilitary rhetoric, describing the efforts to turn the desert into a flourishing settlement as a war that consists of successive battles ultimately leading to a victory. In and of itself, the foundation of the State of Israel marked a victory over exile and the desert: the settlement overcame the desert, the counter-place, and transformed that space back into a Jewish homeland. The conception of an inherent struggle between the settlement and the desert, the cultivated and the wild, was clearly articulated in the theme around which Israel's first international exhibit was organized in 1953.[57] "The Conquest of the Desert" celebrated the achievements of the Zionist settlement using its advanced technological knowledge to make the desert bloom and thereby incorporate it into the Settlement domain.

Mythical Space and "Net Landscape": The Romantic View of the Counter-Place

The settlement narrative highlighted the negative view of the desert as a territory characterized by hostility toward the Zionist settlement, yet this construction of space does not encompass the range of meanings attributed to the desert as a symbolic counter-place in Hebrew culture. The *Yishuv* culture also articulated another, competing conception of the desert that enhanced its positive qualities as a mythical space representing nature, freedom, transformation, and rebirth. Like the negative view of the desert, this approach too assumed a fundamental opposition between the desert and the settlement, although in constructing their meanings as conflicting symbolic landscapes this view assigned positive rather than negative value to the "counter-place."

It was from out of this context that the desert drew its romantic and national appeal as representing nature in its primal and purest form. Desert thus provided the antithesis of civilization and technological progress. It was a mythical landscape that withstood the pressure of history and modernity and preserved its primal beauty and integrity. More than other regions it appeared unspoiled by humans and could therefore provide a sense of permanence and authenticity. The Israeli poet Yehuda Amichai, who belonged to that generation of Hebrew youth of the *Yishuv* period who became desert lovers, devoted a book of poetry to the desert. In one of these short poems he described the primal quality of the desert landscape, "an open land, without subconscious, without wrapping, without food coloring, pure landscape, net landscape."[58]

The desert was closely connected to European images of the Holy Land and hence was more likely to evoke biblical images. In the *Yishuv's* emergent Hebrew culture, the desert served as a mythical space imbued with memories of ancient forefathers and hence a territory that made it possible for Jews to reconnect with that past.[59] Hebrew teachers regarded traveling in the land as critical for the education of the youth and believed that these experiences would help them reclaim their authentic native identities. Traveling, and especially the physical experience of hiking, were considered an effective venue for mnemonic socialization and a means to instill in youth the love of the homeland and intimate knowledge of its landscape.[60] Personal accounts of touring the land often referred to the experience of imagining the ancient Hebrews at the very site they had lived and acted, and this theme appeared in literary texts. The Second Aliya member, Rachel Yanait (Ben-Zvi) imagined Bar Kokhba when she traveled through Betar, the last stronghold of his revolt against the Romans in the second century.[61] The poet David Shimoni saw the images of the Maccabees and heard the sounds of their followers when visiting the site of their ancient home village, Modiin.[62]

More than the urban environment, the ruins of the past and the open vistas of the desert made it easier for travelers to remember the past and imagine the ancient forefathers walking in the same space. Excursions to the desert therefore presented an even more attractive destination for young people.[63] Thus, Nogah Hareuveni described the hikers' response to the unique forms of the desert landscape: "The wonder that these chasms inspire has always made me imagine what our forefathers (...) Standing in the very same places, awestruck by the same scenery, I see how these landscapes reflected in the words bequeathed to us in the Bible."[64] A Hebrew settlement novel about the early days of a new desert settlement articulates the sense of a renewed bond with the ancient Hebrews. When a Bedouin warns a young Jewish settler about the harsh conditions in that land, the latter dismisses this warning: "We are not foreign to this desert," he replies. "In the desert our nation was born. We return to it now." At another point, one of the settlers tells his friend: "Have you ever thought about it that most of the Jewish holidays were formed in the desert and that exile only distorted them? That's it. Here, in the desert, we should give them back their contents."[65]

A popular song called upon Hebrew youth, "Go, go to the desert," thus echoing God's command to Abraham to set out on his journey to the Promised Land. Indeed, trips to the desert had a particularly strong appeal for the new generation of Hebrew youth of the pre-state and early state periods. In contrast to the founders' generation, for whom work and construction represented the shedding of their exilic identity, Hebrew youth born or raised in the *Yishuv* society saw the desert as the territory that allowed them to affirm their native Hebrew identity. In the late *Yishuv* period, desert trips were also used as a cover for paramilitary practices (away from the British Mandatory police) or for intelligence and mapping purposes. Located on the outskirts of the Zionist settlement and far from the center, the desert presented multiple challenges. Hiking in the desert took several days, required strenuous efforts, and involved real dangers. The desert terrain was an unfamiliar territory that was difficult to reach; finding one's way in it required avoiding treacherous tracks, surviving with a limited supply of water and food, and risking arrest by the British police or attack by Bedouins. Some of these trips indeed ended in death when young people lost their way, became dehydrated, stumbled and fell into abysses, were killed by armed Bedouins, or died from accidental explosions of arms that the youth had smuggled with them.[66]

Field trips to the Judean desert brought with them "an unusual combination of danger and a sense of freedom that one lacks in the settled land: On the one hand the real and imagined dangers (...). On the other, though, there was a sense of absolute freedom: No law, no police, nobody would reach you here and tell you what to do. You can go wherever the wind carries you."[67] As Hebrew youth became more familiar with the desert terrain, they no longer had to rely on Bedouin guidance, and this independence enhanced their sense of mastery and freedom: "It is difficult to describe the feeling of victory we had when we became free of the Bedouin guides and began to find the way on our own. We stopped behaving like lost sheep which are led by an omniscient shepherd and became the masters of the desert (...)."[68] At times these trips were taken in defiance of authority, be it the British Mandatory Authority, their parents, or the commanders of the Jewish Haganah underground.[69]

Young people also saw their expeditions into the counter-place as a means of gaining control over that terrain. Thus, as the above quotes reveal, accounts of the hikes in the desert terrain often drew on the rhetoric of struggle and articulated the desire to conquer the desert or reported on feeling a sense of victory in achieving this goal. "Conquering" in this context meant facing the risks and challenges of trekking in the desert and overcoming them. Meir Har-Zion, whose daring and adventurous spirit as a commando soldier earned him a legendary hero's reputation, describes a life-risking route he once took, sneaking out of a youth camp by himself to reach the top of Masada in 1951. When he finally arrived at his destination, after a near-death experience, he was fascinated by the wild view of the landscape but also filled with "crazy happiness mixed with the joy of victory."[70]

Modern field trips to the desert can also be seen as rites of passage in which young people test their ability to become both adults and natives. The desert's location, away from the settled and cultivated center of the country, made it possible for youth to reenact the scheme of transitional rites noted by Arnold Van Gennep, and following him, Victor Turner: rites of passage typically involve venturing outside of the home sphere into an alternative space in which the familiar social order is suspended to allow for the experience of change; the process is completed by the return to the home base in a transformed identity and status.[71] By taking these trips into the desert and facing the challenges they presented, Hebrew youth reshaped their identities as natives and reaffirmed their membership in the *Yishuv* society. The excursion to the counter-place thus served as a way to re-experience the significance of the civilized space as the homeland.[72]

These experiences young people also echoed an ancient memory of the desert as a mythical space of revelation and transformation, a memory encoded in sacred texts and rituals. Moses saw the burning bush in the desert and accepted the mission of rescuing his oppressed brethren from Egypt; and the desert was where the Israelites were transformed from a community of slaves into a free nation and received the tenets of their Jewish faith as a community of believers.[73] In the modern age, the desert provided the space outside of the home territory where Hebrew youth reclaimed their ancient roots and were transformed as natives. During the War of Independence, the Palmach underground commando unit, operating out of jeep vehicles in the Negev, was named "The Negev Beasts" (*Hayot ha-negev*), thus implying that its members were not only the true natives of the region but could also be seen as part of the wild nature of the desert landscape.

Concluding Remarks: Cultural Ambivalence and Space Categories

The desert and the settlement were constructed as symbolic space categories that are both oppositional and complementary, yet their interpretation encompassed contradictory meanings that reveal a more complex approach to space than the examination of the settlement ethos alone might have revealed. The Zionist national vision of reclaiming the Jewish homeland drew on the conflation of space and history in Jewish collective memory. The "desert" and the "settlement" therefore emerged as space categories imbued with national meanings. Desert represented the negative impact of exile on the Jewish homeland, and the emergent Jewish society of Palestine regarded the space around it as a symbolic desert. Settlement, on the other hand, embodied Jews' national life in their own country and was associated with both the Jewish national past in Antiquity and the modern Zionist return to the homeland. Since the Hebrew term *yishuv* constitutes both a verb and a noun, the concept encompassed both the process of colonizing the land and the outcome of this process: hence the reference to the society as the *Yishuv*. The perception of space as desert was tied to the perception of the Orient that presented its inhabitants as dormant, passive, and resistant to change. By contrast, the Zionist settlement was seen as representing a deep commitment to change, construction and progress.

Yet, as a counter-space to the *yishuv*, the desert also held great fascination for members of the new Jewish society. As a site of nature, its primeval landscape and unique geological forms held the mystic of the wilderness. The desert provided a space outside of and away from society, offering its members a refuge from the social order and its constraints and the possibility of experiencing freedom. The desert was seen as land that stood beyond time and projected eternity, but it was also linked to ancient roots and ancestral images that stretched back to mythical beginnings. Having served as the space of national rebirth in Antiquity it became the space to which the Hebrew youth returned in order to reclaim their collective identity as natives and carry that identity back to the settlement. The desert could function as a refuge and a site of inspiration, revelation, and transformation precisely because of its status as the counter-place vis-à-vis the *Yishuv*.

As symbolic landscapes, the settlement and the desert were essentially dynamic categories, constantly in flux, and subject to negotiation and change. The continuing tensions between them constructed their boundaries, yet those remained both fluid and vulnerable, creating a situation that was inherently destabilizing. As a 1936 geography textbook noted, "The desert is not a fixed territory since an industrious people will build and expand its settlement and turn the desolate desert into a flourishing garden."[74] In other words, as the settlement expands by making the desert bloom, the space defined as desert clearly shrinks; and alternatively, when land ceases to be cultivated or a settlement is deserted or destroyed, the desert expands its scope by reclaiming this space. The struggle over space between these two forces also implies an interdependence in creating and shifting the visible and invisible divisions between these categories. Recounting his daily walk into the desert that lies around his hometown of Arad, Israeli writer Amos Oz captures the essence of the relations of these two space categories and the line that separates them: "The desert begins where my street ends (…)."[75]

The cultural ambivalence that underlies Israeli attitudes toward the desert was not fully acknowledged in the Hebrew culture which placed a stronger emphasis on the settlement ethos and highlighted a negative perception of the desert as hostile space. Yet the admiration of the desert marked a persisting undercurrent of a nostalgic search for Middle Eastern roots and was manifest in the desire to escape to the "non-place" from an increasingly complex and pressing political reality. In the decades following the foundation of the state, space categories would continue to be transformed and marked by conflicting attitudes and agendas. Given the rapid development of the center of the country, and to a lesser degree its area in the north, "desert" became, after 1948, increasingly identified with the sparsely populated southern and eastern regions of the Negev, the Arava, and the Judean deserts. The polemics about the Jewish settlement drive in the territories beyond the 1967 borders, the rise of an individualistic ethos and "alternative culture," the emergence of a strong environmentalist movement, the continuing efforts to settle the Negev desert and strengthen "the periphery," and the development of desert tourism that draws on the romantic view of that space—all these reveal Israelis' ongoing engagement in a wide range of issues that underlie the relations between the settlement and the desert and have made more overt the cultural ambivalence that surrounds them.[76]

Notes

1   Amos Oz, *Kol ha-tikvot: Mahshavot al zehut israelit* [All our hopes: Essays on the Israeli condition] (Jerusalem: Keter, 1998), 251. All translations from Hebrew are mine unless otherwise noted.

2   I would like to thank David Tartakover for giving me permission to reprint the postcard with Pesach Ir-Shai's painting from his collection. Ir-Shai's picture was also included in the Israel Museum's exhibit catalog *Kadimah: Ha-mizrah be-omanut Israel* [To the east: Orientalism in the arts in Israel], eds. Yigal Zalmona and Tamar Manor-Friedman (Jerusalem: Israel Museum, 1998), 20.

3   For the discussion of The Place in Hebrew culture, see Zali Gurevitch and Gideon Aran, "Al ha-makom" [On the place], *Alpayim* 4 (1991): 9–44, and "The Land of Israel: Myth and Phenomenon," in *Reshaping The Past: Jewish History and the Historians*, ed. Jonathan Frankel, *Studies in Contemporary Jewry* 10 (1994): 195–210.

4   Zionist historiography marks "the *yishuv* period" as beginning with the First Aliya [Zionist immigration to Palestine] and the *yishuv hadash* (the new settlement) as opposed to the small *yishuv yashan* (the old settlement) of mostly Ultra Orthodox and Orthodox Jews who had lived in Palestine prior to the Zionist immigration. Yehoshua Kaniel notes that the terms *yishuv yashan* and *yishuv hadash* emerged in the late 1880s obscuring lines of continuity between them. See *Hemshekh u-temurah: Ha-yishuv ha-yashan veha-yishuv he-hadash bi-tekufat ha-Aliya ha-rishonah veha-sheniyah* [Continuity and change: Old Yishuv and new Yishuv during the First and Second Aliya] (Jerusalem: Yad Ben-Zvi Press, 1981), 21–34.

5   See Yehuda Gur [Gurzovsky], *Milon ivri* [Hebrew dictionary] (Tel Aviv: Dvir, 1955), 379, and Avraham Even-Shoshan, *Ha-milon he-hadash* [The new dictionary] (Jerusalem: Kiryat Sefer, 1988), II, 512. Even-Shoshan also notes that the definition of *yishuv* implies "a place of culture and orderly social life," while Gur's definitions include an explicit reference to the *yishuv* as civilized place in contrast to the desert.

6   On the distinction between place and space, see Yi-Fu Tuan, *Space and Place* (Minneapolis: University of Minnesota Press, 1977) and Donald W. Meinig, "Introduction," in *The Interpretation of Ordinary Landscapes* (New York: Oxford University Press, 1979). On the view of the Oriental landscape as an ahistorical concept, see Edward Said, *Orientalism* (New York: Vintage Books, 1979), 96; for its impact on Hebrew culture and art, see Ariel Hirschfeld, "Kadima: Al tefisat ha-mizrah ba-tarbut ha-israelit" [To the east: On the construction of the east in Israeli culture], and Yigal Zalmona, "Mizraha, Mizraha" [To the East! To the East!] in *Kadimah*, 11–31 and 47–93 respectively.

7   Even-Shoshan, *Ha-milon he-hadash*, II, 630; for a fuller discussion of the biblical use of the term 'midbar', see Nogah Hareuveni, *Desert and Shepherd in Our Biblical Heritage* (Neot Kedumim: The Biblical Landscape Reserve in Israel, 1999), 26–27.

8   E. V. Walter similarly addresses the concept "expressive space," in *Placeways: A Theory of the Human Environment* (Chapel Hill: North Carolina Press, 1988), 215. W.J.T. Mitchell discusses the concept of landscape as a symbolic construct and explores the impact of cultural images on politics: "Holy Landscape: Israel, Palestine, and the American Wilderness," *Critical Inquiry* 26 (Winter 2000): 193–223.

9   René Descartes, "Principles of Philosophy," *Descartes' Philosophical Writings*, abridged version, trans. G.E.M. Anscombe and Peter Geach (Indianapolis: Bobbs-Merrill, 1971), 181–238. I would like to thank Michael R. Curry for bringing this text to my attention.

10 In his desert poetry book, Yehuda Amichai writes: "We sometimes say:'/But, on the other side…'/Here is the other side." "Le'itim" [Oft], *Nof glui einayim* [Open eyed land] (Tel Aviv: Schocken, 1992), n.p. (English and German translations appear side by side with the

original Hebrew). Note that the Hebrew poem further draws on the double meaning of "on the other side" as "on the other hand."

11   On Zionist collective memory and the perception of exile as a symbolic void, see Yael Zerubavel, *Recovered Roots: Collective Memory and the Making of Israeli National Tradition* (Chicago: University of Chicago Press, 1995), 13–33. See also Amnon Raz-Krakotzkin, "Galut betokh ribonut: Likrat bikoret shelilat ha-galut ba-tarbut ha-israelit" [Exile within sovereignty: Towards a critic of the 'negation of exile' in Israeli culture], *Teoria u-vikoret* 4 (Fall 1993): 23–55.

12   Ze'ev Yavetz, "Shut ba-aretz" [Travel in the land] (1891); repr. in *E'evra na ba-aretz* [Wandering in the land: Travels by members of the First Aliya], ed. Yaffa Berlovitz (Tel Aviv: Defense Ministry Press, 1992), 115.

13   Yehudit Harari, "Bein ha-keramim" [Among the vineyards] (1947; repr., ibid.), 288; emphasis added.

14   Nechama Pohachevsky, "Me-rishon le-ziyon ve-ad marge ayun" [From Rishon Le-ziyon to Marge Ayun] (1908; repr. ibid.), 314–15.

15   Yehoshua Barzilai, "Be-sha'arei yerushalayim" [In the gates of Jerusalem] (1891; repr., ibid.), 82; see also Moshe Smilansky describing travelers' sadness at witnessing the "desolated land" in "Be-ikvot benot ruhaniya" [In the footsteps of the daughters of Ruhaniya] (1894; repr., ibid.), 187.

16   Eliezer Smoli, *Anshei bereshit* [The frontiersmen] (1933; repr., Tel Aviv: Am Oved, 1973); English translation by Murray Roston (Tel Aviv: Masada, 1964). The quote here is based on my translation of the original Hebrew (1973, 184) since Roston's translation (1964, 214) obscures the author's use of the terms 'midbar' and 'shemama'. Note that emphasis added.

17   For other examples of the use of both 'desert' and 'desolate land' as interchangeable concepts, see Yisrael Betser, "Ha-zerif me-ever la-yarden" [The shack in the Transjordan], in *Kovetz ha-shomer* [The guard anthology] (Tel Aviv: The Labor Archives, 1937), 131; Moshe Unger, "Hag ha-mayim" [The water festival], in *Mo'adim le-simha: Mahzor sipurim u-mahazot le-khol mo'adei ha-shana* [Holidays: A cycle of stories and plays for the holidays] (Tel Aviv: Shlomo, no year), 63–69; Natan Alterman's poem "Shir ha-kevish" [The song of the road] written in the 1930s, relates to the workers' ongoing struggle against the desert and glorifies their achievements (see Yoram Tehar-Lev and Mordechai Naor, *Shiru, habitu u-re'u: Ha-sipurim me-ahorei ha-shirim* [Sing, look and watch: The stories behind the songs] (Tel Aviv: Defense Ministry Press, 1992), 20–21. See also Yehoshua Kaniel's comments that members of the old Jewish *yishuv* in Palestine perceived the landscape as symbolically empty to indicate the absence of Jews in it: *Ba-ma'avar: Ha-yehudim be-eretz-israel ba-me'ah ha-19* [In translation: The Jews of Eretz Israel in the nineteenth century] (Jerusalem: Yad Ben-Zvi Press, 2000), 15, 26, 212–14. For the application of the Hebrew term 'shemama' (desolation) to different landscapes, see Yoram Bargal, "Dimuyei nof eretz-israel be-ta'amulat ha-keren ha-kayemet le-israel bi-tekufat ha-yishuv" [Landscape images of the Land of Israel in the Jewish National Fund's propaganda during the *Yishuv* period], thematic issue on "Landscape and nature in the Land of Israel throughout its history," *Motar* 11 (2003/4): 19–28, esp. 21–22.

18   From the diary of the First Aliya settler, Hayim Hissin (March 14, 1882), quoted in Shulamit Laskov, *Ketavim le-toldot hibat-ziyon ve-yishuv eretz-israel* [History of Hibbat Zion and the settlement of the Land of Israel] (Tel Aviv: Hakkibutz Hameuchad, 1982), I, 426. I would like to thank Alan Dowty for bringing this, and other references on this issue, to my attention.

19   *Kibush ha-shemama, catalog le-ta'arukha bein le'umit vi-yerid amim* [The conquest of the desert: A catalogue for an international exhibition and fair], a bilingual catalogue, Jerusalem,

September 2 to October 14 (Jerusalem, 1953; no publisher indicated), the English Section, 7.

20   "Be-Oholei ha-nodedim" [In the Nomads' Tents] in Nahum Gavrieli, ed., *Yediat ha-moledet* [Knowing the homeland] (Tel Aviv: Omanut, 1934), I: 138, 150.

21   S. Shaked, *Zot moladeti* [This is my homeland] (Tel Aviv: Yesod, 1962), 78, quoted in Yoram Bargal, *Moledet ve-geografia be-meah shenot hinukh ziyoni* [Moledet and geography in a century of Zionist education] (Tel Aviv: Am Oved, 1993), 145. Bargal observes that the contrast between the fertility of the land during Antiquity and its vast desolation following Jews' exile from it became more pronounced in Hebrew geography textbooks published after World War I and persisted until the 1960s (ibid., 139). The above quotes demonstrate that this theme appears earlier in the writings of Zionist settlers and visitors to Palestine.

22   Eviatar Zerubavel, *Time Maps: Collective Memory and the Social Shape of the Past* (Chicago: University of Chicago Press, 2003), 16–18.

23   A.D. Gordon, "Avodatenu me-ata" [Our labour from now on] *Ha-uma veha-avoda* [Nation and labor] [Jerusalem, The Jewish Agency, 1952], 244.

24   *Kibush ha-har* [The conquest of the mountain] (Jerusalem: Jewish National Fund, 1956), 6–7.

25   M. Meirovitch, "Kometz ha-zera'im ha-rishon" [The first seeds], in Nahum Gavrieli and Baruch Avivi, eds., *Mikra la-yeled li-shemat ha-limudim ha-revit'it* [Textbook for the Child for the Fourth Grade] (Tel Aviv: Yavneh, 1957), 117.

26   David Ben-Gurion and Izhak Ben-Zvi, *Eretz-israel be-avar uva-hove* [The Land of Israel in the past and in the present] (New York, 1918, in Yiddish; repr. in an abridged Hebrew translation, Jerusalem: Yad Ben-Zvi Press, 1979), 228. See also Rabbi Eliyahu Gutmacher in a letter quoted in *Sefer ha-mo'adim* [The book of holidays], ed. Yom-Tov Levinsky (Tel Aviv: Dvir, 1956), VIII: 330.

27   Haim Tehar-Lev, *Kol min he-harim: Hizayon li-yeladim* [A voice from the mountains: A children's play] (Jerusalem: Sifriat Adama, 1959), 9.

28   Moshe Smilansky made the following observation on the eve of World War I: "From the beginning of the Zionist idea, the Zionist propaganda described the country to which we come as a deserted and a desolate land that longs for its redeemers"; quoted in Yaacov Haroey, "Yahasei yehudim-aravim be-moshevot ha-aliya ha-rishona" [Jewish-Arab relations in the First Aliya colonies], in *Sefer ha-aliya ha-rishona* [The First Aliya book], ed. Mordechai Eliav (Jerusalem: Yad Ben-Zvi Press, 1981), I: 266. Studies of early Zionist photography and films offer similar observations: Ruth Oren, "Havnayat makom: Ta'amula u-merhav utopi be-zilum ha-nof ha-ziyoni, 1898–1948" [The construction of place: Propaganda and utopian space in Zionist landscape photography 1898–1948], *Devarim Aherim* 2 (Fall 1997): 18–19; Ella Shohat, *Israeli Cinema: East/West and the Politic of Representation* (Austin: University of Texas Press, 1989), 15–53 and see also Hillel Tryester, "'The Land of Promise' (1935): A Case Study in Zionist Film Propaganda," *Historical Journal of Film, Radio and Television* 15 (no. 2): 190.

29   A letter from Yehiel Halperin addressing the writing of a monograph on the settlement of Merhavia, the first settlement in the Yisrael Valley, Central Zionist Archives, KKLS 2/2487, quoted in Bargal, "Dimuyei nof eretz-israel", 25.

30   N.H. Karpivner, "Arba ha-ma'ayanot" [The four springs], in *Mikra'ot israel li-shenat ha-limudim ha-revi'it* [Mikra'ot Yisrael: Textbook for the fourth grade], eds. Z. Ariel, M. Blich, and N. Persky (Jerusalem: Masada, 1958), 163–64.

31   Fania Bergstein, "Rigvei ha-galil" [The soil of Galilee], in *Haruzim adumim* [Red beads] (Tel Aviv: Hakibbutz Hameuchad, 1955), 164–70.

32 Zalman David Levontin, *Le-eretz avoteinu* [To our ancestors' land] (1885; repr. in *E'ebra na ba-aretz*), 34–35, 42.

33 Suleiman Mani notes the abundance of trees in "Masa le-aza" [Travel to Gaza] (1885; repr. in ibid.), 50–51; Barzilai, "Be-sha'arei yerushalayim" [In the gates of Jerusalem], 60–69; Yavetz observes the Arabs' orchards in "Shut ba-aretz," 108.

34 Ben-Gurion and Ben-Zvi, *Eretz-israel be-avar uva-hove*, 151–55, 210.

35 Harari, "Bein ha-keramim," 287; Pohachevsky, "Me-rishon le-ziyon ve-ad marge ayun," 307.

36 Ahad Ha'am, "Sakh ha'kol" [In sum], *Kol kitvei ahad ha'am* [Ahad Ha'am's collected works] (Tel Aviv, 1954), 427. Israel Bartal notes that Zionist anthologies did not ignore the Arab population but rather associated them with the decline and destruction narratives. See Bartal, "Kibbutz galuyot textu'ali: Ha-antologia ha-ziyonit be-sherut ha-uma" [Textual metling pot: The Zionist anthology in the nation's service], in *Nof moladeto: Mekhkarim be-ge'ografia shel eretz-israel uve-toldoteiha* [The motherland's landscape: Studies in the geography and history of the Land of Israel], eds. Yossi Ben Arzi, Israel Bartal and Elchanan Reiner (Jerusalem: Magnes, 2000), 482.

37 For a further discussion of the island as a space metaphor for the *yishuv*, see my paper "The 'Desert' and the 'Island': Space Metaphors in Modern Israeli Culture" at the annual meeting of the Association for Jewish Studies, December 2006, in San Diego, California. On the symbolic meaning of the sea, see Hanan Hever, *El ha-hof ha-mekuve: Ha-yam ba-tarbut ha-ivrit uva-sifrut ha-ivrit ha-modernit* [Toward the longed-for shore: The sea in Hebrew culture and modern Hebrew literature] (Jerusalem: Van Leer Institute and Hakibbutz Hameuchad, 2007).

38 For a detailed discussion on the role of the sea in Jewish awareness, see Gil Herbert's article in this volume.

39 Said, *Orientalism*; Zalmona and Manor-Friedman, *Kadima*; Derek J. Penslar, "Zionism, Colonialism, and Postcolonialism," in Anita Shapira and Derek J. Penslar, eds., *Israeli Historical Revisionism From Left to Right* (London: Frank Cass, 2003), 84–98.

40 Moshe Smilansky, "Be-taiha" [In Taiha], repr. in *Sifri he-hadash li-shenat ha-limudim ha-hamishit* [My new reader for the fifth grade] (Tel Aviv: Yavneh, 1950), 21. See also Bargal, *Moledet ve-ge'ografia*, 160.

41 Yosef Weitz, *Ha-galila!* [To the Galilee!] (Jerusalem: Jewish National Fund, 1937), 3.

42 Yosef Klausner, quoted in Yosef Gorny, *Ha-she'ela ha-aravit veha-be'aya ha-yehudit* [The Arab question and the Jewish problem] (Tel Aviv: Am Oved, 1985), 56.

43 Hemda Ben-Yehuda, "Ish tahat gafno ve-tahat te'enato" [Each one under his vine and fig tree] (1944; repr. in *E'ebra na ba-aretz*), 165.

44 Y. Grazovsky describes the differences between the Hebrew and Arab villages in "Tiyul be-harim" [A journey in the mountains] (1938), repr. in *Mi-dan ve'ad be'er sheva: Sefer le-limud moledet be-kitot dalet-hei* [From Dan to Beer Sheva to the fourth and fifth grades], eds. A. Gondelman and Y. Gefen (Tel Aviv: Am Oved, 1945), 16–17.

45 Yosef Weitz, "Herzliya," ibid., 105–106. Many of the texts included in this textbook describe the settlement process and its dramatic impact on the country's landscape.

46 Eliezer Smoli, *Anshei bereshit* [The frontiersmen of Israel], 20 [translation from the English edition]; emphasis added.

47 Ibid., 17. Pohachevsky describes the crowded and dirty streets of old Tiberias and the "desolate desert" stretching in between it and the new Hebrew colony, Rosh Pina, to which she refers as a "Garden of Eden" ["Me-rishon le-ziyon ve-ad marge ayun," 307].

48 S. Yizhar, *Be-fa'atei negev* [On the edge of the Negev] (1945; repr., Tel Aviv: Hakibbutz Hameuchad, 1978), 134–35.

49   For J.B. Jackson's remarks on the symbolism of introducing straight lines into the American landscape in "The Order of a Landscape," in *The Interpretation of Ordinary Landscapes*, ed. Donald W. Meinig (NY & Oxford: Oxford University Press, 1979), 158–60. On the tension between Western and Middle Eastern conceptions of space see also Aziza Kazum, "Tarbut ma'aravit, tiyug etni u-segirut hevratit: Ha-reka le-i shivyon etni be-israel" [Western culture, ethnic labeling and social reticence: The background to lack of equality in Israel], *Israeli Sociology* I (1999): 390–91. Isaiah's prophecy associates the vision of redemption with the symbolism of the straight line: "A voice cries in the wilderness, prepare ye the way of the Lord, *make straight in the desert* a highway for our God" (Isaiah 40:3; emphasis added). Harold Fisch notes that the creation of a straight line in the open space challenges the vastness of the desert. See *Remembered Future: A Study in Literary Mythology* (Bloomington: Indiana University Press, 1984), 133.

50   M. Chuzmer, *Ge'ografia* [Geography] (1918), quoted in Bargal, *Moledet ve-ge'ografia*, 160; emphasis added.

51   Ibid., 159–61; Maoz Azaryahu, *Tel aviv, ha'ir ha-a'mitit: Mitografia historit* [Tel Aviv the real city: A historical mythography] (Sde Boker: Ben-Gurion Institute and Ben-Gurion University Press, 2005), 61–66; Barbara E. Mann, *A Place in History: Modernism, Tel Aviv, and the Creation of Jewish Urban Space* (Stanford: Stanford University Press, 2006); Anat Helman, *Or va-yam hekifuha: Tarbut tel avivit bi-tekufat ha-mandat* [Light and sea surrounded it: The culture of Tel Aviv during the Mandatory Period] (Haifa: Haifa University Press, 2007).

52   Aharon Ever Hadani, *Ha-mifal ba-arava* [The Arava Project] (Tel Aviv: Mitspe, 1931), I: 55, 87.

53   Smoli devotes an entire chapter to the settler family's first attempts to plow a new field and describes how the new trees were planted in straight lines in *Anshei bereshit*; Yehuda Raab, a First Aliya settler and one of the founders of the Petach Tikvah colony, describes the moving ritual of plowing the first furrow in his autobiography, entitled *Ha-telem ha-rishon* [The first furrow] (Tel Aviv: The Zionist's Library, 1957); see also Shaul Katz, "Ha-telem ha-rishon: Idiologia, hityashvut ve-hakla'ut be-fetah tikva ba-asor ha-shanim ha-rishonot le-kiyuma" [The first furrow: Ideology, settlement and agriculture in Petach Tikvah in the first decade], *Katedra* 23 (1982): 57–122. Shlomo Schwartz describes how new immigrants were taught how to plant trees in straight lines to create new forests: *Anashim hadashim be-harim ha-gevohim* [New people in the high mountains] (Tel Aviv: Hakibbutz Hameuchad, 1953), 63, 97, 118.

54   Ze'ev Yavetz notes "the two lines of snow-white houses with their red roofs that can be seen among the trees that are planted in a straight line in front of them," in "Shut ba-aretz," 117; similarly, Dvorah Eilon-Sireni describes how she recognized a Hebrew settlement when she saw the straight lines of trees, in "Ba-negev" [In the Negev] in *Sifri ha-hadash la-kita ha-hamishit*, 41.

55   See the images in the exhibit catalogues *Kadima:* and *Ta'amula ve-hazon; Omanut sovietit ve-israeli 1930–1955* [To the east and propaganda and vision in Soviet and Israeli art] (Jerusalem: Israel Museum, 1997). See also Rachel Arbel, ed. *Kahol ve-lavan bi-zeva'im: Dimuyim hazuti'im shel ha-ziyonut 1897–1947* [Blue and white in colors: Zionism's visual images, 1897–1947] (Tel Aviv: The Diaspora Museum and Am Oved, 1996).

56   Yael Zerubavel, "The Conquest of the Desert and the Settlement Ethos," in A. Paul Hare, ed., *The desert experience in Israel: Settlement and science in the Negev* (Lanham, MD: The University Press of America, 2008).

57   The official translation of the Hebrew title of the exhibit *kibbush ha-shemama* is "The Conquest of the Desert." For further discussion of the exhibit, see ibid.

58   Amichai, "Nof" [Open landscape], in *Nof gelui einayim* [English translation provided in the original], n.p.

59 On the use of the land as a symbolic bridge to the ancient past, see Yael Zerubavel, "Antiquity and the Renewal Paradigm: Strategies of Representation and Mnemonic Practices in Israeli culture," in *On Memory: An Interdisciplinary Approach*, ed. Doron Mendes (Bern: Peter Lang, 2007), 331–48.

60 On mnemonic socialization as a process of transmitting the memory of the past to members of society, see Eviatar Zerubavel, *Time Maps*, 2–5. On the educational value of hiking, see Rachel Elboim-Dror, *Ha-hinukh ha-ivri be-eretz-israel* [Hebrew education in Palestine] (Jerusalem: Yad Ben-Zvi Press, vol. 1, 1986; vol. II, 1990), I:389, II:346; Yaffa Berlovitz, *Lehamzi eretz, lehamzi am: Sifrut ha-aliya ha-rishona* [Inventing a Land, Inventing a People: Literature of the First Aliya] (Tel Aviv: Hakibbutz Hameuchad, 1996), 183–86; Mordechai Naor, ed., *Tenu'ot ha-no'ar, 1920–1960* [Youth movements, 1920–1960] (Jerusalem: Yad Ben-Zvi Press, 1989), 246–56; Shaul Katz, "The Israeli Teacher-Guide: The Emergence and Perpetuation of a Role," *Annals of Tourism Research* 12 (1985): 49–72; Gurevitch and Aran, "Al ha-makom," 30–34; Bargal, *Moledet ve-ge'ografia*, 37–3.

61 Rachel Yanait Ben-Zvi described the euphoric feeling and the sense of establishing a direct bond with their ancestors that she and her friends experienced while walking in the fields. When they returned home by horse and wagon crowded with Arabs, their spirits plummeted as the immediate reality sank in, undermining the earlier vision. *Anu Olim* [We climb up] (Tel Aviv: Am Oved, 1962), 22.

62 David Shimoni, "Le-kivrot ha-makabim" [To the Maccabees' tombs], repr. in *For Hanukkah: An Anthology* (Tel Aviv: The Histadruth Youth Center, 1935), 19–21 and in *Orot: Homer la-hanuka* [Lights: Materials for Hanukkah] (Jerusalem: Jewish National Fund, 1949), 18–19. See also Smilansky, "*Be-ikvot benot ruhaniya*," 187.

63 On Israeli backpackers in the Far East and elsewhere, see Erik Cohen's article in this volume.

64 Hareuveni, *Desert and Shepherd in Our Biblical Heritage*, 129. Interestingly, the English translation changed here the collective pronoun 'we', which the author uses to describe this experience in the original Hebrew text, to the first singular 'I'. See *Midbar ve-ro'e be-mosheret Israel* (Lod: Neot Kedumim, 1991), 115.

65 Alexander and Yonat Sened, *Adama lelo zel* [A land without a shade] (Tel Aviv: Hakibbutz Hameuchad, [1951], 1977), 105, 14.

66 See Yehoshua Yellin's story about an 1871 excursion to the desert in: "Ha-to'im ba-midbar" [Those lost in the desert] (1924), repr. in *E'ebra na ba-aretz*, 11–23; for Hebrew youth's desert experiences in the later *Yishuv* period, see Shmaryahu Gutman, "Siah sayarim" [Hikers' conversation] in *Yam ha-melah u-midbar yehuda, 1900–1967* [The Dead Sea and the Judean Desert, 1900–1967], ed. Norchechai Naor (Jerusalem: Yad Ben-Zvi Press, 1990), 236–55, and Azaria Alon, "Me haya midbar yehuda avureinu?" [What the Judean Desert meant to us?] Ibid., 270–75. Meir Har-Zion recounts in detail his various adventures in the desert and the dangers he faced there as a youth, in *Pirkei Yoman* [Diary chapters] (Tel Aviv: Epstein, 1969).

67 Alon, "Me haya midbar yehuda avureinu?," 271.

68 Ibid., 274.

69 Rafi Tahon, *Yam ha-melah saviv-saviv* [Around the Dead Sea], *The Dead Sea and the Judean Desert*, 246; Guttman, "Siah sayarim"; Meir Har-Zion, *Pirkei yoman*, 62, 80, 115. The stories often describe how they manipulated security regulations to express their adventurous spirit and defiance of order in these excursions.

70 Ibid., 78–79.

71 Arnold Van Gennep, *The Rites of Passage* (1908) (Chicago: University of Chicago press, 1960) 114; Victor Turner, *The Ritual Process* (New York, 1969), 80–118.

72   See also Tuan, *Space and Place* on the significance of the home territory and the reaffirmation of its meaning through exploring space outside of its domain.

73   Yosef Hayim Yerushalmi, *Zakhor: Jewish History and Jewish Memory* (Seattle: University of Washington Press, 1982), 11–12.

74   Avraham Yaacov Braver, *Hora'at moledet be-veit ha-sefer ha'amami* [Homeland instruction at the primary school] (Tel Aviv, Dvir, 1930), 2–5, quoted in Bargal, *Moledet ve-ge'ografia*, 129.

75   See note 1.

76   For a more detailed discussion of these issues, see: Yael Zerubavel, *Desert in the Promised Land: Nationalism, Memory, and Symbolic Landscapes* (Chicago: University of Chicago Press, forthcoming).

11 Jews and the Big City

Explorations on an Urban State of Mind

Joachim Schlör

View into the rotunda at the Kaisergalerie, Unter den Linden, Friedrichstrasse, Berlin, 1880s

"Jews have always been an extraordinarily urban people." With a general statement like that, many universities offer courses that deal with the Jewish encounter with the city. Then the description becomes more specific, concentrating on one of the many aspects of this relationship: "This seminar explores various aspects of the Jewish encounter with the city, examining the ways that Jewish culture has been shaped by and has helped to shape urban culture."[1] Usually these courses range widely across and among different fields of Jewish history — "Jewish urban experience from ancient times to the present; public space and private; the city and the sacred; Jewish ghettos and quarters; or the struggle over modern Jerusalem"[2] — but can also focus on more specific fields such as migration. "This course is intended to engage the students in a comparative analysis of the changes that urbanism entailed for Jewish immigrants coming to a city such as New York; the nature of Jewish interaction with the city and with other groups in the city (…),[3] or literature; "This course examines poems, novels, memoirs, and short stories by Jewish writers who have worked in a wide variety of urban and cosmopolitan settings over the course of the last century.[4]

In my own course at the University of Southampton, "Modern Jewish Culture and the Big City," students were fascinated to see how a basic model — the structure of a (traditional) Jewish community — could be applied to so many different geographical and political settings throughout the centuries; to learn how specific practices and rituals pertaining to Jewish law (creating or abolishing an *eruv*,[5] the Sabbath "border"; spending the week of *Sukkot* in a hut on the balcony or on the roof of one's house) *take place* in the built environment of a city; and to study how centuries of urban living produced a body of texts (descriptions of cities; travel reports; novels; sermons; rabbinical responsae) which indeed can be described as signifiers of an "urban state of mind." A reading of Georg Simmel's *Metropolis and Mental Life* seemed to work well in the context of debates about the challenge of modernization and the "prize" of urban life, for the communities as well as for individual Jews.

On the Road to a New Jewish Urban Historiography

Against this background, the idea of this essay is to ask whether (or not) the city in general, and the modern "Big City" in particular, can be described as a place in which traditions, experiences, and images of the — assumed — special relationship between modern Jewish culture and the spatial conditions of modernity and modern life come together, in a concentrated, dense, perhaps even in an essential form. Is the city, especially the Big City of modern times — Berlin, London, New York City — a special kind of *Jewish* space, more so than the countryside, the village, or the small town? And is it a Jewish *space* that combines elements of temporality and transience with elements of duration and substance, such as architecture, forms of settlement and the like? Is the Big City a physical place *and* a space of reference at the same time—not only for the Jews, but maybe more so than for others? The greater part of the material I use for this essay concerns the situation of German Jews after 1933: when their lives were threatened, but also their form of living: as an urban community which had grown in and with German cities, especially Berlin, and had to try and save (and transform) this urban experience in the places of their emigration.

The subject seems to cross over many fields of Jewish studies and there have been a number of general publications in recent years that should be mentioned, starting with Steven Zipperstein's critical reading of books about Jewish historiography and the modern city (1987),[6] Ezra Mendelssohn's edited volume *People of the City: Jews and the Urban Challenge* (1999),[7] a French volume edited by Chantal Bordes-Benayoun on *Les juifs et la ville* (2000)[8] and also, if I may, my own habilitation work on the debates about the notion of a "special" relationship between modern Jewish and modern urban cultures.[9] And of course there are numerous studies about specific cities, the processes of urbanization, the city-country relation and a wide range of other topics. Still, there is a more general dimension to this area, something like an open—but uncharted—road to questions of perception, identity, and imaginations of "self" and "other," as the sentence, "Jews have always been an extraordinarily urban people," testifies. The Frankel Institute for Advanced Judaic Studies announced "Jews and the City" as its "Theme for 2007–2008" with the following text:

> Jews have lived in cities for millennia and scholars have consistently studied Jewish urban life. Much scholarship on Jewish religion still assumes that Jews are not particularly oriented to spatial forms and representations, preferring temporality or textuality to spatiality. One form of Jewish social organization associated with cities, namely, "the ghetto," entered urban studies as an analytical concept to diagnose the attributes and socio-pathology associated with enforced residential segregation. The phenomenon is most familiar with regard to the mandatory residence within specifically Jewish precincts in medieval and early modern times, whether as ghetto in Christendom, or mellah in Muslim lands. A focus on Jews and the city asks how the physical geography of built environments—bridges, walls, and invisible boundary markers—structures and reflects inter-group relations.[10]

The notion of "the city" invites researchers from practically every field, from more traditional studies in religion and history to new programs in literary criticism, cultural studies and urban geography:

> "It raises questions about the relation of text to space, of representation to practice, of prayer to built environment, of difference to holiness, of creative constructions to physical ones. It invites examination of fruitful intersections of gender and sexuality, commerce

and entertainment, politics and public culture, labor and domesticity, class and religion, as mediated through urban spaces, as well as inter-ethnic relations, cultural-brokering, identity-formation and ethnicity. The commonality is urban space and Jews as one group among many who reside in cities."[11]

This last point also makes the topic useful for studies in Jewish/non-Jewish relations and for the examination of Jewish culture as "one moment in a larger culture, an organ in a larger organism," as David Biale has put it.[12] My own interests lie at the intersection of studies of modern Jewish history and culture where studies of urban history and culture meet. I have written about Berlin,[13] Tel Aviv,[14] and Odessa,[15] and I think that indeed, "urban history [presents] a great forum of the historical sciences, a 'central place' at which an unusually large variety of disciplines, interests and tendencies could converge."[16]

A Special Relationship: Jews and Space

Eli Barnavi, in his *Historical Atlas of the Jewish People*, has tried to explain the basic tension in the notion of "Jewish space":

> Stereotypes are sometimes grounded in reality. The word 'Jew', whether it is used pejoratively or not, immediately evokes the association of *mobility*, a propensity to wander, to move from one place to another. (...) The Jewish perception of space is marked by unique characteristics: it comprises a notion of multiple spaces—rather than one of a single space; and between these spaces—a void. In other words, the Jewish spatial experience is *differential* and *discontinuous*. Although this applies to some extent to mankind in general (a one-spatial man does not exist and has never existed), Jewish history has extended this existential condition to stereotypical dimensions.[17]

This "special relationship" between the Jewish experience(s) and the idea of space and place has touched on many different notions and forms of space throughout history, from God's request to Abraham, "Go unto the land that I will show you," to the forty-year-long wandering in the desert after Exodus, from the erection of the Temple in the city of Jerusalem to the destruction of the Temple in the city of Jerusalem, from Ezekiel's prophecy of/in exile brought to those who were in Babylonian captivity, on the river Khebar, in a place called "Tell-abib," to the foundation of the first all-Jewish city of Tel Aviv in 1909, from Jerusalem to Alexandria to Córdoba and Seville to Constantinopel and Hamburg and New York, from Jerusalem to Rome, from Jerusalem to Cologne and Speyer and Ratisbon and Cracow and Vilna and Odessa and to Jerusalem and Tel Aviv: many of the central "spaces" connected to Jewish history and culture—are cities.

> My heart is in the East [Jerusalem] my body in the extreme West [Spain]"—in this famous verse, Judah Halevi, the greatest poet of the Jewish Golden Age in Spain, epitomizes the

perception of multiple spaces and their discontinuity. A space of the heart, a space of the body, and in between them a void (albeit traversed by the poet himself on his journey to the Holy Land). Yet Jewish spatial experience goes beyond the polarity of the Land of Israel/ Diaspora: it is not necessarily rent between two poles, for besides the experience of two places separated by a gulf, it also incorporates other spaces for which the Jew has a variety of mental stances, including indifference. This uniqueness of Jewish spatial experience has been a constant factor in Jewish history, both when dominated by religion and when molded by Zionism or modern secular ideologies. Jewish consciousness constantly shifts between awareness of physical spaces (the birthplace, for example) to spaces of reference (the ancestral homeland, Hebrew, etc.), a shift which actually *constitutes* the Jewish spatial experience.[18]

Those insights are anything but long-accepted in the academic realm. When in May 2003 a conference on "Jewish Conceptions and Practices of Space" took place at Stanford University, it was

> born of the conviction that the field of Jewish Studies generally remains in the grip of the notion that time exerts more control over Jewish life than space, a notion exemplified by Heschel's famous assertion that Jews lived and created almost exclusively in the dimension of time.[19]

In the introduction to a special edition of *Jewish Social Studies*, which came out of the conference, Charlotte Elisheva Fonrobert and Vered Shemtov argued that there is enough material to show that "the identification of concrete 'place,' both as evidence and as an analytical tool, does of course heavily inform particular areas of study (e.g., the Temple, the State of Israel) of the ancient and modern periods," but still, "at present it cannot be said that the discipline of Jewish Studies systematically trains its students or encourages its scholars to identify Jewish conceptions and practices of space."[20]

Actually, there are many starting points from which to begin this institutionalization. There are a wide range of topics related to place and space in every aspect of Jewish studies as an academic field, in religion— the Talmud "identifies and debates what are clearly defining characteristics of the conception and practice of space and (concrete) place: distance, measurement, size, juxtaposition"—as well as in history or literature.[21] Space, Fonrobert and Shemtov maintain, is a category of evidence and at the same time an analytic tool because the individual and collective "creation and experience of place" has been central and integral to the practice of Jewish life, and, "Jewish texts of all periods contain countless descriptions of concrete places where Jews comfortably practiced their rituals, struggled to do so, or were prevented from doing so. The history of Jewish text and ritual is inseparable from the history and nature of such places." Jewish studies can learn from neighboring disciplines such as geography and anthropology "that neither individual nor community can experience time in this world without claiming, occupying, naming, shaping, negotiating, and losing 'real' space."[22] Research into the wider fields of "Space and Religion" or "The Location of Religion" has been very fruitful, and Jewish studies should make more use of the works of Jonathan Z. Smith and Kim Knott[23] and enter into a dialogue with neighboring research about Christianity and Islam — and space. A discussion of the spatial dimensions of religious practice and experience could also help to build a bridge between the more traditional fields of Judaic studies (Bible, Talmud, Rabbinical Literature) and newer approaches from cultural studies. I have tried to show this with an overview of "Jewish Forms of Settlement,"[24] and it is also the basic assumption for the "Jews and the City" project at the Frankel Institute:

> Scholars of religion have argued that there are distinctive forms of urban religion that inscribe elements of the sacred on cities. Religious figures have shaped cities into sacred spaces while commerce and entertainment, politics and street performances have produced religious change. Cultural studies critics have examined key urban types, such as the flaneur, as well as invisible cities, sites of memory, representations of the city in varied media, ruins and slums. Urban studies has paid attention to the impact of the built environment on a wide range of cultural behaviors from intimate family formations and individual self-fashioning to institutional structures and public performances of ethnic identities. Urban and cultural geographers have mapped the transformations of space into place.[25]

Even if we wish to deal only with the Big City of modern times, traditions and associations of a much older history will inform our task.

With the rise of the Israelites as a political and military factor in the Middle East, with the establishment of the Kingdom under David and the erection of the First Temple under Solomon, the city of Jerusalem became the main focus of "Jewish" urban history. The destroyed Jerusalem was the point of reference for the exiles in Babylon, even though they were so far from it. Again, after the destruction of the Second Temple and the beginning of "Diaspora," Jerusalem remained the focus of reference, as Sidra deKoven Ezrahi has shown in her wonderful book on *Exile and Homecoming*.[26] The idea of Jerusalem has been transformed and applied to cities all over the world:

> Concepts of sacred space, examined effectively within such cities as Jerusalem, can be applied to diasporic cities. The sacred city of Jerusalem served as a touchstone of Jewish religious imagination regarding the organization and meaning of sacred space. Jews living far from the land of Israel invoked the name of 'Jerusalem' to apply to their cities to indicate how much at home they were and to claim that Jewish piety and learning flourished there. Both types of 'Jerusalem' involve mentalities of spatial orientation and symbolic meaning.[27]

References to Jerusalem — or the idea of a "Jerusalem" wherever it might be: real or imagined, heavenly or on the earth, made of stone or of text, in Lithuania or in Israel — can be found, and analyzed, throughout history.

The additional value—of meaning, imagery, and association—in the notion of "Jews as city-dwellers par excellence" (Karl Kautsky[28]) can already be found in sources from the medieval period. When the bishop of Speyer invites Jews to settle and pay taxes and develop his town, he uses the formula: "In the name of the holy and indivisible trinity. I, Ruodgerus [Rüdiger], called Huozmann, Bishop of Speyer, in my attempt *to turn the town Speyer into a real city*, thought that I would increase the honor of our place by having Jews settle here."[29] His was obviously mostly an economic motive, but in the course of the following centuries an image of Jewish urban culture (in so many different forms and epochs and, at the same time, very consistent) was created—and, of course, also challenged and disputed—that, to paraphrase Peter Gay (as quoted below), seemed to say something apt about "the City and the Jews." But what?

The Image: "A Jew, Somebody Who Lives in a City"

The memoir collection in New York's Leo Baeck Institute contains a great number of reports from German Jews who emigrated after 1933. I don't know who Gabrielle Kaufmann was and how she managed to escape from Germany. In her "Letter from a German Attorney now living in Palestine," she writes (in the wonderful English of the German immigrant):

> Each country sees something different in a 'Jew'. For the English the Jew is a kind of savage Patriarch from the East, who lives in Whitechapel; for the Russians the Jews are a mass of handicraft workers, small trade-dealers, difficult to embody in a social order; for the French he is a citizen who through religion has a peculiar place within the nation; for the Germans of 1936 the Jews are the 'Elders of Zion'. Conscious or unconscious, the European comprehends under a Jew, somebody who lives in the city, a physician, lawyer, owner of a department store, agent, generally black-haired and small, intellectual, bargaining, or both together. Those who do not fit into this picture through appearance, habit of living, those who are blond, tall, who work in the country, will be regarded as exceptions. And now the sensation of this Jewish country: *there are only exceptions.*[30]

This is, obviously, written with the intention of presenting a differentiated image of the Land of Israel and its Jewish population, the *yishuv*, sometime in the 1930s. On a more abstract level, the text can be read as a good example for the constructed character of such images about Jews in general, like a plea, as it were, for getting beyond prejudices and seeing—and accepting—people as they "really" are. In the context of this project on *Jewish Topographies*, I would like to isolate one of the constructed images and repeat it here: "Conscious or unconscious, the European comprehends under a Jew, somebody who lives in the city." This formula, "who lives in the city," seems to describe a state of affairs and even a state of mind and does not (only) refer to statistical evidence or the occupational structure of a given Jewish population. It is, in a very short and condensed version, the summary of a long debate about the relationship between modern Jewish and modern urban cultures.

How can we read Frau Kaufmann's notion? The sentence has two actors, "the European" and "the Jew." The first actor "comprehends" something, or perhaps better put, has a certain image, a certain idea about the other—namely that this other can best be described or understood as "somebody who lives in the city" (like fish in the water or Italians in Italy). In general, stereotypes tend to tell us much more about the person who has stereotypes or feels or uses them than about the stereotypically described object. So we should start our deconstruction of this phrase with the question of why on earth this "European" seems to think, first, that a Jew is somebody who lives in a city when, for example, there are very obviously many Jews who do not live in cities; and second, that it should be possible to characterize anyone by saying that this person lives in a city when there are very obviously so many other aspects of life that seem more important and better able to characterize a person or, indeed, a whole culture or nation or religion.

As historians or sociologists we would then go on to deal with part two: we know the historical background, the "reason" for the fact that, yes, Jewish life in Germany, in Austria, in Hungary, but also in France and England, at a given period of time, was indeed mostly an urban existence. But the fact that the overall and general developments of modernization and urbanization between, say, 1830 and 1900, affected the Jewish minorities in some European countries more quickly and intensively than it did the population as a whole doesn't tell us anything about the "character" or the "essence" of a Jewish person or indeed Judaism in general.

One could even go on and ask if others share Frau Kaufmann's observations. Does she just give us an impression or an idea of what she has in mind, or does she correctly refer to views held more generally? Do we find the images she describes in other sources as well? Yes, we do. We find these stereotypes in so many different sources that, for my feeling, a rational "deconstruction" of the image is insufficient. I am by far not the first one to have this feeling; in a way I come back to Peter Gay's observations about the "Berlin-Jewish spirit." In his article, which he calls "A Dogma in Search of some Doubt," Gay tells us that in 1929 when the Ullstein publishing house launched a new Journal called *Tempo*, Berliners soon nicknamed it "Jüdische Hast." And Gay comments: "The nickname seems to say something apt about Berlin, about its Jews, and about the two together—but what?"[31]

The image of "the" modern Jew as "city-dweller par excellence" has often been used in an anti-Semitic context. The article "Großstadt" (Big city) from the notorious anti-Semitic lexicon *Sigilla Veri of Semi-Kürschner*[32] summarizes and illustrates the pre-existing combination of anti-urbanism and anti-Semitism most clearly:

The Jews flourish and thrive in the big cities, while the Aryans that are lured to go there are extinguished in huge numbers. The internally rotten and contaminated cities are extended Ghettos, places that are naturally not detrimental for the Jews' health; all Jews are somehow brother murderers for one might say that Cain is the father of the big city, since it was him who according to the first book of Moses went away from the lord and founded a city.[33]

We have heard so much about the negative effects of urban life, of Jewish traditions and even the principal threat on Jewish life in general; that it is quite charming to hear of the contrary from time to time. In 1933, a time of immediate danger for the Jews in Germany, Ludwig Feuchtwanger writes in his *Bayrische Israelitische Gemeindezeitung*:

> In general people make the mistake to view the influence of the big city on the religious life and on Jewish life in general too negatively. (…) In a narrower circle Jewish life is indeed possible also in London, Berlin, New York, maybe it can even be cherished in a stronger and deeper, more concentrated way. A while ago Ludwig Heitmann, a reverend from Hamburg, has described—in the journal 'Morgen', December 1929—rather pessimistically the consequences of big city development: a growing disengagement of all forms of life from deeper commitments; a growing outer congealment of all of life's functions; an ongoing inner social isolation and loneliness; the feeling of fear among those "freely floating in space". We have to ask, though, if this urban objectification and disenchantment does of necessity weaken the disposition for a fulfilled, active Jewish life, both in the sense of Jewish religion and Jewish culture. There are signs that we can answer this in the negative. The Jew in the big city, guided by young and strong leaders and by a good, courageous Jewish press, which does not avoid the important things in a wrong neutrality, can regain joy in things Jewish (a new learning of religious life or an engagement in community activities), especially when trying to get away from the sober and cruel reality of urban life, and thereby save himself from a grey and dark daily life.[34]

Furthermore, anyone who wants to understand contemporary Jews, has to try and understand them from the point of the city, and in the city:

> The economical, socio-political, and spiritual-religious fate of Jewry in today's world, especially of the Jews in Germany can be understood most profoundly in the light of their being completely urbanised and inhabitants of the metropolis.[35]

City life can be fascinating, attractive—and Jewish. Feuchtwanger maintains that "the modern Jew" in particular needs modern schools and the best education, needs proximity to theaters, concerts, and lectures, needs, in other words, an urban life. In the German language at least, "modernity"— as opposed to modernization, which indicates a process and the notion of a breaking-up, of discussion, of hope for the future, *and* nostalgic longing for the past—bears more meaning than just the definition of a period in history. When we hear the word *Modernität* or, for that matter, *Urbanität*, we also hear a similar subtext as in "somebody who lives in a city": a state of affairs, a state of mind, a character, an essence. What happened when this state of mind was questioned and challenged by the Nazi rise to power in Germany 1933?

The "Case of the Jews" or the Urban Utopia of Moritz Goldstein

I would like to tell the story of a manuscript written by Moritz Goldstein. The elements on the front page include: a title — "Die Sache der Juden" ("The Case of the Jews"); a motto — verse 15 of chapter 36 of Ezekiel, the prophet: "Neither will I cause men to hear in thee the shame of the heathen any more, neither shalt thou bear the reproach of the people any more, neither shalt thou cause thy nations to fall any more, saith the Lord GOD." *(Ezekiel, 36, 15;* a date — 1938. and an archive stamp — American Guild for German Cultural Freedom, 90 Versey Street, New York, New York. By going through these elements and trying to understand their meaning and function, I hope to arrive at a point from where I can better understand a sentence such as "Jews have always been an extraordinarily urban people."

Let us begin with the author. Who was Moritz Goldstein? Born on March 27, 1880, Goldstein died in 1977 in New York. Shortly before his death, at the age of 97, he finished the proofreading of his memoirs, *Berliner Jahre: Erinnerungen 1880–1933,*[36] in which he tells us how, as a small boy, he could look down from his parents' apartment onto the hustle of Berlin's "Kaisergalerie" (see opening page of this article). His father had immigrated to Berlin from a town in Upper Silesia and made a career that brought him to the post of a managing director of this very symbol of modernity and urban culture in Berlin: the Passage or Kaisergalerie, corner Unter den Linden and Friedrichstraße. Moritz Goldstein: Berliner — someone who lived in the city. He went to school in Berlin and briefly served in the military, during which time he experienced anti-Semitism for the first time. When someone asked him what he would like to do for a living, he answered: "Ich möchte Kulturgeschichte studieren." Instead, he settled for German Literature, studied with Erich Schmidt, and wrote his doctoral dissertation on a subject he didn't like very much. Still, Schmidt's guidance and help brought him a job, and this job, to make a longer story short, brought about one of the most important texts, *Deutsch-jüdischer Parnaß*, and — following its publication in the journal *Kunstwart* — one of the most important debates about the self-image, the self-understanding of German Jews shortly before World War I. Editing a series of German classics, "Bong's Klassiker-Bibliothek," Goldstein described an experience that he put into the words most of us know: "We Jews are administering the spiritual property of a nation which denies our right and ability to do so."[37] In a way, we can say that Goldstein in this essay summarized the ambivalent German-Jewish experience with modernization and modernity.

On the one hand, the process of emancipation and assimilation had brought his father to the "Kaisergalerie" and the son to the job at Bong's publishing house. But on the other hand, growing anti-Semitism and the general feeling of alienation gave him the impression that he was not fully accepted in German society, while at the same time he had lost his ties to Jewish tradition and community. In his essay he argued for a new renaissance of Jewish and especially Hebrew literature and art; in his real life, he did nothing to promote such a movement but instead worked as a journalist with the *Vossische Zeitung,* where after some years he became quite famous as the paper's legal correspondent, writing under the name of Inquit.

In 1933, Goldstein, his wife Toni, and their son Thomas emigrated to Italy. Having to leave that country in 1938, they then went to England via France.

1938. The date, in handwriting, was put on the page later. It indicates the time during which Goldstein wrote this manuscript, on a typewriter that did not have German "umlauts." 1938 is a decisive year; I don't have to elaborate on that. The persecution of German Jews reached a dramatic peak with the pogroms of November, and shortly after that, emigration from Germany was made more and more difficult until it became nearly impossible in the following year. Goldstein, having reached an insecure asylum in Manchester, wants to do something, to create a plan to save the Jews of Europe from immediate danger. He sits down to write a very long manuscript that contains four chapters dealing with nearly every aspect of the Jewish experience in Germany, especially the dangers of nationalism, and one final chapter in which he develops his plan. He calls the manuscript "Die Sache der Juden," and this is of course Inquit, the legal correspondent, speaking: The *Case of the Jews* is brought before the public. At least that was the idea, because the manuscript was never to be published.

In this context, Moritz Goldstein has to take on a new role, maybe a new identity. And this is where Ezekiel comes in. I quote some of the first sentences of "Die Sache der Juden", just to show that Goldstein turns into a prophet:

> O I have seen them, on that first day of April, 1933, when German Jewry should be pilloried in front of the world, in the form of a boycott against their shops. In the afternoon during rush hour I stood at the corner of Jerusalem and Leipziger streets in Berlin, sent there by the Vossische Zeitung, to see for myself—I was then still allowed to do that—and I let them pass me by, the German citizens of all ranks, with kith and kin, and I looked into their faces and saw how they grinned and showed each other the inscriptions which for example Tietz department store had painted on its windows, an arrow marked 'One way street to Jerusalem'—more did their wit not pull off.

Ezekiel, having already been brought into Babylonian captivity in the year 598 BCE, is a prophet in exile, and he has been called the "prophet of exile."

Jerusalem has been destroyed, but God does not dwell among those who have sinned and stayed in Jerusalem. The divine presence has migrated from Jerusalem and come to stay with those in captivity, in exile. The only true Israel is in Babylon, in exile. One day it will be back in the Land of Israel, but not for the time being.

For Goldstein, the "Jerusalem" of the German-Jewish "symbiosis" has been destroyed. The belief in asssimilation has proven false. Dissimilation is the fate of the Jews, and worse will come. One day, the solution will be to go and settle in the Land of Israel—but not for the time being. The British restrict immigration, the Arab population is hostile. And, what is maybe more important, it would take too much time to build up a Jewish State in Palestine. So, here is the plan:

> You cannot *make* a land, with a population rooted to the soil. A land has to grow. There may be better or worse conditions for its development, factors which facilitate or obstruct its growth. But in general one has to wait for it. Growth needs time, long and silent time.
>
> But you can make a city.[38]

Goldstein's proposal is to build a gigantic city, which will be able to take in Jews from Europe in a short period of time, give them work and housing, as well as the opportunity to prepare themselves for a return to the Land of Israel, whenever that time comes.

> Because the settlement of Jews, of hundreds of thousands, of millions of Jews on the land would take too much time, build them a city, have them build a city, temporarily, as an intermediate solution. There is no place in the world for a large Jewish settlement on the countryside (i.e., a state). Or maybe there is place, but all places belong to states, and the states do not give [them] away. But if there may be no space for a country of the Jews, there still might be space enough for a city of the Jews. Even for a big [city], a world city, a gigantic city, a city of millions.

Like any other utopian project, this city does not yet have its place. Goldstein cannot and will not say where it should be built; he wants readers to imagine that the plan can work and that it can be successful. We come slowly back to people who live in cities:

> There is so much complaint about the wrong occupational structure of the Jews, where agricultural production is lacking nearly completely or has been lacking until now. I think this is a lopsided way to put it. The occupational structure of the Jews was not wrong, it was completely right because they could just make a living under the circumstances they were forced to live in. It would be wrong if one would take the Jews, as they are now, and set them down on a soil where they would have to help themselves, cultivate grain and win ore and coal from a mine. This is exactly one of the reasons why Jewish colonisation is so difficult. But these same Jews, transplanted into a city, have the right occupational structure: they have learned what is demanded in a city.

One of the most interesting elements in Goldstein's utopian project is, in my view, that the city he describes is a city built by Jews, but not exclusively for them. The urban experience of Berlin is one of failed integration and assimilation, but the consequence, for him, cannot be nationalism or exclusivity. When I have tried to read and understand Goldstein's idea, I followed the academic and public debates about the connection between modern Jewish and modern urban existence in different fields: the image is of course topical in anti-Semitism and its mixture of hostility against Jews and hostility against cities. The terms "Verjudung" and "Verstädterung" are practically synonymous for some authors, Spengler, for example, as well as in the whole range of practices such as "Rassenhygiene" and "Geopsychology." Anti-urbanism is a fixed element in the overall nationalist and reactionary discourse in Germany, and the same is true for anti-Semitism. The image of the city-dweller as a weak assimilationist is being discussed inside the Zionist movement, first on a rather theoretical level—but soon enough also in a very practical context, i.e., the Polish immigration of 1924/25 and the German immigration from 1933 on, both of which bring to Palestine people who used to live in cities and have no intention of becoming "Bauerntölpel," country bumpkins, as Isaac Deutscher put it.

It is an interesting experience to discuss questions of Zionism and of the practical state-building process in Palestine alongside the notions of urban culture, urbanism, and anti-urbanism, and to include sources and scholarly literature from the fields of religion, geography, anthropology, and folklore studies. In a sense, what we have to do if we want to understand Goldstein's project, is "Kulturgeschichte studieren." And whereas the different aspects of the history of hostility against the big city have been studied by quite a number of scholars, the history of "city-love" or "city-respect" has not yet been written.

The historiography of "the relationship between city and Jews" should develop further and try to get into an interdisciplinary exchange with migration studies and studies on transnationalism and cultural transfer. Often enough

> this relationship has been described as an exceptional, a special or even a unique one. Such views can be found along the entire political spectrum, reaching from the extreme right to the left. In numerous anti-Semitic writings, the Jews were accused of being the major source of the degenerative influence of the urban environment, for it was they, the ultimate city people, who were benefiting from the rise of urban capitalism while at the same time being immune to the city's negative influences.[39]

Shulamit Volkov has shown how stereotypes against urban Jews were part of an anti-Semitic nexus juxtaposing anti-urban and anti-modern patterns.[40] I would argue, with David Weinberg, that we should pursue studies about the meeting points of modern Jewish and modern urban life: "It was also in urban society that nineteenth- and twentieth-century Jews made their livelihood, established modern communal and religious institutions, and created distinctively new cultural forms."[41]

The Big City is the place where the spatial dimension of Jewish history and culture finds its modern form and expression.

I think this is the direction in which our studies on "Jews and the City" should go: comparative, transnational, focussing on interethnic relations rather than isolating the assumed and oftentimes misinterpreted *Jewishness* of spaces, and in close contact with new fields of cultural studies such as performance studies:[42] *How* does one live an urban life? In Berlin, in London, in New York— but also, today, in Jerusalem and Tel Aviv? For a long time, studies about Tel Aviv, for example, have relied on "interior" sources only, the "city without history" seeming to explain and speak for itself. Today, with a growing non-Jewish (and also non-Arab) population, it will be necessary to study the formerly "all-Jewish city" ("la ville cent-pour-cent juive") as a place where different urban experiences of the Jewish Diaspora meet and leave their mark; but in the meantime, it is also a place of immigration from Thailand and the Philippines, from Nigeria, from Romania and Russia. In a certain way, the Tel Aviv of today is "co-constructed"[43] as a world city by all its inhabitants, not only by Jews. When we accept the idea that "urbanity," or urban culture, can be anything but exclusive, then maybe this new kind of interaction and co-construction makes Tel Aviv more of a "Jewish (big) city" than it was before.

Notes

1   Beth S. Wenger, "Jews and the City" (course description, University of Pennsylvania, Spring 2003), http://www.sas.upenn.edu/urban/syllabi/Spring%2003/urbs227.pdf.
2   University of Maryland, Meyerhoff Center for Jewish Studies: "The Jew and the City through the centuries" (course description), http://www.jewishstudies.umd.edu/courses/200.html (accessed August 9, 2007).
3   University College London, Department for Hebrew and Jewish Studies, Undergraduate Prospectus 2007.
4   Bruce Thompson, "Diaspora, Urbanism, Ethnicity" (course description at UC Santa Cruz) http://reg.ucsc.edu/soc/aci/winter2006/ltmo.html (accessed August 9, 2007).
5   For a detailed discussion of the eruv, see Manuel Herz's article in this volume.
6   Steven Zipperstein, "Jewish Historiography and the Modern City: Recent Writing on European Jewry," in *Jewish History* 2, no. 1 (1987): 73–88.
7   "People of the City: Jews and the Urban Challenge," ed. Ezra Mendelsohn, *Studies in Contemporary Jewry* XV (1999).
8   Chantal Bordes-Benayoun, ed., *Les juifs et la ville* (Toulouse: Presses universitaires du Mirail, 2000).
9   Joachim Schlör, *Das Ich der Stadt: Debatten über Judentum und Urbanität, 1822–1938* (Göttingen: Vandenhoeck & Ruprecht 2005).

10 University of Illinois, "Fellowship Opportunities," https://www.grad.uiuc.edu/fellowship/listing/2943 (February 27, 2007).

11 Ibid.

12 David Biale, "Confessions of an Historian of Jewish Culture," in *Jewish Social Studies* 1 (1994): 40–51; here 45.

13 Joachim Schlör, "Bilder Berlins als 'jüdischer Stadt': Ein Essay zur Wahrnehmungsgeschichte der deutschen Metropole," in *Archiv für Sozialgeschichte* 37 (1997): 207–229.

14 Joachim Schlör, *Tel-Aviv: From Dream to City* (London: Reaktion Books, 1999).

15 Joachim Schlör, "Sieben Werst von der Hölle. Jüdisches Leben in Odessa," in *Odessa Odessa: Die Stadt und ihr Traum; Eine universale Liebeserklärung aus Berlin,* ed. Shelly Kupferberg (Berlin: Elefanten Press 1999).

16 Derek Fraser and Anthony Sutcliffe, eds., *The Pursuit of Urban History* (Baltimore: Edward Arnold, 1983), XI, IX.

17 Eli Barnavi, "Introduction," in id., *The Historical Atlas of the Jewish People: From the Times of the Patriarchs to the Present* (New York: Hutchinson, 1992).

18 Ibid.

19 Charlotte Elisheva Fonrobert and Vered Shemtov, "Introduction: Jewish Conception and Practices of Space," in "Jewish Conceptions and Practices of Space" (Special Issue), *Jewish Social Studies* 11, no. 3 (2005): 1–8; here 3–4; the reference to Heschel refers to Abraham Joshua Heschel, *The Sabbath: Its Meaning for Modern Man* (New York: Farrar, Straus and Giroux, 1951).

20 Fonrobert and Shemtov, *Introduction* (as are the following quotations).

21 For the field of literature, cf. Barbara E. Mann's special edition of *Prooftexts* 26 (2006) and her introduction "Literary Mappings of the Jewish City: Other Languages, Other Terrains," 1–5. Other articles cover cities such as Vienna, Moscow, Lodz, Cairo, Bagdad, and Hanna Soker-Schwager's beautiful text, "Godless City," on (Ya'akov) "Shabtai's Tel Aviv and the Secular Zionist Project," 240–281.

22 Mary Minty, written communication, quoted in Fonrobert and Shemtov, "Introduction," 4n14.

23 Jonathan Z. Smith, *To Take Place: Toward Theory in Ritual,* Chicago Studies in the History of Judaism (Chicago: Chicago University Press, 1992); Kim Knott, *The Location of Religion: A Spatial Analysis of the Left Hand* (Oakville, CT: Equinox Pub., 2005).

24 Joachim Schlör, "Jüdische Siedlungsformen: Überlegungen zu ihrer Bedeutung," in Elke-Vera Kotowski, Julius H. Schoeps, Hiltrud Wallenborn, eds., *Handbuch zur Geschichte der Juden in Europa,* Vol. 1: "Länder und Regionen," Vol. 2: "Religion, Kultur, Alltag" (Darmstadt: Primus 2001), vol. 2, 29–47.

25 University of Illinois, "Fellowship Opportunities."

26 Sidra DeKoven Ezrahi, *Booking Passage: Exile and Homecoming in the Modern Jewish Imagination* (Berkeley: University of California Press, 2000).

27 University of Illinois, "Fellowship Opportunities."

28 Karl Kautsky, *Rasse und Judentum,* Ergänzungshefte zur Neuen Zeit, Nr. 20, ausgegeben am 30. Oktober 1914 (Stuttgart: J.H.W. Dietz 1914), 56.

29 "In nomine sancte et individue trinitatis. Ego Ruodgerus qui et Huozmannus cognomine, Nemetensis qualiscumque episcopus, cum ex *Spirensi villa urbem facerem,* putavi milies amplificare honorem loci nostri, si et iudeos colligerem." "Bischof Rüdiger von Speyer verbrieft den in Altspeyer aufgenommenen Juden bestimmte Freiheiten," September 13, 1084, in Alfred Hilgard, *Urkunden zur Geschichte der Stadt Speyer* (Strassburg: K. J. Trübner, 1885), no. 11, 11–12. (trans. and emphasis added by the author), http://www.historia-iudaica.org/dt_q1084.html (accessed August 17, 2007).

30   LBI New York, Collection A 28/3, Box 2, Folder 4: Gabrielle Kaufmann, Correspondance from Israel (italics by the author).

31   Peter Gay, "The Berlin-Jewish Spirit: A Dogma in Search of Some Doubts," in *Freud, Jews and Other Germans: Masters and Victims in Modernist Culture* (New York: Oxford University Press, 1972), 169–188; here 170.

32   The name refers to *Kürschners Gelehrtenkalender* a periodically published encyclopaedia of scholars and academics in Germany.

33   Erich Ekkehard and Philipp Stauff, eds., *Sigilla Veri: (Ph. Stauff's Semi-Kürschner) Lexikon der Juden, -Genossen und -Gegner aller Zeiten und Zonen, insbesondere Deutschlands, der Lehren, Gebräuche, Kunstgriffe und Statistiken der Juden sowie ihrer Gaunersprache, Trugnamen, Geheimbünde, usw.*, Vol. 2, 2nd ed. (Erfurt: U. Bodung, 1929), 826. Trans. Tobias Metzler.

34   Ludwig Feuchtwanger, "Die Grossstadt als jüdisches Schicksal: Eine statistische Studie anlässlich der Volks-, Berufs- und Betriebszaehlung vom 16. Juni 1933," in *Bayrische Israelitische Gemeindezeitung* 12 (1933): 181–82.

35   Ibid., 178.

36   Moritz Goldstein, *Berliner Jahre 1880–1933*, Dortmunder Beiträge zur Zeitungsforschung (München: Verlag Dokumentation, 1977).

37   Moritz Goldstein, "Deutsch-jüdischer Parnass," in *Kunstwart: Halbmonatsschau für Ausdruckskultur auf allen Gebieten*, ed. F. Avenarius (first march issue, 1912): 281–294; here 284.

38   Moritz Goldstein, "Die Sache der Juden" (manuscript). Elisabeth Albanis and Till Schicketanz (University of Mainz) prepare an edition of Goldstein's writings, including this manuscript.

39   Tobias Metzler, one of my doctoral students, who writes a comparative study about the Jewish communities of London, Berlin, and Paris, sums up his dissertation project with these words.

40   Shulamit Volkov, "Anti-Semitism as a Cultural Code: Reflections on the History and Historiography of Anti-Semitism in Imperial Germany," in *Leo Baeck Institute Year Book* 23 (1978): 25–46, Shulamit Volkov, "Readjusting Cultural Codes: Reflections on Anti-Semitism and Anti-Zionism," in *Journal of Israeli History* 25, no. 1 (2006): 51–62.

41   David Weinberg, "Jews and the Urban Experience: Introduction," *Judaism* 49, no. 3 (2000): 278.

42   Tobias Metzler's dissertation project "explores some of these 'new cultural forms' urban Jews created. It investigates paradigmatic elements in the formation of urban Jewish cultures in the three European capital cities London, Berlin, and Paris during a period reaching from the final decades of the nineteenth up until the middle of the twentieth century. In doing so, it employs a broad notion of cultures, incorporating both traditional concepts of culture, depicting the city as the place of a distinct cultural infrastructure, institutions and 'monuments' as well as elements of culture as patterns of action, emphasising the diversity of Jewish attempts to take up the challenge of the urban environment and creating Jewish spaces and places in its midst."

43   Steven Aschheim has used this term to describe German-Jewish history beyond the notions of assimilation and integration. Steven E. Aschheim, "German History and German Jewry: Boundaries, Junctions, and Interdependence," in *LBI Year Book* XLIII (1998): 315–323.

Part IV

Exploring and Mapping
Jewish Space

12 Travel and Local History as a National Mission

Polish Jews and the Landkentenish Movement in the 1920s and 1930s

Samuel Kassow

לעמבערגער רעראקצי׳ע קאלעגיום: ה. ריקער און י. פינעליס

ישראל אשענדארף / לעמבערג

יארעמטשע. וואסערפאל.

דער מעגטש פאטאגראפירט זיך ביים וואסערפאל אין טאל,
אויסן בארג צווישן די פעלזן, ער וויל צינויפבינדן זיין פארגענ־
גלעכקייט מיט וואסער און בארג.

דער וואסערפאל ווערט קיינמאל נישט מיד אין זיין באװע־
גונג: וואַרפט זיך קאמפפסלוסטיג אויף די פעלזן, שרומט פאר כעס,
פארבלינעמט זיך פאר פרייד און ברילט פאר נצחון.

דער בארג.ווערט קיינמאל נישט אויסגעריצט.אין זיין פאר־
גליווערהונג: זאלן רעגנס גיסן, זאלן שנייען פאלן, זאלן ווילדערװען
שטורעמס – רויק שטייט ער, מעכטיק אין זיין שווייגן, א דענקמאל
פון אייגענער גבורה.

שרייט דער מענטש: גיב מיר דיין אימפעט, וואסערפאל!
בעט זיך דער מענטש: גיב מיר דיין רו, בארג!

נאר דער וואסערפאל ווייסט נישט פון שעגקען און אוועקגעבן.
ער איז געהודינט אלין צו פארגינסט, מיטדרייסן און מיטנעמען.

און דער בארג פרעגט: וואס זוכסטו אויף מיינע מעכטיקע,
ריזיקע פלייצעס? רו? – דיר גיט נאר די א קליין בערגעלע ערד.

יארעמטשע 1938

Introduction

In June 1938, a time of growing worry for Polish Jewry, the *Jewish Society for Landkentenish/Krajoznawstwo* (ZTK)[1] published the June issue of its regular journal, which appeared in two languages, Polish and Yiddish. That month the main theme of the journal was kayaking, and members of the kayaking section filled the issue with articles that described their travels along Poland's rivers and lakes. A certain Y. Maynemer compared kayaking to a marriage, where one needed to know and rely on one's partner.[2] Y. Tseylan described his picturesque voyage down the Vistula, the beautiful scenery and the campsites along the river banks in the evening. Sometimes, though, the mood changed as the kayakers visited Jewish townlets (*shtetlekh*) along the way and talked to the local Jews about their struggles to make a living. But then the depression lifted, at least for a while as the group resumed their trip down the river.

> We return full of impressions. This wonderful excursion down the Vistula has given us profound memories and has deepened and heightened our appreciation of this world. The beautiful scenery and intense emotions will affect us forever.[3]

Another traveler also noticed the dichotomy between the exaltation of a kayaking trip through the Augustow lakes region and the grim mood of the local Jews that they met along the way. His group encountered a good deal of hostility from non-Jews as they passed through bigger towns.

> But these 'trifles' did not spoil our upbeat mood. The optimism (*bitokhn*) that is rooted in us, the belief that times will get better more than made up for everything. Now we are impatiently waiting for the chance to get back (into the boat) and to travel along Poland's rivers and lakes.[4]

The back pages of the issue gave the usual summary of ongoing Landkentenish activities. The Lodz chapter reported that in the past five months it had sponsored 15 trips and organized lectures on such diverse topics as Yiddish folk idioms, the Yiddish theater, Jewish humor, mountaineering in the Alps, and the writings of Andrzej Strug. There were courses in Yiddish, English, and French. The skiing section reported that it had arranged classes in physical training and sponsored a ski trip to Zakopane.[5] But there were also clouds on the horizon. In the past the section had organized many excursions to state and municipal enterprises and industrial plants. But now the group found it harder to get permission to visit these sites.

The Vilna section reported that it had organized about a dozen walking tours of the city on Saturday afternoons, with such disparate destinations as the old Jewish cemetery, the Tobacco monopoly headquarters, the Lithuanian Museum, old caves, the legendary Romm Press (a leading publisher of holy Jewish books), and an art exhibit.[6]

In the Polish-language section of the journal an editorial note informed readers that just as the staff was preparing this particular issue, the Polish Kayaking League had begun discussing a motion to expel all Jews from the organization. And thus, the editors continued, the decision to prepare a kayaking issue changed from an "intention" to an "obligation."

One does not usually associate Polish Jewry with kayaking, hiking, or skiing. Nor does one usually think that these sports belong in the same organizational framework as language courses, tours of factories, or trips to art museums. But this seemingly eclectic range of activities all fell into the purview of the Jewish Society for Landkentenish/Krajoznawstwo. And in the process, Polish Jews showed that they lived in Poland as equal citizens and not as guests.

The Lankentenish idea, devoted as it was to fostering Jewish pride and self-awareness, derived from two major sources. The first was the Polish ideal of Krajoznawstwo. Indeed Landkentenish was a direct Yiddish translation of the Polish Krajoznawstwo, which meant literally "knowing the land". Before 1918, when they regained their independence, "knowing the land" gave Poles an important means of defending their national identity.

The second source of the Landkentenish idea was the Jewish cultural revolution of the late nineteenth and early twentieth centuries, symbolized by the Yiddish writer Yitzhak Leybush Peretz, the writer and folklorist S. Ansky, and the historian Simon Dubnow. This cultural revolution, with its emphasis on collecting (in Yiddish, *zamling*) documents, artifacts, songs, and folklore, nurtured the ideal of Landkentenish by highlighting the central role of the people, rather than traditional religious texts, in the survival of the Jewish nation and in the determination of its future. They also claimed new status for East European Jewry as the major center of the Jewish people rather than the intellectual backwater described by Heinrich Graetz and other scholars. Its cultural distinctiveness, they argued, had developed over centuries, and was rooted not only in national memory but also in the experience of sharing land and space with their mainly Slavic neighbors, including the Poles.

Founded in Warsaw in 1926, the *Jewish Society for Landkentenish/ Krajoznawstwo* became one of the leading cultural organizations of prewar Polish Jewry.[7] The society brought together intellectuals and laypeople in a collaborative effort to study Jewish history, material culture, architecture, and folklore through tourism, lectures, photography, and recreation. The Society—and especially its historians, architects, and photographers—sought to foster *doikayt* (hereness), a deep sense of rootedness to the Polish lands where Jews had lived for hundreds of years. During the course of its existence (1926–1939) the society faced formidable obstacles. Given the economic problems of Polish Jewry, would it be able to attract members who could afford the cost of excursions, ski trips, or photography courses? Would it be able to make Yiddish culture, such an important part of the Landkentenish idea, attractive to a constituency that included many acculturated Jews who were teachers, white-collar workers, and professionals? Would its leaders be able to instill in potential members an awareness of the fine balance between recreational and educational tourism, between purposeful study and leisure? How would the society counter a view, widely held in more traditional sectors of Jewish society, which saw hiking, mountain climbing, and skiing as forms of *batlones*, a frivolous use of time?[8]

Yet, by 1939, the Society had achieved a great deal. A study of its activities serves as a healthy reminder that scholars must broaden their perspective on interwar Polish Jewry, go beyond politics and ideology, and pay more attention to questions of leisure, popular culture, and tourism as they seek to understand Europe's largest Jewish community before the Holocaust as it fought for its dignity, rights, and culture.[9]

Krajoznawstwo: Knowing the Polish Homeland

During the period of the partitions, and especially in the decades preceding World War I, many Poles saw Krajoznawstwo as a way to protect Polish identity by telling Poles to "know the land" through hiking and travel, through the study of geography, flora and fauna, regional folklore and history. A major landmark of the movement was the publication, between 1880 and 1902, of the *Słownik Geograficznego Królewstwa Polskiego i innych ziem słowiańskich* (Geographical dictionary of the Kingdom of Poland and other Slavic Lands) under the editorship of Filip Sulimierski, Bronisław Chlebowski, Władyslaw Walewski, and Józef Krzywicki. The *Słownik* included a wealth of information on virtually all the lands of the former Polish-Lithuanian Commonwealth. It outlined the history of villages and towns, described lakes, rivers, and mountains, and provided statistical information and details on climate. Many amateurs, scattered throughout the lands of the former Commonwealth, contributed to the Dictionary. The project sent a clear message that a suppressed people could use memory, geography, history, and folklore to stake its claims to land. This was not just the province of elite scholars but an entire movement that included both professionals and laypeople in a common national effort.[10]

The goals of the Society were wide-ranging and somewhat diffuse. According to one historian of the society, its goals were to:

> a) organize excursions led by qualified specialists to investigate different aspects of krajoznawstwo. b) to collect material on folklore, archeology, art and industry c) to open libraries and publishing houses d). to document the landscape through photography e) to organize the protection of artistic and natural sites f) to organize conferences and lectures g) to organize exhibitions h) to publish tourist guidebooks and monographs on different regions.[11]

As we will see later on, the Jewish Landkentenish society after its founding in 1926, enunciated goals and a statement of purpose that were strikingly similar.

When Poland regained its independence in 1918, the entire Polish Krajoznawstwo movement dedicated itself to the heady task of recreating a single nation from territories that had been divided among three empires. Krajoznawstwo would bring Poles, long separated, together and enable them to reconnect with the old-new state. The movement in the interwar period cultivated a finely balanced synthesis between loyalty to a new centralized state and devotion to the "small fatherlands"—the traditional regions, each distinguished by landscape, dress, and local culture. This balance was reflected in the resolutions of the First All Polish Congress of Krajoznawstwo, which the PTK organized in Poznan in 1929. There the Congress defined Krajoznawstwo as a "social movement" that needed to transform feelings into deeds. Love of the land and devotion to the people had to be translated into a far-reaching program of education "in the countryside and in the cities," a program that had to "take into account regional differences."[12]

Thus Krajoznawstwo functioned on many different levels. Edward Wieczorek noted that it could be defined functionally as a striving to know one's native land; institutionally as a social movement that brought together experts and laypeople for this purpose; culturally and sociologically as a movement that occupied a place in its people's cultural history through such media as journals, guidebooks, films, and recordings of folklore.

The Polish Krajoznawstwo movement developed during a period when tourism throughout Europe and the United States was undergoing significant transformation. Tourism all over the developed world was becoming democratized; it was no longer regarded as being just for the rich. By the same token, in many different countries there were attempts to lend meaning and significance to tourism that went beyond mere leisure.

> In recent decades scholarly attempts to study and evaluate the place of tourism in popular culture have suggested ways in which tourism in the late 19th and 20th centuries helped develop new notions of national identity, overcome the alienating effects of modern

industrial urban society and engender a deeper sense of a national community by linking it not only to abstract ideals but also to concrete landscapes and historical sites.[13]

In some ways these insights contribute to a better understanding of the significance of Krajoznawstwo for both Jews and Poles in interwar Poland. As will be seen, the leaders of the movement (for both peoples) regarded Krajoznawstwo as a way to bridge the gaps between the rural and the urban, the world of natural wonders like the Tatras and the more prosaic attractions of the local radio station or electric plant, and to create a modern citizen, physically fit and intellectually curious. Far from being a passive tourist, this new citizen was actively involved in getting to know the country and always ready to transmit that knowledge to others. Poles and Jews certainly took pride in a long history. Krajoznawstwo could supplement this sense of history and serve as a catalyst to renegotiate the sense of national identity in a new Europe marked by rapid political and economic change.

In the 1920s and 1930s, Poles saw Krajoznawstwo as a way to reunite their long-lost country and to bridge the psychological gaps that divided Poles from the Russian, Prussian, and Austrian partitions. The new State did all it could, given economic constraints, to make its mark on the land and to "reclaim" space. The PTK issued new tourist guidebooks that tried to instill a new sense of ownership of and pride in the nation's natural wonders, in its urban centers and rural heritage and in its regional cultures. There was a palpable sense of achievement as the new Polish State, even in the face of economic difficulties, built new sections of Warsaw, new railways to link formerly separated areas, erected a new port in Gdynia, razed old Russian landmarks and monuments, and established new national parks.

But for Jews, the Polish Krajoznawstwo movement was a double-edged sword. To claim lands as Polish was often a tricky business when forty percent of the population of the new State belonged to national minorities. Many of the towns listed in the *Slownik Geograficzny* had Jewish majorities, and especially east of the Bug River, many of the regions had large Ukrainian, Belarussian, and Lithuanian populations that had their own national claims.

With regard to the Jews, other problems arose. The decades that preceded World War I saw a marked deterioration of Polish-Jewish relations in all three partitions. Many Jews increasingly saw themselves as part of a modern secular Jewish nation rather than as a religious group. More and more Poles, on the other hand, especially those who followed the teachings of Roman Dmowski (the chief ideologue and cofounder of the National Democratic Party commonly called the *Endecja*), regarded the Jews as an alien element that had to be eliminated through emigration if the Poles were to develop their own middle class and become a normal nation. But by the early twentieth century, even many Polish liberals, who had been ready to accept Jews into the Polish nation if they assimilated into Polish culture, felt betrayed by the rise of a new Jewish nationalism that saw the Jews as a separate nation.[14] To be sure, for a time, Polish positivism had recognized the need to learn about the Jews and their folklore, and a leading positivist ethnographic journal, *Wisla*, included Jewish material. As the great Polish writer Boleslaw Prus remarked, "how ridiculous that we (…) do not study the customs, religion and life of almost one million of our fellow citizens, who will sooner or later be fused with us into a uniform society."[15] But for Prus, Alexander Swiętochowski, and other positivists, to learn about the Jews did not imply that they were ready to accept the presence of a separate Jewish nation in Poland. Knowing the Jews was intended to facilitate communication with the Jewish masses in order to ease their eventual assimilation. When most Jews made it clear that they would not become "Poles of the Mosaic faith," toleration and liberalism often gave way to anger and a readiness to accept nationalist assertions that the Jews were indeed a foreign body implanted on the Polish nation.

Krajoznawstwo for Jews

The rise of a modern Jewish nationalism took many forms. Some Jews turned to Zionism, others to revolutionary socialism expressed in such movements as the Bund. But politics was only one aspect of this search for national self-assertion. History, literature, and folklore would play their part in the new crusade to build a new Jewish nation, and it would only be a matter of time before these trends fused in a determination to build a Jewish Krajoznawstwo—Landkentenish—that focused on many of the same lands and the same landscapes so beloved by the Poles (as well as by the Ukrainians, and the Belarussians, and the Lithuanians, and the Germans).

As Jews looked at the guidebooks published by the PTK, they discovered that the same books that provided thrilling descriptions of the Tatras or the architectural wonders of Vilna or Zamość ignored Jewish synagogues and antiquities, centuries-old evidence of the Jewish presence in Poland.

When Polish guidebooks did discuss Jewish sites, the results were not always positive. For example, in a comprehensive guidebook to Wilno (Wilno in Polish, Vilne in Yiddish, Vilnius in Lithuanian)[16] published in 1923 by the Wilno section of the PTK in 1923, we read, in a description of the old Jewish section of Wilno, that

> even though the quarter had certain charms: this (positive impression) is weakened by the typically oriental slovenliness of the inhabitants of this unhygienic quarter and by an intolerable stench which makes it impossible, on hot summer days, for a cultured European to visit these sites.[17]

The historical section of this PTK guide to the city totally ignored its rich Jewish history and the reasons why East European Jewry called it the Jerusalem of Lithuania. The guide described Wilno not as a multinational city but as a Polish city that had long suffered under savage Russian repression. This stance becomes particularly evident in a passage recalling April 19, 1919, the day the Polish forces entered the Bolshevik-held city. While the PTK guide raved about the "wave of happiness [that] swept over the population" and saw it as "the finest moment in the many centuries of Wilno's history,"[18] the city's Jewish population (which at that time was about forty percent) experienced this event quite differently, as the incoming Polish army launched a savage pogrom. According to the Anglo-American Morgenthau Commission, it took 55 Jewish lives and resulted in the arrest of hundreds of Jews and extensive damage to Jewish property.[19] The point here is not to revisit the charges and countercharges that swirled around the events of April, 1919, but to illustrate why Jews felt unwelcome and alienated in many branches of the PTK.[20]

A comparison between the Vilna guides issued by the Polish PTK and the Jewish ZTK is quite illuminating. The former, already cited, only described one Jewish site, the Old City Synagogue, and perfunctorily mentioned the old Jewish cemetery and the Choral synagogue built in 1892. The guide to Vilna issued by the Jewish ZTK, written by Zalmen Szyk, was slated to run as two volumes, although only the first volume, published in 1939, actually appeared.[21] Szyk devoted hundreds of pages to the city's famous synagogues, study houses, and libraries, described in great detail the major role that Vilna/Vilne played in the development of modern Yiddish and Hebrew literature and cited the many poems and literary works devoted to Vilna in various languages. Szyk's guidebook included information about the Poles, the Lithuanians, and the Belarussians, and the importance of Vilna for their cultures. The book contained rich information about geography, Vilna's environs, and the region's unique characteristics. The Polish guide used Krajoznawstwo to create the image of a city that belonged to one people. Szyk used Landkentenish to describe a multiethnic tapestry — even as he described Jews as settled inhabitants, not aliens or guests.[22]

Despite growing tensions between Jews and Poles, most Polish Jews still saw Poland as their home and still nourished hopes that their situation would improve. Yes, the anti-Jewish violence of 1918–1921 and the growing anti-Semitism in the country, especially after the death of Józef Pilsudski in 1935, caused widespread bitterness and disappointment. This was a period when the old nineteenth-century ideals of assimilation, the notion held by some that a Jew could become a "Pole of the Mosaic persuasion," gave way to the realization that Poles and Jews were two separate nationalities. Yet paradoxically, the terminal decline of the assimilationist ideology in the

1920s and 1930s was accompanied by unprecedented acculturation. More Jews than ever before spoke Polish as their first language. Serious Jewish dailies such as *Chwiła, Nasz Przegląd* and *Nowy Dziennik* appeared in the Polish language. By the 1930s, a majority of Jewish children were getting their primary education in Polish state schools.

In short, while most Polish Jews did not regard themselves as Poles they nonetheless increasingly embraced Polish culture and harbored the hope that eventually they would find their place as equal and loyal citizens of the Polish state. In the meantime, they had to stake their claim to equality by showing that Poland was their home. Landkentenish was an important tool for accomplishing this.

The Intellectual Roots of Jewish Landkentenish

Three individuals, as mentioned above, stand out in the cultural revolution that indirectly fostered the Jewish Landkentenish movement: the Yiddish writer Y. L. Peretz, the writer and folklorist S. Ansky (Solomon Rappaport), and the historian Simon Dubnow. Several factors linked the three. First, they had all experienced, each in his own way, a return to a strong identification with Jewishness after various attempts to either assimilate or to embrace a universal, European culture. Second, they all tried to find a new form of Jewishness that could navigate between a religious tradition, in which they no longer believed, and assimilation, which they rejected. Third, while they rejected neither Zionism nor Hebrew as a language, they all believed strongly in the affirmation of the Jewish folk in Eastern Europe and its Yiddish culture.

This provided a foundation for a resurgence of interest in what would become the pillars of Landkentenish: *zamling*, or collecting materials to document the history, folklore, and material culture of East European Jewry. Another foundation of the Landkentenish movement was inherited from the Haskala (the Jewish enlightenment): a new interest in the physical regeneration and "productivization" of the Jewish people. This legitimized not only educational reforms and calls for a return to the land, but also advocated tourism and physical activity as ways to promote healthy individuals and a healthy people. Indeed, the Haskala fostered a new awareness that the Jew was not just a member of a collective group but also an individual in his own right, entitled to happiness and to the enjoyment of the benefits of modern culture.

While Peretz, Ansky, and Dubnow built on many of the premises of the Haskala, they nonetheless all agreed that some of its assumptions needed rethinking. The Haskala had hoped that education and acculturation in the course of the nineteenth century would lead to a decline in anti-Semitism. Jews would give up the negative and backward aspects of their culture—Yiddish, traditional dress, customs that set them off from their gentile neighbors— but hold on to what was positive in their Jewishness: religious monotheism, ethical liberalism, a knowledge of Hebrew to complement fluency in Polish, German, or Russian. In the words of the Hebrew poet Yehuda Leyb Gordon, Jews would, "be a man in the street and a Jew at home."

The *maskilim* (supporters of the Haskala) viewed the Jewish masses as backward, needing enlightened intellectuals who would break the hold of obscurantist rabbis and lead them forward. What Peretz, Ansky, and Dubnow all questioned was the Haskala's assumption that the relationship between the intelligentsia and the folk was a one-way street. They each began to realize (Ansky and Peretz more than Dubnow) that the folk had much to teach the intelligentsia and that its traditions were vital pillars of national identity. Even as they affirmed many aspects of the Haskala, each in his own way paved the way for a more positive attitude toward folk culture and toward the folk language, Yiddish, and thus legitimized a new movement to study the history and folklore of the Jewish masses in Eastern Europe.

Yitzhak Leybush Peretz (1852–1915), one of the three major classic Yiddish writers, started out writing in Polish and was not only steeped in Polish culture but also quite well acquainted with the world of Polish ethnography and Krajoznawstwo. He had been a prosperous lawyer in the provincial and picturesque town of Zamość, but a political denunciation led to his disbarment and to a personal crisis. Peretz, who until the late 1880s had written in Polish and Hebrew, composed his first Yiddish work, *Monish,* in 1888.[23]

In time, Peretz became not only an important Yiddish writer but also one of the major theorists of "Yiddishism." Yiddish literature and Yiddish theater would draw on Jewish religious texts, traditional folklore, and the Hasidic heritage to create a new secular culture that would be both Jewish and universal. "Through Peretz," as Mark Kiel has observed, "Jewish folklore became everything its opponents feared most: a source of modern national pride and a means of preserving tradition, the folk's defining character, in a new secular key."[24]

Literature and theater, according to Peretz, could bring the Jews into the world of European culture—not as beggars, but as equals. Instead of emigrating from Europe, the modern Jews would create a culture of their own that would be both national and cosmopolitan. They had to be proud of their Yiddish language and they had to fight for their rights as Jews. In turn, their neighbors would come to appreciate Jewish culture and accept the rights of the Jews to be themselves.

But if Jewish tradition was to serve as a base for this new secular culture, it had to be studied, not in order to preserve a religious way of life but to preserve an integral link between past, present, and future. For Peretz, the priority was to collect (*zaml/tsu zamlen*) folksongs, folk sayings, and documents: "Collect, transcribe and inscribe. Come together and learn to read, sing together, recite, enjoy, create the atmosphere for the artistic (...) Later genius will come and create."[25]

Like Peretz, S. Ansky (Shloyme Zanvl Rappoport, 1863–1920) also came to Yiddish literature by a roundabout route.[26] Born in Vitebsk, he started writing in Yiddish and spent some time as a tutor in small Jewish towns. Frustrated and disappointed in Jewish society, Rappoport joined the Russian revolutionary movement and went to live among ordinary Russian workers and peasants. He became a Russian writer—and for a brief moment, even contemplated conversion to Russian Orthodoxy. Although he eventually did not convert, he remained suspended, in his own words, "between two worlds."

In the stormy years just before and during the Revolution of 1905, Ansky slowly returned to his Jewish roots. One major factor was the impact that Peretz's writing made on him. As David Roskies has pointed out, "for the first time he discovered a modern European sensibility expressing itself in Yiddish."[27] Once again Ansky started to write in Yiddish.

One of Ansky's major contributions to the cultural revival of East European Jewry was his appreciation of the importance of Jewish folklore. Like Peretz, Ansky called on the Jewish intelligentsia to recognize the enormous reservoir of spiritual energy that lay untapped and unnoticed in the hundreds of *shtetlekh* of the Pale. Ansky put the study of Jewish folklore at the center of his cultural agenda. In 1912, Ansky organized an extensive folklore expedition that combed the Pale of Settlement for sources. In the space of two years it collected

> 2000 photographs, 1,800 folktales and legends, 1500 folk songs and mysteries (biblical Purim plays), 500 cylinders of Jewish folk music, 1000 melodies to songs and *niggunim* without words, countless proverbs and folk beliefs, 100 historical documents, 500 manuscripts and 700 objects acquired for the sum of six thousand rubles.[28]

This treasure trove of material was not only meant to record Jewish life. It could also give Yiddish writers valuable source material to produce a new culture that could realize Peretz's vision of a sensibility that was both Jewish and European. A new modernist culture could use Jewish themes and produce works that would not embarrass the most demanding literary critic. This is exactly what Ansky himself achieved in his great play, *The Dybbuk*.

Simon Dubnow (1860–1941) advanced a bold plan to use the study of Jewish history to reduce conflict and promote harmony between "The Old and the New Judaism." Dubnow came from an observant family in the town of Mstislav, and like Ansky and Peretz, he rebelled against Orthodox tradition. But by the time he had reached his mid-twenties Dubnow found himself intellectually stymied. He had rejected the world of Jewish tradition but where else could he go? What could he believe in? Then one day he had an epiphany. History could replace religion. History could become the religion of the secular Jews. If they could not follow the laws of the Torah, they could still remain true to their people by studying its past. Furthermore, just like Peretz and Ansky, he came to see that there was really no contradiction between the "Jewish" and the "universal." The study of Jewish history could bring into harmony his Jewishness and his allegiance to universal, progressive values.[29]

But how does a people without a state and without a territorial base study history? East European Jews had no archives or universities. Valuable Jewish documents and chronicles lay scattered about in private homes, attics, and cellars. This, Dubnow, warned was not only a scandal but also posed a national danger. When future historians wrote about Jews, what sources would they use—the documents left by anti-Semitic bureaucrats or sources written by Jews themselves?

In 1891, Dubnow issued a call to Russian Jews to collect documents that future Jewish historians could use. No people could afford to ignore its own history—or to leave the writing of it to others. This appeal had an enormous impact. The call to *zaml*, to collect, touched a chord. In St. Petersburg, Russian-speaking Jewish lawyers and doctors, alienated from the synagogue but looking for alternative expressions of Jewishness, responded to Dubnow's call with alacrity. In 1908, along with Dubnow and Ansky, they founded a historical-ethnographic society in St. Petersburg to collect and publish documents on Jewish history. This society published an important historical journal, *Evreiskaia Starina,* which included landmark articles about Jewish history, folklore, and material culture.

Like Peretz and Ansky, Dubnow envisioned a new secular Jewishness that would take in the best of Jewish tradition. Armed with this culture, the Jews would find their place in the liberal democratic Russia that would replace the doomed autocracy. What Dubnow foresaw was a new Europe based on different nationalities—including Jews—enjoying national-cultural autonomy.

The traumatic events of World War I and its aftermath lent a new urgency to the project of national regeneration outlined by Peretz, Ansky, and Dubnow. In 1915, as hundreds of thousands of Jews became penniless refugees, Peretz, Ansky, and the writer Jacob Dineson issued a joint appeal for Jews to gather sources in order to write their own history. If they did not do so, their enemies certainly would.[30]

One year later, in 1916, a group of Vilna Jewish intellectuals published the first of the Vilna *zamlbikher*, an almanac of Jewish life in wartime that largely followed the template of Landkentenish and zamling.[31] Articles and reports brought together the past and the present: studies of traditional synagogue architecture alongside reports on new schools and soup kitchens, articles on Jewish social psychology alongside compilations of jokes and folklore, studies of new secular schools next to accounts of the Jewish book trade and publishing. This collective effort brought together religious and secular Jews, Hebraists and Yiddishists, Zionists and Bundists. The *zamlbikher*, which at first glance seemed to be little more than a collection of miscellanea, in fact became a critical weapon of national self-defense. Implicit in the *zamlbikher* was the belief that the emerging Jewish nation in Eastern Europe was a work-in-progress, the sum total of what the Jews, as a people, did, a modern nation and not just a religious group.[32]

Many of the Vilna intellectuals who participated in the Vilna *zamlbikher* also played a leading role in the founding of the YIVO (the Yiddish Scientific Institute) in Vilna in 1925.[33] The YIVO had two main goals: to promote scholarly research in the Yiddish language and to use the insights of interdisciplinary scholarship in Yiddish to bolster the morale and cultural vitality of a beleaguered people. If ordinary Jews knew more about themselves and their language, the YIVO argued, they would have more self-respect and more determination to fight for their rights. To capture the totality of the Jewish experience in Eastern Europe, the YIVO consciously tried to bridge the boundaries between history, sociology, linguistics, psychology, and anthropology, and to develop new, interdisciplinary methodologies.

Charting a Course

The YIVO and the ZTK would have an almost symbiotic relationship. To fill its libraries and archives, the YIVO relied on networks of *zamlers* throughout Poland (and other countries). In its instructions to the *zamlers*, in its conferences and in its exhibits, the YIVO served as the natural ally and the source of inspiration for the Landkentenish idea in Jewish Poland, and the journals of the ZTK reprinted YIVO materials and questionnaires. Scholars, writers, and community leaders associated with the YIVO would provide critical intellectual guidance for the ZTK.

In an article published in the society's organ, *Landkentenish,* in 1933, the historian Emanuel Ringelblum, who played a major role in the YIVO, justified the need for a separate Jewish Landkentenish society. He argued that "the centuries of urban life, the remoteness from nature, life within the narrow, stifling confines of the ghetto have caused the Jew to feel distant and estranged from the beauty and glory of nature."[34] Properly organized, tourism could bring about not only individual regeneration but also a healthy national revival. Landkentenish could become a solid pillar of an emerging secular Yiddish culture and help turn back the tide of assimilation. But Ringelblum, like other leaders of the Landkentenish movement, feared that the movement could easily take the path of least resistance and degenerate into a mere tourist agency. What was needed was a "Jewish Landkentenish Society," not a "Landkentenish society for Jews."[35]

There were other reasons, Ringelblum reminded his readers, for a separate Jewish Landkentenish society. Polish societies were at best indifferent to Jewish history or the study of Jewish architectural objects. But even if Poles were better disposed, there would still remain a key difference between Polish and Jewish tourism. In most countries, including Poland, tourists could already use comprehensive guidebooks to museums, cities and architectural treasures that had been prepared by professional scholars. Jewish tourists, however, unlike their Polish counterparts, had to do their own spadework. Next to nothing was known about local Jewish history, cemeteries, old synagogues, or regional folklore. There were no guidebooks, few guides, and no points of reference. The Jewish Landkentenish movement had to join forces with the YIVO and combine recreation with serious *zamling*. "Landkentenish," Ringelblum wrote, "really means not only learning (*derkenen*) a city's past, its monuments and buildings; it also means getting to know the people (*folk*) with its centuries old folklore and creative traditions."[36]

As an example of what Landkentenish could become, Ringelblum singled out a local history of Pruzhany, produced by a dedicated group of teachers and students from the town's Yiddish school.[37] It grew out of a project that students and teachers had prepared for a national exhibit organized by the CYSHO (the Central Yiddish School Organization). The completed volume contained a wealth of information not only on the town but also on the region: its geography, economic structure, Jewish labor, Jewish communal and social institutions, and local architecture.

As important as the result was the process: the Pruzhany project showed that Jewish history could be researched and written by ordinary people. Not just scholars in the big cities but also ordinary students and teachers in small towns could come together and study the Jewish past.

What made Ringelblum especially happy was the basic approach of the Pruzhany group. From the very beginning of the project, the Pruzhany collective refused to view Jewish history as a separate and isolated discipline. They treated the Jewish presence in Pruzhany from a comparative perspective, with Jews perceived as an integral part of the region's social and economic system. Ringelblum wrote appreciatively:

> We feel that the writers see proud working Jews. They don't feel that they're guests in Pruzhany. Rather they regard themselves as long established veterans who have put down deep roots in the local area thanks to their work and toil.[38]

In other words, Pruzhany showed lankentenish in action. It brought people together. It changed their consciousness. They "put their town on the map," and in the process, learned more about themselves.

Other leaders of the Landkentenish movement stressed the role of tourism that was no longer associated with luxury hotels and fancy railroad sleeping cars. All over Europe, tourism was increasing and its democratization, as well as the greater accessibility of physical sports, would have important cultural and psychological effects. The Yiddish writer Michal Bursztyn thus noted:

> Hiking through the countryside, spending time in the villages, in the mountains and at the seashore instills in the urban dweller the psychological need to live a healthier life. For us Jews this can become a catalyst in our drive to become a more productive element and to return to the land. And then slogans stop being empty words and instead give way to deeds (…). And we are seeing a new phenomenon among Jews: our merchants and white collar workers, to some extent artisans and workers as well, who all follow a sedentary life style, are beginning to understand that it is healthy to escape the shop and the office, to take a hiking stick (not a wandering stick!) in hand, don a back pack and hike through the country. One relishes natural beauty and gets strength to continue the hard struggle for existence.[39]

For Bursztyn, the benefits of Landkentenish were not only psychological but also political. Jewish rights were under constant attack, and Jews were accustomed to responding by pointing to documents that allegedly promised them equality. But in fact, Landkentenish and tourism would be more effective in the long run.

> How long have we Jews been on the Polish land? Eight hundred, nine hundred years? Our ties to the earth are more important (for defending our rights) than any documents (*yikhus briv*). All along the Polish rivers lie towns and settlements with a rooted and diverse Jewish life that stretches back hundreds of years. The shtetl with its little forest and little stream, surrounded by neighboring villages, signified the direct, unmediated ties to nature, as far as that was possible, given the specific circumstances under which we lived. The contact with the earth is the sine qua non for the continuation of our existence. That contact strengthens us and roots us (*farbirgert unz*) in this country and gives our neighbors the sense that we are equal. And if we can't achieve our goal of settling Jews on the land right now, then at least [landkentenish can prepare the psychological groundwork for the future].[40]

Bursztyn, who would become one of the most important writers for the society's journals and one of the major leaders of the Jewish Landkentenish movement, completed an important novel published in 1937. Set in a shtetl near Warsaw, *Bay di taykhn fun Mazovie* (Along the rivers of Mazowsze) put many of the ideals and arguments of the Landkentenish movement into a literary framework. Unlike many Jewish literary portrayals of the shtetl, this one contained many gentile characters—Germans as well as Poles—and also gave an account of growing anti-Semitism. In other words, Bursztyn was describing Landkentenish, not nostalgia, the shtetl as it really was, not its idealized version. The novel's town, like many others in Poland in the late 1930s, experienced a pogrom. (To get by Polish censorship, the chapter on the pogrom was set in a Chinese town whose name resembled Smolin, the fictional shtetl described in the book.) After the pogrom, even as many Jews lost hope, the main character, a tough old gardener named Hersh Lustig, reflected a stolid determination to overcome all obstacles and live his life in his native town. At the end of the novel, as the shtetl Jews buried their victims, Lustig swore that "'the town will be rebuilt. I, Hersh Lustig, give you my word.' His face was blazed with courage in the red sunset."[41] For Bursztyn, Poland was not a land of exile. In good times and bad, it was home.

The ZTK attracted not only historians like Ringelblum and writers like Bursztyn but also important community activists like Dr. Leon Wulman. Wulman headed one of the most important organizations of Polish Jewry, the TOZ (*Towarzystwo Ochrony Zdrowia Ludności Żydowskiej/Society for the Protection of Health*) that dedicated itself to improving health care for Polish Jewry and to disseminating the idea of "social medicine": proper diet, nutrition, clinics for poor children, etc. His work in the TOZ, which gave him deep insight into the myriad problems—social, physical, cultural, and psychological—that faced Polish Jewry, made him all the more conscious of the need for Landkentenish. In a 1933 article, Wulman praised the work of the Polish PTK and urged Jews to follow its example and get to know their country:[42]

> Our society has the goal of getting to know the country where we work and live, where our parents and ancestors built a life over the course of centuries and participated in the political, economic and social life of (Poland).[43]

But, Wulman emphasized, the ZTK also had tasks that only it, as a Jewish society could carry out.

> Our society is Jewish. It must be Jewish in form and in content. (...) We not only want to learn about this land (...) but also to learn about our own part in the life of this land, our history here, our labors, our customs, our antiquities, life forms, cultural treasures etc. We want to know Jewish life in Poland through and through, we want to investigate it and understand it; this is a task that only we, relying only on ourselves, can do.[44]

The ZTK, Wulman continued, could succeed only if it were a truly open and democratic organization, linked with the broad masses of Polish Jewry. The ZTK, Wulman warned, was not a philanthropic organization, dependent on the leadership of a few wealthy donors. Nor was it a political party. "We are a collective voluntary society (gezelshaft)," Wulman wrote. The ZTK, he implied, had to do all it could to avoid the factionalism and infighting that marked so much of Polish Jewish life and reach out to all who shared its ideals.

Landkentenish in Action, 1926–1939

The formal founding of the ZTK took place in Warsaw in 1926. The Warsaw section began with two hundred members and soon organized various sections devoted to tourism and excursions, publications, protection of Jewish antiquities, medical, legal, library, industry-technology, pedagogy, and fundraising initiatives. From the very beginning the society emphasized that it recruited members from all political persuasions and was determined to remain a nonpolitical organization.[45]

From the very beginning, excursions played a major role in the activities of the society. In 1929, the Warsaw branch alone organized 123 excursions with 2,953 participants, and in 1936 there were 228 excursions with 5,304 participants. Local excursions visited sites within the city: industrial plants, public institutions, utilities such as electric and water works, Okecie airport, the Polish national radio station, the training site for police dogs, and a state vodka factory. In 1936 the society sponsored forty-two trips to the Warsaw suburbs, where a guesthouse in the wooded area of Milosna was especially popular. There were twenty-seven longer trips within Poland, including Vilna-Troki, Kazimierz on the Vistula, the Kampinos Forest, the mountain town Zakopane, the port of Gdynia, and other areas. In addition, the society sponsored four foreign trips. Twenty-seven members participated in a 15-day trip to Paris, while sixty-four traveled on 3 different visits to Yugoslavia.

The ZTK rapidly expanded throughout Poland. By 1938 it boasted fifteen branches with 3,000 dues-paying members.[46] In the previous year it had organized sixteen hundred excursions with 35,000 participants, twenty-eight courses that attracted 300 students, two hundred lectures with an audience of 12,000, thirty summer and winter colonies and camps with 6,000 participants, and five Landkentenish exhibitions.[47]

The ZTK also maintained rented "colonies," including Zakopane, Kazimierz on the Vistula, the lakeside resort of Druskenniki, and the Baltic resort of Karwia. Members could sign up for two- or four-week stays. By the standards of the day, the ZTK's colonies were not cheap. But they had little trouble attracting visitors and helped keep the national society solvent.

The colonies sponsored a full range of summer and winter activities. The society's colony at Zakopane organized downhill skiing, cross-country ski treks, hikes in the Tatras and, of course, lectures in the evenings. Michal Bursztyn, describing the colony on the Baltic seashore at Karwia, tried to explain how important it was for the society to keep it from becoming just another beach club. There should be lectures for the visitors on topics such as the Baltic, the dunes, the peculiar topography of the sea region between Danzig and Karwia, and about the ethnography of the area, especially the culture of the Kashubs.

Beginning in 1928, the society began to organize courses for tour leaders, as it frequently encountered difficulties in finding qualified people to lead excursions (not only for its own constituency but also for organized student trips since Jewish school principals all over Poland turned to the ZTK for assistance) and to run the colonies. These courses included lectures in local history, map and compass reading, and photography. In an article in the society's journal, one of the leaders of the ZTK, Dr. H. Seidengart, remarked that it was as important for a good trip leader to know photography as it was to be able to read a map since part of the purpose of Landkentenish was to leave a documentary record of each trip.[48]

The society organized hundreds of lectures for the general public. An important aspect of ZTK activities were lectures on the history of Jewish art and architecture by specialists such as Szymon Zajczyk and Meir (Majer) Balaban. Also popular were lectures about visits to such "exotic" regions as Africa or India. Some of the lectures were held in Esperanto, with a translation into Polish. Many sections of the ZTK supported Esperanto as a way of breaking down national barriers.

Publications played a major role in the activities of the ZTK. The first publication of the ZTK was *Land un Lebn* (Land and life), of which three numbers appeared in 1928. In 1930 a regular journal entitled *Wiadomości ZTK* (ZTK news) began to appear. This journal was greatly expanded in 1935 and appeared in a bilingual format. The first issue of the Yiddish language *Landkentenish: tsaytshrift far fragn fun lankentenish un turistik, geshikhte fun yidishe yishuvim, folklor un etnografiye* (Knowing the land: magazine for questions on turism, the history of Jewish communities, folklore and ethnography) appeared in 1935 and continued until 1939. Both journals became a major forum for articles about the history of Jewish art and architecture in Poland.[49] Special issues of the ZTK journals were devoted, as we saw earlier, to kayaking or skiing. In 1939, special issues appeared that were devoted to particular cities such as Lwow and Grodno.

The ZTK journals also printed many articles on regional history by the younger generation of Polish Jewish historians. Especially noteworthy were two articles on the theme of regionalism by Philip Friedman. In a wide-ranging discussion, Friedman saw the growing interest in regionalism as a backlash against hyper-centralization and the excessive influence of homogenized, mass culture that threatened to vitiate the diverse local cultures that had given Europe its special characteristics. Supporters of regionalism, Friedman pointed out, did not question patriotism or national unity but rather sought to combat the ongoing threat of an artificial, monotone conformity that was an unfortunate by-product of modernization. "The more diverse, the more colorful, the more creative the parts, the stronger will be the whole."[50] In Poland itself, Friedman emphasized, regional factors were playing a greater role in education, in tourism, and even in fashion. Polish Jews, Friedman urged, should follow this example. As for Polish Jews, they too were to note and study their own regional differences.

Problems and Controversies in the ZTK

It was inevitable that the ZTK would also encounter problems and difficulties. A major blow came in 1934 when the Polish Ministry of Communications withdrew the deep discounts on train tickets that members of tourist societies in Poland received. The ministry eventually restored the discounts to the PTK and even to the German *Beskiden-Verein*, but not to the ZTK. This caused a great deal of bitterness and consternation and became a major topic at national meetings of the ZTK.

The ZTK also discovered that it was easier to proclaim the ideals of the Landkentenish movement than it was to actually realize them. Many articles complained that too many visitors to the colonies came to relax, hike, and ski: lectures about Landkentenish, or serious treks to learn about local customs, were not what they had in mind. Over time this disquiet about the proper balance between Landkentenish and mere tourism became one of the most prominent themes in the society's publications.

Another problem that the society had to face was the role of Yiddish. On the one hand, the very nature of Landkentenish dictated that the Yiddish language and its culture play a prominent role in all of the society's activities. After all, how could one have a Landkentenish society that ignored the language of the folk? The society issued its publications in both Polish and Yiddish. It worked closely with the YIVO and disseminated YIVO questionnaires aimed at stimulating research into Jewish ethnography, local history, the photographing of local landmarks and architectural preservation.

But it appeared that far too many members remained indifferent to Yiddish culture. In a far-ranging article on the ZTK's colonies, Michal Bursztyn pointed out that the visitors were primarily teachers, white-collar workers, students, and professionals. About ten percent had completed higher education. Few were ordinary workers.[51] The social composition of the colonies meant that a great proportion of the visitors, perhaps most, were Polish-speaking rather than Yiddish-speaking and a careful reading of the ZTK's publications shows ongoing concern about how to stimulate interest in Yiddish culture.

How can one evaluate the work of the ZTK? This article can only touch on some of the major issues and much research needs to be done. In a time of rising discrimination and in difficult economic circumstances, the ZTK, which constantly stressed that the Polish landscape was not foreign territory but "home," worked against tough odds. At a time of ongoing politicization and radicalization of large sectors of Polish Jewry, the ZTK went against the trend, advocating activities that many observers saw as basically unserious. The historian Philip Friedman noted that the Landkentenish idea came under fire from many quarters. Zionists attacked it because it seemed to affirm Poland rather than Palestine as home. Communists and radical leftists criticized its lack of political fervor.[52] But these criticisms did not deter Landkentenish activists, who shared different political beliefs, from working together. And this Landkentenish ideal inspired its members even after the start of World War II. Emanuel Ringelblum founded the Oyneg Shabes archive in the Warsaw Ghetto. Joseph Zelkowicz, a leader of the Lodz society, wrote important reportages on the Lodz Ghetto.[53] Michal Bursztyn arrived as a refugee in the Kovno Ghetto, where he wrote short stories about the ghetto experience. Bursztyn, Zelkowicz, and Ringelblum all perished in 1944.

The ZTK, through its publications, left a record, partly of unfulfilled hopes, partly of real if unheroic achievements. If not all its members became Yiddishists, the fact remains that the ZTK was one of the few arenas in Polish Jewish life that tried to be truly bilingual and to show respect to both Polish- and Yiddish-speaking Jews.

In the late 1930s a hike in Zakopane, a kayak trip on the Vistula, or a visit to the local airport might be considered nothing out of the ordinary. Or on the contrary, one might see this as evidence of a quiet determination of Polish Jews to come forward and say that despite everything, they were entitled to enjoy equal rights and be treated with human dignity in a country where they had lived for eight hundred years. And they hoped to enjoy those rights not only as Polish citizens but also as Jews.

*This is a facsimile of the December 1938 issue of *Landkentenish*. The lead article, by Israel Ashendorf is entitled: *Jaremcze* [Waterfall].

A person likes to be photographed at a waterfall in a valley, or on a mountain among the boulders. He wants to link his mortality to the water and the hills. The waterfall never tires in its activity; it hurls itself aggressively against the rocks, foams in anger, is intoxicated from joy and sparkles with a sense of victory.

The mountain never abandons its stolid immobility. Let rains pour, let the snows fall, let the storms rage—the mountain stands quietly, powerful in its silence, a monument to its own heroism.

And man implores: give me your power, waterfall! Give me your calm, mountain!

But the waterfall knows nothing of bestowing and giving away. It is used to pour everything, to carry along, to take away.

And the mountain asks: What are you looking for, on top of my mighty, giant shoulders? Calm?—You will find calm underneath a little mound of earth at once.

Notes

1   The Society used both its Yiddish and its Polish names. In Polish it was the *Żydowskie towarzystwo krajoznawcze* or ZTK. In Yiddish it was called the "Yidisher gezelshaft far landkentenish." Both, the Yiddish and the Polish term can be translated with "Knowledge of the Land" or "Local history and geography," resp. "Heimatforschung" or "Landeskunde" in German.

2   "The kayak is the symbolic ring that brings together two people, just as it is the case (*lehavdil*: to make a separation) with a married couple. But in the kayak people have to get along even better than is the case of two loving partners in a marriage." Y. Maynemer, "Masn turistik un individuele turistik" [Mass tourism and individual tourism], *Landkentenish* 28, No.2 (June 1938): 1.

3   Y. Tseylan, "Mitn Shifl oyf der Vaysl" [With a boat on the Vistula], *Landkentenish* 28, No.2 (June 1938): 3–4; here 4.

4   Tsel-Kuf, "Mit kayakn oyf Wigry un Oygustover Ozeres" [With kajaks on the lakes of the Wigry and Augustow areas], *Landkentenish* 28, No.2 (June 1938): 5–7; here 7.

5   "Fun der landkentnerisher bavegung", *Landkentenish* 28, No.2 (June 1938): 8.

6   Ibid., 9.

7   There is very little scholarship on Landkentenish in any language. David G. Roskies, "Landkentenish: Yiddish Belles Lettres in the Warsaw Ghetto," in *Holocaust Chronicles: Individualizing the Holocaust through Diaries and Other Contemporaneous Personal Accounts*, ed. Robert M. Shapiro (Hoboken, NJ: Ktav, 1999) focuses more on literature and the Warsaw ghetto than it does on the interwar Landkentenish society. See also passing remarks in Roskies' *The Jewish Search for a Usable Past* (Bloomington, IN: Indiana University Press), 26–27; 54–55.

8   This issue raises concerns that go far beyond the theme of Jewish Landkentenish and that in part reflect wider tensions and polarities associated with attitudes toward and perceptions of vacation, leisure time, and tourism in modern Europe. One basic tension has derived from the perceived opposition of "travel" and "tourism." As Ellen Furlough points out, writers like André Siegfried, Paul Fussell, and Daniel Boorstin lamented the decline of "travel"—cultured, intelligent, individualistic—and its replacement by what they regarded as pre-packaged, commodified, herd-like mass tourism. See Ellen Furlough, "Making Mass Vacations: Tourism and Consumer Culture in France, 1930s–1970s," *Comparative Studies in Society and History* 40 (1998): 247–86. In her study of the history of vacationing in the US, Cindy Aaron stressed the ongoing confrontation in American culture between a negative view of leisure, condemned as idleness and a more positive view, which saw it as restoring the mind and body to health, see Cindy S. Aron, *Working at Play: A History of Vacations in the United States* (New York: Oxford University Press, 1999).

9   The idea of tourism and travel as a basic element of modern citizenship achieved a new resonance in the 1930s as both Nazi Germany and the Stalinist regime trumpeted their successes in giving the masses opportunities for travel and leisure that had previously been reserved for a privileged few.

10   For a survey of the development of the Krajoznawstwo movement from the 1870s on, see Edward Wieczorek, "Krajoznawstwo nauką w służbie badania przestrzeni" [Knowing the land: a scientific discipline that facilitates the study of space], http://khit.pttk.pl/index. php?co=tx_krajoznawstwo (accessed June 15, 2007).

11   Ibid.

12   Ibid.

13   John Sears argues that the development of landscape tourism in the nineteenth century, in a country that lacked the long historical traditions of Europe, was vital to "America's invention of itself as a culture." But as Sears points out, tourists visited not only the scenic mountains and waterfalls but also more prosaic destinations—towns, factories, and asylums—that served as a reminder of America's special ability to both produce and to engage in enlightened social reform. John Sears, *Sacred Places: American Tourist Attractions in the Nineteenth Century* (New York: Oxford University Press, 1989), 4. Cf. also Dean MacCannell, *The Tourist: A New Theory of the Leisure Class* (New York: Schocken Books, 1976), and Michael Schudson, "Review Essay: On Tourism and Modern Culture," *American Journal of Sociology* 84, No.5 (1979): 1251.

14   See Alina Cała, *Asymilacja Żydów w Królewstwie Polskim, 1864–1897* [The assimilation of the Jews in the Polish Kingdom, 1864–1897] (Warsaw, 1989), 216–268; Theodore Weeks, *From Assimilation to Antisemitism: The Jewish Question in Poland, 1850–1914* (DeKalb, Ill.: Northern Illinois University Press, 2006).

15   Quoted in Norman Davies, *God's Playground: A History of Poland*, Vol.2 (New York: Columbia University Press, 1982), 257. In turn I found this quote in Mark Kiel, "Vox Populi, Vox Dei: The Centrality of Peretz in Jewish Folkloristics," *Polin* 7 (1992): 88–120; here 112.

16   In addition to the common toponym Vilna, the Polish resp. the Yiddish toponyms are employed here when focusing on a certain ethnic dimension of the city, rather than on the city as a multiethnic entity.

17   Juljusz Kłos, *Wilno: przewodnik krajoznawczy* [literally: "Wilno: a knowing the land guide," or, more figuratively, "Wilno: a scholarly tourist guide"] (Wilno: Wydawn, 1923), 218.

18   Ibid., 45, emphasis added.

19   A good recent treatment of these events can be found in Przemysław Rożański, "Wilno: 19– 21 Kwietnia, 1919 roku" [Vilnius, April 19–21, 1919], *Kwartalnik Historii Żydów* 217 (2006), 13–34.

20   The PTK had many branches and it would be going too far to level blanket charges of anti-Semitism. A leader of the Warsaw ZTK, Aleksander Dubrowicz, noted, for example, that when the ZTK began to organize courses for their leaders in 1928, it received help and support from the Warsaw branch of the PTK. But it certainly can be said that even when the PTK and its branches did not actively exclude Jews, they certainly showed little interest in Jewish history and material culture. See Dr. H. Seidengart, "Dziesięć lat krajoznawstwa" [Ten years of knowing the land], *Krajoznawstwo-wiadomości ZTK* [Knowing the land: News of the ZTK], March, 1937: 14–30; here 17.

21   The manuscript for the second volume was lost in World War II.

22   Zalmen Szyk, *Toyznt yor Vilne* [Vilna: A thousand years], Vol.1 (Vilnius: Gezelshaft far land-kentnish in Poyln, Vilner opteylung, 1939).

23   On Peretz see Ruth R. Wisse, *I.L. Peretz and the Making of Modern Jewish Culture* (Seattle: University of Washington Press, 1991).

24   Mark Kiel, "Peretz in Jewish Folkloristics," 93.

25   Ibid., 103.

26 On Ansky, see the excellent collection of articles by Gabriella Safran and Steven J. Zipperstein, eds., *The Worlds of S. An-sky: A Russian Jewish Intellectual at the Turn of the Century* (Stanford: Stanford University Press, 2006).

27 S. Ansky, *The Dybbuk and other Writings*, ed. David Roskies (New York: Schocken Books, 1992), xviii.

28 Ibid., 13.

29 A good summary of Dubnow's life and thought is contained in Sophie Dubnov-Erlich, *The Life and Work of S.M. Dubnov: Diaspora Nationalism and Jewish History*, ed. Jeffrey Shandler (Bloomington: Indiana University Press, 1991); see also Anke Hilbrenner, *Diaspora-Nationalismus: Zur Geschichtskonstruktion Simon Dubnows* (Göttingen: Vandenhoeck & Ruprecht, 2006).

30 Reprinted in David Roskies, *The Literature of Destruction: Jewish Responses to Catastrophe* (Philadelphia: Jewish Publication Society, 1988), 209–210.

31 Tsemakh Shabad and Moshe Shalit, eds., *Vilner Zamlbukh* [Anthology of Vilna], Vol.I (Vilna, 1916), Vol.2 (Vilna 1918).

32 See Samuel Kassow, "Jewish Communal Politics in Transition: the Vilna Kehille, 1919–1920," *The YIVO Annual* 20 (1991): 61–93; Anne Lipphardt, "Vilne, Vilne unzer heymshtot: Imagining Jewish Vilna in New York," in Marina Dmitrieva and Heidemarie Petersen, eds., *Jüdische Kultur(en) im Neuen Europa: Wilna, 1918–1939* (Wiesbaden: Harrassowitz, 2004), 85–97; here 87–88.

33 See Cecile Kuznitz, "The Origins of Yiddish Scholarship and the YIVO Institute for Yiddish Research" (Doctoral Diss., Stanford University, 2000).

34 Quoted in Roskies, "Landkentenish," in *Holocaust Chronicles*, ed. Robert Shapiro, 11–29; here 11.

35 Emanuel Ringelblum, "Fun der reaktsiye" [From the editors], *Landkentenish* 1 (1933): 3–8; here 3.

36 Ibid., 5.

37 G. Urinski, M. Volansky, N. Zuckerman, eds., *Pinkes fun der shot Pruzhene* [Memorial Book of the town of Pruzany] (Pruzany, 1930); Emanuel Ringelblum, "An interesanter onheyb" [An interesting beginning], *Literarishe Bleter* 27 (1931): 533–534.

38 Ibid., 533.

39 M. Bursztyn, "A nayer factor in yidishn lebn" [A new factor in Jewish life], *Landkentenish* 1 (1933): 9–13; here 11.

40 Ibid., 11–12.

41 Michal Bursztyn, *Bay di taykhn fun Mazovie* [On the riverbanks of Masovia] (Buenos Aires: Yosef Lifshits-fond fun der literatur-gezelshaft baym YIVO, 1970), 210; the original edition was published by the Jewish PEN club in Warsaw in 1937.

42 Leon Wulman, "Di ideologishe yesoydes fun unzer program" [The ideological foundations of our program], *Yedies fun yidisher gezelshaft far landkentenish* (May/June 1933): 14–15.

43 Ibid., 14–15.

44 Ibid., 15.

45 Dr. H. Seidengart, "Dziesięć lat krajoznawstwa," 14.

46 By comparison, in that same year the PTK had 93 branches with 11,570 members.

47 "Po zjeździe" [After the conference] *Krajoznawstwo-wiadomości ZTK* (April 1938): 1–5; here 3.

48 Dr. H. Seidengart, "Dziesięć lat krajoznawstwa," 18.

49 Especially noteworthy are the articles by Zajczyk on the restoration of the old synagogue of Adelsk as an example how a little money and qualified experts enabled a dedicated team—together with the support of the local Polish authorities—to rescue and restore a

wooden synagogue from the eighteenth century, as well as his call for a documentation project for the synagogues all over Poland; an article by H. Altman on the positive attitude shown by Poles to their folk art vis-à-vis the lack of interest shown by Jews in Jewish art and material culture, and by Yiddish literary critic Nakhman Mayzl on Zakopane as a theme in Yiddish literature; Sh. Zajczyk, "Di Shul in Odelsk" [The synagogue in Odelsk], *Landkentenish* 1 (1933): 75; id., "Poznajmy dawną sztukę Żydów polskich" [Let us learn the old art of Polish Jewry"], *Krajoznawstwo-wiadomości ZTK* (April, 1935): 3–6; H. Altman "Vegn yidisher folkskunst" [On Jewish folkart], *Landkentenish* (June, 1936): 5–8; Nakhman Mayzl, "Zakopane Motivn in der Yidisher Literatur" [Zakopane motives in Yiddish literature], *Landkentenish* (April, 1935): 9–10.

50 See Philip Friedman, "Regionalizm," *Yoyvl Numer Landkentenish* (April, 1937): 3–7.

51 Michal Bursztyn, "Landkenerishe arbet oyf di kolonyes un vanderlagern" [Landkentenish work in the colonies and during the hiking tours], *Landkentenish* (June 1935): 1–5. A survey of members of the Warsaw branch in 1930 showed that 40 percent of the members were white-collar and office workers, 11 percent members of the free professions, 10 percent teachers, 8 percent students, and only 7 percent were workers. It should be pointed out that workers' parties like the Bund and the Left Poalei Tsiyon had their own sports clubs and hiking groups.

52 "Regionalizm," *Landkentenish* (June, 1937): 2.

53 On Ringelblum and Zelkowicz during the Holocaust, see Kenneth Helphand's article in this volume.

13 Taking Distance

Israeli Backpackers and Their Society

Erik Cohen

Tourist hot-spot in the Thamel neighborhood of Katmandu; the Hebrew version of the sign reads "40 Rupie per kilo", 2007

In his epochal *Exkurs über den Fremden* (originally published 1908; translated into English as *The Stranger*, 1950), Georg Simmel characterized the stranger as a person in an incongruent position:[1] he (or she) is spatially ("topographically") close to the locals, but socially remote from them— sharing with them only some general human traits, but none of the more particular, intimate ones.[2] While such a position may be uncomfortable, as Simmel, an acculturated German Jew, knew from his own predicament, it has an intellectual advantage: sitting on the fence between the inside and outside of the society of sojourn, the stranger may perceive it more "objectively" than do the ordinary locals. While this claim might be contested in this age rife with a postmodern insistence on a multiplicity of truths, it is still possible to argue that one can gain fresh insights, and more independent judgments regarding oneself, one's identity, and one's society by taking some distance, or temporarily "estranging" oneself geographically as well as psychologically from one's normal, everyday existence. Simmel's approach is particularly germane in the context of this anthology, which approaches contemporary Judaism from a spatial perspective. However, we should note that Simmel does not give us a clue as to how the stranger achieves such a privileged perspective on the society in which he sojourns. In real life, sojourning strangers, such as middleman minorities,[3] usually live in groups rather than as isolated individuals. I submit that they form their perspective on the host society in the course of interaction in the relative security of the protective shield provided by their local ethnic communities.

In this chapter, I shall explore the possibilities inherent in the trips contemporary Israeli backpackers take to places that are geographically and socioculturally remote from their country of origin, for a more personal distance taking—one that might lead them to novel insights into their identity and place in Israeli society, and possibly also to new perspectives on that society, with its manifold, long-standing problems. Special attention will be paid to the social frameworks within which these backpackers sojourn on their trip, and to the possible role these frameworks play in that process.

My interest in this topic is linked with a more general concern: the quest for novel ways to achieve a breakthrough of the dead-end in which Israel found itself, following the exhaustion of the Zionist program on the one hand,[4] and the impasse in Israel's efforts to be integrated into its regional context on the other.

The question of the relationship between tourism and sociocultural change is a moot one. Dean MacCannell, in his influential book, *The Tourist* (1976), saw in tourism the opposite of revolution: "(…) tourism and revolution (…) seemed to me to name the two poles of modern consciousness—a willingness to accept, even venerate things as they are on the one hand, a desire to transform things on the other."[5] By implication, there should be little transformational potential in tourism. MacCannell's tourist, if alienated from his society, will just seek authenticity and real life elsewhere; but he is not concerned with changing his society of origin.

There is, however, also a commonsense belief that "travel broadens the mind," or, in its colloquial Hebrew version, "opens the head," and thus changes the attitudes, views, and even the conduct of travelers; the historical influence of travel on the travelers' societies of origin is beyond doubt. Contemporary mass tourists, spending much of their trip complacently ensconced in an environmental bubble adapted to their accustomed needs and preferences, are unlikely to return home with new ideas or initiatives. Backpackers, however, seem to differ from conventional tourists. They are often portrayed as engaged in a contemporary form of a rite of passage.[6] They are believed to be in a state of flux, or liminality,[7] and to travel on quests to achieve independence, form their personal identities, and consider their futures in their home societies. They thus seem to be more impressionable, more open and susceptible to experiences and influences on their trip than conventional tourists. It is therefore reasonable to expect that they might employ their encounters with strange people and cultures to gain a fresh perspective on themselves and on their societies of origin. That expectation is particularly pertinent for Israeli backpackers. They come from a society characterized by frequent crises and perpetual turbulence, to which they will return at the conclusion of their trip, and during their sojourn abroad, can hardly avoid concern over their own and their society's future.

The question therefore emerges, do Israeli backpackers tend to take a distance, or "estrange" themselves in the course of their trip, in order to gain a new perspective on themselves and on Israel in light of their experiences abroad? If so, do they constitute one avenue by which a stagnating society could be revived?

Despite my long-standing interest in backpackers, I have not conducted independent fieldwork on Israeli backpackers. This article therefore draws upon the work of other researchers. However, though a growing body of studies on Israeli backpackers exists,[8] those questions have not yet been explicitly explored in these studies. I shall therefore attempt to extricate whatever evidence exists regarding those questions from research that was primarily concerned with other issues.

Israel and Its Backpackers

Israeli Jewish society is culturally wide open, deeply engaged economically, culturally, and scientifically with the world system, and in many areas is in the forefront of globalization processes. Contrariwise, however, it is also socially closed, parochially self-concerned, perennially involved in internal controversies, and has little awareness of how the opinions and experiences of others might be relevant to its own concerns. There is a strong pressure to conform to its basic ideological tenets (e.g., to be a "Zionist," whatever that may mean nowadays). Moreover, Israel is a geographically isolated state, with little intercourse even with those neighboring states, like Egypt and Jordan, with which it is formally at peace. A glaring gap has developed between the internal and external perspectives on Israel, manifested dramatically in the contrasting views on Israel's recent war on Hezbollah in Lebanon, held by the Israeli political establishment and by foreign governments and media.

Many Israelis feel that their country is a "pressure cooker,"[9] in the double sense of permanent internal tension and turbulence, and being closed to the outside. From early on, travel abroad has therefore been a principal means of escape from daily reality for growing numbers of Israelis. Owing to the country's regional isolation, they traveled by air to destinations beyond the neighboring countries, predominantly to Europe and to the United States; recently, trips to the "Far East" have become fashionable. Backpacker tourism is one mode in which Israelis travel; it expanded from a trickle of individuals in the 1960s into a rapidly growing stream in the last two decades.[10] Directed at the outset primarily at Europe, by the mid-1980s the stream had turned increasingly to Asia and to a smaller extent to South America.[11] "Escape," not just from the myriad pressures inherent in life in Israel, but also from some pressures related to personal circumstances, has been a major motive for the growing popularity of backpacking. Many early backpackers came from *kibbutzim* (collective settlements), escaping the stifling enclosure of communal life. Many second-generation *kibbutz* youths used the year off, granted them by their communities after discharge from the army, to backpack abroad. This practice gradually spread into Jewish urban youth, who began to take off increasingly on a "long trip after military service,"[12] to work off the strains and constraints imposed by increasingly unpopular military duties, and to enjoy a period of freedom and reflection.

Young army leavers constitute the bulk of Israeli backpackers; but they are joined by older individuals, who are primarily seeking relief from personal life crises or who are disenchanted with life in Israel.[13] Different age groups entertain different modes of engagement with Israel and their place in it, as we shall see.

Israeli Backpacker Enclaves

During the period of rapid expansion of Israeli backpacking, backpacking on the global scale also underwent a rapid transformation. Much of the recent literature on backpackers has focused on the progressive institutionalization of the phenomenon.[14] This resulted in the emergence of a growing chasm between the backpacker ideology of freedom and independent, individual travel, juxtaposed to the mode of travel of conventional tourists,[15] and the actual incorporation of backpacking into the tourist system.[16] Moreover, studies of contemporary Western backpackers indicate that in contrast to the earlier drifters,[17] many are conformist rather than alienated from their societies of origin.

Though embracing an ideology of "getting in with the natives," and avowedly aiming "to cover as much space as possible" on their trip,[18] contemporary backpackers actually tend to spend much of their time with similar-minded youths in backpacker enclaves rather than with the locals.[19] In the enclaves, which exude "an atmosphere of their own," they are provided with various "backpacker-related services,"[20] such as inexpensive guest houses and hostels, restaurants, coffee shops, bars, and forms of entertainment, catering to the tastes of contemporary Western youth. While this inward orientation is frequently judged to be inferior to the exposure of the early drifters to the strangeness of the host society, its possible constitutive function in gaining a new perspective on oneself and one's society, from a remote place that is free of constraints, has generally been overlooked by existing studies.

At first glance, Jewish Israeli backpackers could be expected to differ significantly from their co-travelers from other countries: they seem to be more enterprising, adventurous, and self-assured, many having acquired the skills for self-dependent action, under unfamiliar and difficult conditions, in the army.[21] Some Israeli backpackers have taken lonely and dangerous trips to remote places; a few even perished on the route. They thus seem to be capable of more direct, individual exposure to the strangeness of host environments, and less dependent upon the backpacker enclaves, than other Western backpackers. However, studies of contemporary Israeli backpackers uniformly confirm the opposite: the tendency of the bulk of Israeli backpackers to ensconce themselves, not just in any backpacker enclave, but in distinctly Israeli ones.[22] The biggest and best known Israeli enclaves are located in Kathmandu in Nepal, on Khao San Road in Bangkok, Thailand, in Goa in India, and in Cuzco, Peru; but many smaller ones are found in several other Asian and South American countries.

Few other nationalities have created such all-embracing backpacker enclaves as Israelis have in India. Darya Maoz' PhD study presents a detailed description of such an Israeli enclave; she reports that in a small Northern Indian village the Hebrew language is heard everywhere. There are signs in Hebrew all over the place advertising a variety of services. The Israeli backpackers read Hebrew books and newspapers, listen to Israeli music, update themselves on Hebrew news (apparently from the internet), and speak Hebrew among themselves and even to some of the locals. They prefer

"Israeli" food, served in local restaurants that feature menus in Hebrew.[23] Idith Heichal entitled her M.A. study of Israeli backpackers in Asia, "Israeli Localism Wanders in the Orient,"[24] a title implying that the Israeli backpackers carry the environmental bubble of their home everywhere they go in Asia. Anteby-Yemini, Bazini, Gerstein and Kling describe in some detail the manner in which Israelis re-create an Israeli "locality" in a foreign environment.[25]

This pronounced tendency of Israeli backpackers to create their own enclaves demands explanation. Doubtlessly it is in part a consequence of practical exigencies: despite coming from a polyglot society of immigrants, the younger generation of Israeli-born youth suffers from serious limitations in the use of foreign languages, even English, the *lingua franca* of backpackers throughout the world. Many encounter serious difficulties even in basic communication on practical matters in English.[26] This deficiency imposes serious limitations on their intercourse, not only with the locals, but also with other backpackers, and induces them to seek the company of other Israelis. That tendency is also influenced by the pronounced "neophobic" (fear of novelty) attitude regarding food,[27] characteristic of Israeli youths, which drives them to sites offering "Israeli" dishes.[28] Indeed, food is the iconic mark of the presence of Israeli backpackers and of Israeli enclaves. In many backpacker enclaves popular with Israelis, such as the town of Pai in northwestern Thailand,[29] there is at least one "Israeli" restaurant.

However, the emergence of Israeli backpacker enclaves is not related merely to functional factors. It has some deeper roots in Israel's Zionist ethos, even if this ethos has been transposed to an alien environment, completely unrelated to Zionist aspirations.[30] Israeli backpacking, though non-hegemonic,[31] nonetheless reflects elements of the Zionist ethos: it is an extension onto a global scale of the *tiyul* (excursion) of the early Zionist youth movements in the Land of Israel.[32]

The language and mode of travel of the young backpackers has also been more immediately influenced by their experiences in the army, especially their service in the occupied Palestinian territories.[33] In the early 1990s, a local Israeli newspaper had already published an article with the ironic title *"Thailand be'yadeinu"* ("Thailand is in our hands," a military term to announce the occupation of a tactical objective).[34] The military approach is also readily apparent in the backpackers' attitudes to the inhabitants of the destinations they frequent, to whom they tend to refer to as *hamekomiim* (the locals), a term used in the Israeli army for the inhabitants of the occupied territories. It is even more directly reflected in the terms in which distinctive sub-groups of Israeli backpackers in India are known: *kovshim* (conquerors) and *mitnahalim* (settlers),[35] terms that are obviously borrowed from the terminology of the occupation. Maoz, however, while describing in detail the style of travel of the two sub-groups, did not relate these appellations to the creation of Israeli enclaves—which appear to be the factual realization of the traveling ethos, implied in the twin metaphors of "conquest" and "settlement."

The Israeli enclaves abroad could be seen as reflecting, on a global scale, a deeply ingrained Zionist precept: the establishment of Jewish settlements in the Land of Israel, even though the enclaves' concrete constitution, ambience, and purpose are far removed from those of the settlements of the early pioneers, or from those of the contemporary "settlers" in the occupied territories.

There exist well-developed communication networks among Israeli backpackers: "(…) [they] have developed communal structures such as meeting places, restaurants, lodging, mutual aid centers, and information points (where one can share the experiences of other travelers)."[36] They are assisted on their trip by files in travel-supply shops where the prospective backpackers tend to purchase their equipment,[37] by visitor books in Israeli embassies,[38] and by word-of-mouth information, flowing from returning, experienced travelers to departing ones. It became quite common in recent years for backpackers to leave for their trip to Thailand or India with a detailed itinerary, according to which they advance from one Israeli enclave to another. Trips thus became highly standardized, constituting a backpacker tourist sub-system, no less cut off from the flow of local life than that of conventional tourists. Even backpackers who leave Israel with the avowed intention of staying away from other Israelis soon succumb to the temptation of the enclaves, which provide them with a familiar kind of shelter and with security in a strange and often confusing environment.[39]

Israeli backpackers are often ambivalent regarding their attachment to the enclaves, being aware that they thereby forfeit a fresh, unmediated experience of the destination, which they had hoped for prior to departure. Israeli researchers noted the irony in the attachment of Israeli backpackers' to their enclaves (e.g., in the title of Heichal's M.A. Thesis), but paid insufficient attention to the possibility that the enclaves play a significant intermediary role in the formation of the travelers' identity and of their relationship to their country of origin.

Several studies pointed out that the enclaves reproduce Israeli locality wherever Israelis arrive in significant numbers.[40] But they tend to disregard the analytical significance of the combination of familiarity and strangeness by which the enclaves are marked: they are, in Victor Turner's terms, a familiar place in a foreign place. They thus resemble in some ways the liminal stage in tribal rituals of passage, in which adolescents leave their familiar place of abode and move to a remote, unfamiliar place in which they create an effervescent *communitas*[41] before returning to their familiar place. Though important differences exist between tribal rites of passage and backpacker journeys,[42] the resemblance between the Israeli backpacker enclaves and the tribal youths' sites of *communitas* is striking: both are sites of collective estrangement, and thus highly conducive to the formation of new identities. To be sure, backpackers do not form a *communitas*, though their sojourn in a strange environment may bring them closer together than they would be in Israel. They are groupings in a liminal state that makes their permanently rotating members susceptible to new perspectives and ideas, about which they can converse more intimately in the foreign environment than they could at home. The remoteness of the enclaves from everyday Israeli reality facilitates free and unrestrained discussion and reconsideration of that reality. I suggest that the enclaves create the conditions under which the backpackers feel secure, free, and unrestrained enough to be able to distance themselves from their society and open up to one another. The enclaves thus serve as a template within which backpackers can evolve, in interaction with their peers, new individual identities and alternative views of their society and of their own place in it. They thus seem to serve as the principal link between the trip and the reconstitution of the Israeli backpackers' personal identities and attitudes toward their society.

Unfortunately, existing studies do not contain enough direct information on the actual content of the conversations the Israeli backpackers conduct in their enclaves to validate these suggestions. However, considerable evidence exists that suggests young Israeli backpackers do tend to construct or reconstruct their identities on these trips. While Anteby-Yemini et al. stress the maintenance of the backpackers' (external) identity,[43] Ayana Haviv offers a detailed description of the tendency of young backpackers to evolve a new identity, which is sharply at odds with that of the older generation of Israeli Jews.[44] However, these studies do not tell us much about how the formation of these new identities is accomplished. The impression one gets is that it is a personal, psychological process. I argue, however, that it is a collectively negotiated issue, and that negotiation takes place in social interaction between the backpackers, especially in the enclaves.

It is important to note that this new identity does not imply an alienation from Israel, though it may reflect considerable alienation from some facets of Israeli society. In fact, the adherence of the youth to Israel, and their pride in being Israelis, is strengthened rather than weakened by the trip.[45] The majority intends to return to Israel after their trip, and they see their futures there. But the youths are critical or resentful of some features of Israeli society, and especially of the army. Their new identity is more individualistic and hedonistic, in contrast to the, at least rhetorically, more collectivistic orientation of the older generation,[46] propagated by the Israeli establishment. This new identity thus contains some striking "postmodern" traits, common among contemporary Western youth.[47] By implication, the young backpackers seek to reconstitute Israel's national identity to suit their needs and aspirations, and thereby implicitly conceive alternative ways to shape the future of Israeli society.

Maoz' recent study of different age groups of Israeli backpackers came up with the significant finding that the older the age group, the more alienated its members are from Israel, critical of the state and its politicians, and inclined to seek an alternative place in which to spend their lives.[48] This finding seems to indicate that alienation is related to life experience. Young Israeli backpackers who have just completed their military service, may be critical of the army but still intend to spend their lives in Israel, though they might seek to reshape it in accordance to their vision. The older ones, distressed by their experiences and personal crises, have often given up on their society, and are seeking an alternative place of abode.

Conclusion

Simmel's idealized stranger is a lonely individual, observing society from its margins. However, sojourning middleman minorities, the principal embodiments of Simmel's stranger in the real world, live in groups, and so do Israeli backpackers. I proposed in this article that these backpackers, taking a geographical as well as spiritual distance from their own society, formulate or reformulate their identities, and take a new perspective on that society in the liminal space of their enclaves, those little islands of familiarity located in the remote recesses of their host countries. These small, familiar places in foreign settings appear to permit a privileged and uncluttered view of the place from which they have departed, but to which they will also return. Outside the "pressure cooker," an unforced reconsideration of the basics of one's life and society becomes possible. Unfortunately, the extant studies tell us little about the re-imagining of Israel by the backpackers. Haviv's description of the Israeli backpackers' ideology comes closest to prefiguring their image of an ideal society. She asserts that their "(…) individual-oriented ideology contrasts strongly with the Israeli older generation's ideals of collectivity"; however, the Israeli backpackers

> "(…) do not divorce themselves from their community. Instead, they transform it into a new kind of national community that allows for individual expression within it. In this way, the opposition between 'individuality' and 'collectivity' is not overly reified by Israeli backpackers; instead, they see the voluntary community being created around them [on the trip] as the ideal atmosphere for promoting individual growth and self discovery."[49]

Haviv's description of the backpackers' image of the "ideal community" of Israelis on their trip implicitly foreshadows their image of the ideal community in their home country. This image seems to reflect a desire for a society that combines individual freedom with social harmony—an ideal set against the internal dissensions and external conflicts prevailing in contemporary Israel. Maoz' recent article reinforces that conclusion: she asserts that for the backpackers the enclaves represent an ideal image of Israel, but, in her view, not so much a model for Israel's future, but rather a romantic replica of its pioneering past.[50]

The backpackers' alternative image of Israel is vague with respect to specific issues. But the free and harmonious ambience of the enclaves, on which the image is modeled, implies a broader desire for social harmony and a generally more tolerant attitude toward the convictions and aspirations of others. Insofar as these attitudes persist upon the backpackers' return to Israel, they will predispose them to favor an amelioration of the intense tensions and conflicts plaguing Israeli society, and a mellowing of all extremism, whether religious, ethnic, or nationalist. However, insofar as the image harks back to a utopian past, it is doubtful that it possesses a more general, realistic potential to revive a complex, but stagnating, society.

Several researchers have reported on the influence exerted by the culture of host societies on the appearance and conduct of Israeli backpackers. The principal argument of this article, that the Israeli backpackers' identities and images of a future Israeli society are primarily shaped in the course of their free and unrestrained interaction in the enclaves—and not just by their contact with foreign surroundings—is as yet merely suggestive since it rests on meager empirical evidence. Neither the processes of social interaction within the enclaves nor the attitudes of returning backpackers to the burning issues of Israeli society have up to now been systematically examined. A thorough study of these issues could significantly advance our understanding of the role of the highly popular practice of backpacking in the dynamics of Israeli society.

Notes

1   I am indebted to Nir Avieli and Darya Maoz for their comments on an earlier version of this article.

2   Georg Simmel, "The Stranger," in *The Sociology of Georg Simmel*, ed. K. Wolff (1908; repr., New York: Free Press, 1950), 402–408.

3   Edna Bonacich, "A Theory of Middleman Minorities," *American Sociological Review* 38 no.5 (1973): 583–94.

4   Erik Cohen, "Israel as a Post-Zionist Society," *Israel Affairs* 1, no. 3 (1995): 203–14.

5   Dean MacCannell, *The Tourist: A New Theory of the Leisure Class* (New York: Schocken Books, 1976), 3.

6   Erik Cohen, "Backpacking: Diversity and Change," in *The Global Nomad*, eds. Greg Richards and Julia Wilson (Clevedon: Channel View Publications, 2004), 43–59.

7   Victor Turner, "Betwixt and Between: The Liminal Period in 'rites de passage,'" in *Betwixt and Between: Patterns of Masculine and Feminine Initiation*, eds. L.C. Mahdi, S. Foster and M. Little (La Salle, IL: Open Court, 1987), 5–22.

8   For example: Idith Heichal, "Lokaliut israelit nodedet be-mizrah" [Israeli localism wanders in the Orient], (master's thesis, Department of Sociology and Anthropology, The Hebrew University of Jerusalem, 2000); Yehuda Jacobson, "Secular Pilgrimage in the Israeli Context: The Journey of Young Israelis to Distant Countries" [in Hebrew] (master's thesis, Tel Aviv University 1987); Darya Maoz, "Hebetei mahzor hayim be-masa shel israelim be-hodu" [Aspects of the life-cycle in the journeys of Israelis in India] (PhD thesis, Hebrew University of Jerusalem 2004, English summary); id., "The Conquerors and the Settlers," in Richards and Wilson, eds., *Global Nomad*, 109–122; id., "Erikson on the Tour," *Tourism Recreation Research* 31, no. 3 (2006): 55–63; id., "Backpackers' Motivations: The Role of Culture and Nationality," *Annals of Tourism Research* 34, no. 1 (2007); Chaim Noy, "'The Trip really Changed Me': Backpackers' Narratives of Self-Change," *Annals of Tourism Research* 31, no.1 (2004): 78–102; Chaim Noy, "Israeli Backpacking since the 1960s: Institutionalization and Its Effects," *Tourism Recreation Research* 31, no.3 (2006): 39–53; Chaim Noy and Erik Cohen, eds., *Israeli Backpackers and Their Society* (Albany NY: State University of New York Press, 2005); Dalit Simchai, *Ha-shvil haze mathil kan* [This trail starts here] (Tel Aviv: Prague Publications, 2000).

9   Ayana Haviv, "Next Year in Kathmandu: Israeli Backpackers and the Formation of a New Israeli Identity," in Noy and Cohen, eds., *Israeli Backpackers*, Chap. 1, 45–88; here 56–8.

10   Noy, "Israeli Backpacking since the 1960s," 39-53.

11   Noy and Cohen, eds., *Israeli Backpackers*, 26–7.

12   Oded Mevorach, *Ha-masa ha-arokh aharei ha-sherut ha-tsv'ai* [The long trip after the military service: Characteristics of the travelers, the effects of the trip, and its meaning] (PhD thesis, Hebrew University of Jerusalem, 1997).

13   Maoz, "Erikson on the Tour."

14   Richards and Wilson, eds., *Global Nomad*.

15   Peter Welk, "The Beaten Track: Anti-Tourism as an Element of Backpacker Identity Construction," in Richards and Wilson, eds., *Global Nomad*, 77–91.

16   Cohen, "Backpacking: Diversity and Change" and Irena Ateljevic and Stephen Doorne, "Theoretical Encounters: A Review of Backpacker Literature," in Richards and Wilson, eds., *Global Nomad*, 60–76.

17   Erik Cohen, "Nomads from Affluence: Notes on the Phenomenon of Drifter Tourism," in id., *Contemporary Tourism: Diversity and Change* (1973; repr., Amsterdam: Elsevier, 2004), 49–63.

18   Richards and Wilson, eds., *Global Nomad*, 260.

19   Erik Cohen, "Pai: A Backpacker Enclave in Transition," *Tourism Recreation Research* 31, no. 1 (2006): 11–27.

20   Richards and Wilson, eds., *Global Nomad*, 260.

21   Mevorach, *Ha-masa ha-arokh*.

22   Noy and Cohen, eds., *Israeli Backpackers*, 14–15.

23   Maoz, "Hebetei mahzor hayim" and Maoz, "Backpackers' Motivations."

24   Heichal, "Lokaliut israelit."

25   Lisa Anteby-Yemini et al., "'Traveling Cultures': Israeli Backpacking and Reconstruction of Home," in Noy and Cohen, eds., *Israeli Backpackers*, 89–109; here 92–6.

26   Ibid., 100.

27   Claude Fischler, "Food, Self, and Identity," *Social Science Information* 27 (1988): 275–92; Erik Cohen and Nir Avieli, "Food in Tourism: Attraction and Impediment," *Annals of Tourism Research* 31, no. 4 (2004): 755–78.

28   Ironically, the so-called "Israeli" food, preferred by the Jewish Israeli backpackers, is not Israeli at all. Most of it is of Arab origin, such as *shakshuka, houmus* and *t'hina*; the rest is European, such as *Schnitzel* and *chips* (French fries).

29   Cohen, "Pai," 11–27.

30   Erik Cohen and Chaim Noy, "Conclusion: The Backpackers and Israeli Society," in Noy and Cohen, eds., *Israeli Backpackers*, 251–64.

31   Ibid., 252–53.

32   Shaul Katz, "The Israeli Teacher-Guide: The Emergence and Perpetuation of a Role," *Annals of Tourism Research* 12 (1985): 49–72.

33   Maoz, "Hebetei mahzor hayim," 287–90.

34   E. Halfon, "Thailand be-yadeinu" [Thailand is in our hands], *Ha'ir*, June 14, 1991, 34–9; Noy and Cohen, eds., *Israeli Backpackers*, 31.

35   Maoz, "Conquerors and the Settlers"; Noy and Cohen, eds., *Israeli Backpackers*, 27–8.

36   Anteby-Yemini et al., "Traveling Cultures," 98.

37   Jacobson, "Secular Pilgrimage."

38   Noy and Cohen, eds., *Israeli Backpackers*, 15.

39   Cf. Anteby-Yemini et al., "Traveling Cultures," 100.

40   Ibid., 92–97; Heichal, "Lokaliut yisraelit."

41   Victor Turner, *The Ritual Process* (Chicago: Aldine, 1969).

42   Cohen, "Backpacking: Diversity and Change," 52–5.

43   Anteby-Yemini et al., "Traveling Cultures," 101–102.

44   Haviv, "Next Year in Kathmandu," 45–88.

45   Ibid., 63.

46   Ibid., 49–53.

47   Ibid., 52.

48   Maoz, "Hebetei mahzor hayim" and id., "Erikson on the Tour."

49   Haviv, "Next Year in Kathmandu," 71.

50   Maoz, "Backpackers' Motivations."

14 Tales of Diaspora in the New Fluid Atlas of Virtual Place

The Internet Project
"The Man Who Swam into History"

Shelley Hornstein

manuscript of the work you hold was almost complete, but I was searching for an introduction that would hold together this collection about my

(From The Man Who Swam Into History by Robert Rosenstone, pp XI-XII)

Introductory page, web project, "The Man Who Swam into History," 2007

> It is a little-noted fact of history that the rivers of Eastern Europe were jammed with swimmers in the last quarter of the nineteenth century. Not one grandfather but a whole generation of grandfathers sidling, walking, waddling, hurrying, moseying, lurching, striding, flinging, leaping, jumping, tiptoeing, plunging, screaming themselves into previously empty waters. They were not yet grandfathers, but somehow the image is of aged men, dressed fully in black, yarmulkes affixed firmly to their scalps, long white beards floating miraculously and gently on the surface as they flash toward far-off shores (…). In later years, none of these grandfathers were ever known to go near the water (…). Decades later they were full of foolish tales, babbled in languages that grandsons neither understood nor cared much about. But each grandfather had this one moment of undisputed triumph that would quietly resonate through future generations. (Robert Rosenstone, *The Man Who Swam Into History*)[1]

In this quintessentially diasporic vignette of the "old country" fused with the "new," author Robert Rosenstone reminds us that personal stories handed down from one generation to the next take on an afterlife that continues to breathe through the lifeblood of progeny. No matter how embroidered, hyperbolic, incongruous, or rearranged the tales may be, vignettes of private life have transitioned to the public realm primarily through literary pathways and books. Only recently, since the advent of the internet and its wildfire ability to reach scores of people at record speeds, have the stories, visualizations, actualizations, and dramatizations of private lives become even more exposed, available, and communal. The space of representation of our private lives is not just a video clip or a simple photograph transferred to a computer monitor. Rather, the new place of visual encounter—indeed the most novel venue of choice for viewing—is specifically, and not unimportantly, virtual space where place, and even the location of place within that space, is the operative concept. It certainly behoves us, therefore, to take a closer look at this relatively unexplored terrain, and particularly at what it offers for the shaping of *topos* and topographies, and even more strategically, Jewish topographies. Just as mapmaking and topographical representations of faraway lands in the sixteenth century, or Maxime du Camp's photographic travelogues of Palestine, Nubia, and Syria, or *La Mission Heliographique's* photographic record of France's patrimony were snapshots of everyday places and life to capture, treasure, and display to the public eye, so too, accounts of private life and personal travelogues through virtual images now find many different forms of representation on the internet. What the internet demonstrates first is that private lives matter and that anyone's history can in some way stand for our own; and second, that the internet is arguably the new atlas for playing out those tales and itineraries. If an atlas is a collection of maps that documents geography, then the world-wide-web is equally a collection of maps that encompass hitherto unknown or uncharted topographies that deserve to be explored, chronicled, and legitimated as (structured) entities. The stories and histories that are featured online demand to be interpreted, yet it is only with new templates that put space and the spatial turn in our immediate viewfinder (or should I say, monitor), that we will be able to articulate critical thought about these new fluid spaces.

Not only do I want to speak of space and the open question of Jewish space, but I also want to peek into and begin to open up an epistemology of the spatial turn in light of the following closely-knit terms integral to this project, namely, *diaspora, travel, topography*, and *architecture*. To be sure, each of these terms deserves a separate chapter, but in the reduction of space and time here, my question, above all, is: how does each of these terms appear to blend together effortlessly—like some swirling concoction—when brought into virtual space, and even virtual Jewish space? What is it about each of these terms that lends itself so agreeably to this comfortable fit? In the era of smart technologies enabling users to access information with sensors and computing devices that are largely interactive and designed to suit various visitor profiles in tandem with, and sometimes complementary to, the standard desktop or laptop computer, we can now exercise our cultural power by creating another experience or imaginative telling of tales. The internet, as arguably the most successful smart technological invention, is also the most comfortable: because of its privacy and responsiveness to our command, it is a place for exploration. One "travels" from site to site, across topographies, transforming topographies, in an architectural fantasy of illusory constructions. The traveling action calls upon a wide range of ambulatory effects that include displacement, disruption, or continuity.

The Man Who Swam Into History: The Web Project

This paper describes a website project I created that springboards and complements a "ficto-memoir" or perhaps a straight up novel entitled, *The Man Who Swam Into History; The (Mostly) True Story of My Jewish Family* <www. mosaica.ca.>[2] Written by California Institute of Technology History Professor, Robert Rosenstone, and published by the University of Texas Press in 2005, the novel recounts the family migrations of three generations, commencing with the picaresque character, Chaim Baer. This fearless character braved the Prut River by swimming across it from Moldavia to Romania in order to escape the military draft, and so commences his diasporic journey.

Rosenstone begins the tale as a historical travelogue in Stepney Green, in London's East End, as the launching pad for his reflections on his family's past. While the novel is a personal story, it really belongs to so many. The website project, unlike the book, provides the visitor or user with an assortment of historical minuets. Each of these, while tied to the past in some way, is tied equally to the present by connecting links. For example, a visitor to the site can explore and link this story to contemporaneous sites today by clicking on an archival photo of a location referred to in the book; subsequently, the user or reader is "transported" to a contemporary website that promotes tourism for that location today. The passage below illustrates the point: the towns of Tetscani and Moinesti are mentioned, but where are these places, I asked myself? Are they the same today? Do they still exist?

> Between the southern bank of the Prut and the Moldavian village of Tetscani there is a blank, more of time than space. Chaim Baer acquires a wife, a business, a house, a vegetable garden, a cow, and then a second cow (…). The family loads its possessions in a cart and leaves Tetscani for Moinesti, some kilometers away. A few months, or a year or two later, they make the same journey in the opposite direction. These periodic moves partake of a mystery. Official government policy prevents Jews from becoming citizens and from owning land. But nothing is said in historical studies, constitutions, available decrees and statues to indicate that there was a time limit on Jews residing in small towns or rural areas (…). The upset and confusion of packing, hauling, setting up in Moinesti were an unsought diversion. There is evidence enough that Chaim Baer was the sort of man who understood the spiritual benefits of an occasional change of scenery.[3]

These questions informed the central quest of the project: we are attempting at all turns to locate or place into a topography the sites described in this and any book. As part of a visual imagining, the act of reading triggers the visual cues as we shape an architectural landscape in our imagination. More to the point, when reading a narrative about the past, and specifically when that past is hinged to places that seem to be almost whimsical, the reader is hard-pressed to make the links to a contemporary, and in fact real, place. Where is the Prut River, exactly, and what has become of this and that shtetl in the Romanian landscape as their names roll off the narrator's tongue?

Traveling in Virtual Space

This website project offers us the possibility of reaching beyond the immediacy of the text, so that while listening and journeying on this narrative tale we are journeying, too, with our mouse: with each click one window balloons atop another, or the spoken text is silenced abruptly and a random association is made to the next screen. To begin, an image animates the screen. It invites us to play with its surface. A refragmentation of images that appears in the book is toppling one another in an invented structure. Each of the images provides some sort of animation to the user, or the performer of space, as Henri Lefebvre would have it.[4]

The layered tower is building, and the geographic and topographic level of stories accumulates. We are linked immediately to Montreal today or any other contemporary and active place referred to in the novel. In this way, any sense of nostalgia is battled to ensure that these cited places of the past represent a currency of today. Elements of playfulness are also woven through this piece thanks to the user's initiation by way of the mouse or keyboard, or countermanded by the timed text programmed into the software, even when the spoken voice-over narrating the scene is interrupted or blended with a piano scherzo or squawking hen. Music, spoken text, or roll-overs that make images disappear and reappear, or create cacophonies of chirping birds, cows mooing, tongue-clicking and snapping cicadas, or keyboard typing and other background sounds that weave the internet experience with the simulated or taped "real" experiences, are activated at varying intervals.

What were my objectives for this site? Why create it at all? How do I create a site that represents my critical engagement but extends the ideas Robert Rosenstone presents in his book? How do I attempt to create a sense of unlimited space in this new site of spatial topography that could be a site of Jewish history and take on the ideas of diaspora? First, let's work from the present backwards. When the "pilot project" was launched on the homepage of the Mosaica website, I assured Robert Rosenstone that he would be the first to see it. Our email conversation captured some of these thoughts and provides a much-needed context for the web project:

From: Shelley
To: Robert Rosenstone
Sent: Wednesday, February 28, 2007
Subject: website

Hi Robert,

(…) there is a counter-intuitive idea behind it all where I do desire for it to be disorienting to a degree and in this way create some sort of feeling of disjointedness—read, diaspora—and a sense of uprootedness or at least unknowingness about the next step along the journey.

Another aspect that was critical for me was to create a web project that picks up those different subject positions so prominent in your book. As a user on the internet, we're always well aware of the multiple users of a site and the different, multiple and perhaps even somewhat endless (or the illusion of endless) journeys we may have on the site. I was also very conscious of not wanting to in any way "illustrate" the book, or even come close to "duplicating" or rather "representing" the book. So everything is about having some touchstone to the book (voice-over, running text, some of your photos) but always a journey that is unknown to anyone, including you. And syrupy returns to the past were also aspects I wanted to avoid, hence the links to URLs or sites of today that are triggered simply by words or phrases in your narrative. These triggered "keywords" were compiled by the many different students who read your book and created ongoing lists of keywords which we then compiled and almost randomly selected [and worked with me]. In this way, the site reflects the many creators.

Yet there are many bits missing (…). And I've struggled with how to create a kind-of "user's guide" (to the perplexed?). But I'm still working on it: I want to give context but not too much. The project is inspired by your work, but it's not a duplicate, of course (…). You're so close to the project that it must feel like a movie-after-the-book, which it is not whatsoever nor attempts to be. But I'm trying to make it something that speaks about diaspora journeys, connections to today – where are those places in Montreal you describe and what do they look like today, for example, or where (…) is the Prut River EXACTLY and what does it look like now? to chase away that maudlin hankering for the past yet make the work stand out as relevant and alive and the story of so many.

Shelley

To: Shelley
Sent: Thursday, March 01, 2007
Subject: Re: website

Dear Shelley,

(…) I understand, though I think there is a fine balance between achieving such a feeling of disorientation and just confusing people. That of course is what I have struggled with in some of my experimental narratives, including the man who swam. You can probably get away with more dissonance on a website than on a page (…). Certainly I agree that you want to avoid syrupy returns to the past and I don't expect the site to illustrate the book or duplicate it.

I agree, give some context but not too much. I see how that would be a real conceptual dilemma. The thing is that you don't want to put off people who would actually like to hear a whole section or many sections read—or even, god bless them, go out and buy the book (…) Let me give one funny reaction [to your web project]. When I first saw it I really resented having my mother's face covered with that shawl. It made me feel, in some odd way, that my mother was being altered! My mother!! Now of course nobody but I and my brothers will know that photo is of her, and since you can't see the face anyway it doesn't really matter. But I am interested by my own reaction. It was like a violation. Perhaps no more a violation than my writing about her. She surely never expected to wind up a heroine in a book or a face in a public place (website) for all the world to see if it wishes. Odd, isn't it? Perhaps the first time the sense of how we recycle words and images and use people in the past for our own needs. It makes me want to write something but I'm not sure how much I have to say beyond the sudden feeling of strangeness. Don't get me wrong. I am not saying don't use the image; I am examining my own reaction in a larger context.

I suppose we all wish our works could be more, but I understand the frustration, the gap between your vision and the reality. Here at the age of 70 I look back at seven books and think was it worth it? Was this all I had to say? What was all the work and fuss about? This only because I am struggling with a new one about Jews and Muslims in tenth century Cordoba. A historical novel, I think, but possibly a piece of musical theater? Anyway I do think one must experiment with new forms of telling and that is what you are doing (…).

Robert

Creating the Project

If my objective for the website project was to create a parallel and complementary project to the book, then a part of the objective prior to confronting the subject matter of visualizing diaspora on the internet was to come to terms with the challenge of complementing a work of literary fiction in visual and audio form. But how does one create a project that cuts across the fear of (diluted and misrepresented) duplication? In what way can one create the book in another form — a challenge met routinely for example, when a book becomes a film? And how would I demonstrate what I took from Rosenstone's own approach to his novel about its inherently diasporic style, if such a thing might be qualified; that is to say that his story of historical peregrinations through a variety of geographic locations, and the plethora of subjects who tell the stories, was what I found to be perfectly amenable to virtual space?

Let me suggest an example in literature. Architectural analogy was a device Marcel Proust used to describe a literary work since he often found it difficult to describe literature by using strictly literary models.[5] Indeed, Proust would turn to a metalanguage and architectural analogies, in particular, for its ability to bridge the constructions and articulations of space in order to "transfer" ideas within lexical space. I like to imagine that Proust would have found comfort in the internet since initially he had proposed labelling each section of his book with architectural terms, such as a porch way or an apse's stained-glass windows.[6] Through Proust's literary transfers I have my inspiration to articulate Rosenstone's work in an internet project, one that transfers ideas from lexical to virtual space without duplicating the initial work of fiction, but instead creating a new but related project.

Rosenstone's story in book format is not recounted as a linear narrative but is told through tales, sequences, and fragments sliced into each other and reassembled from three generations in five countries across two continents and over a century. Told as a web project for the digital and visual realms, Rosenstone's textual and narrative chronology is further fractured, perhaps exaggerating or creating sensations of frustrations of place, yet complements and stretches the story he unfolds. Indeed, where Rosenstone succeeds at decentering any possibility of a protagonist with multiple subject positions, I took my cue. How perfectly apt his literary strategy is for a virtual world! The context of place in virtual space is challenged and reconceived: volumes and flatness, depth and breadth, marked and unmarked boundaries or edges, frames of all given and hitherto unknown spaces are called into question and taken up anew. In this way, the project attempts to activate multiple subject positions by altering the computer user's otherwise passive voice into active iterations by what could amount to endless potential visits to, and manipulations of, the site. It is hoped that this pilot web art project invites users to build on the idea of biography and history by creating pathways through juxtaposed and intersecting spatial and temporal frames. Unlike a film-after-the-book, the web-user — as the player or hero/heroine — activates new formulations of the experience of migration. New narratives and architectures of what we can now describe as a transnational place are created that are irresolute, erratic, and part of a journey of spontaneity. In so doing, the user expands the spatio-temporal dimensions of a *would-be* diaspora and encourages a re-evaluation of any notion of Jewish place. By using the internet, each player begins to imagine, through personal stories, what it means to be out of place, in place, and displaced. This enables a negotiation of architectonically constructed virtual space to create layered and diverse narratives of place.

Virtual Diaspora

While initially conceived within the specificities of the internet, these places also raise the important concept of travel in the real world, a concept not widely used to discuss diasporic experiences. Yet diasporic journeys are indeed a form of travel. Such a form of travel need not necessarily be tied to the optimism and beauty of travel but can be tied to the hesitancy, fears, and even the rejection of travel when considered as a displacement of oneself or the act of being uprooted from home. However, if we are to imagine diasporic travel as another type of experience, following Jonathan and Daniel Boyarin and others, we can see it rather as celebratory and political, or as an affirming of the power of place and topographies of clusters that allowed for the preservation and indeed the enriching of a Jewish cultural identity.[7] The internet is one pathway that, through the travel metaphor, can enhance the experience or imaginative telling of diasporic tales. Just as Google Earth and other "cybermaps" are testaments to the desire to travel virtually, this project, and others like it, open up the possibility for new visual perspectives to give us another way of knowing home and place.[8]

For both diasporic journeys and the internet, material presence is forever shifting and quested.

> If your eyesight is very good, if your mind is suitably concentrated, if the angle of the afternoon sun on the serrated surface of the Pacific is just right, if the clouds are gone and the breeze calm and the beach almost empty and only a few white sails scud slowly across your retina, then it is possible to stand on the bluff in Santa Monica beneath the palms and watch the snow fall in Montreal.[9]

The lack of a geographical topography that is measurable, calculable, and able to be *sited* and mapped, is a constant, so it seems almost natural for the internet to be a place of topographical real estate, or a place where there is always an attempt to shift dialectically between the need to anchor the unanchorable and always shifting ground, a ground that seems without borders and only points to places elsewhere while conveying the illusion of presence. By bringing the two terms 'diaspora' and 'internet' into a bonded relationship, each user becomes an *agent provocateur*: diaspora and internet are complex and weighty words that represent profound worlds of meaning that seem to be—at least at first blush—at odds with each other, or even utterly unrelated. Yet upon closer reflection, the internet has always attempted to create a sense of "home" on journeys, random and otherwise, through terminology such as "navigation" and "surfing"—and even in this project, by "swimming"—and other smooth forms of mobility in space. It is no small coincidence that the lexicon of the technology environment also turns to architecture (as in *web architecture*) as the label of choice to describe the entire range of creations (and constructions) for the World Wide Web. Architecture, in the real and the virtual worlds, is a framing device, boundary marker, or building mechanism that relates to constructions of frames, shapes and spaces of the internet and the virtual space it opens up to users. A vocabulary of home and place figure prominently in internet language: "home" or "homepage," "navigation," and even the term "cyberspace" contain conceptualizations of place in a virtual environment. If anything, the terms 'diaspora' and 'internet' are in fact more closely related than we think, both sparking images of itineraries without end and presenting the constant vacillation of ideas about what a fixed place and home might be.

Because Rosenstone's book led me to think about the idea that home can have multiple locations, it has also raised the prickly issue of what the notion of a Jewish space might mean. In a special issue of *Jewish Social Studies* on Jewish Conceptions and Practices of Space, Charlotte Elisheva Fonrobert and Vered Shemtov argue for a balance to occur so that space is considered equally along with time in Jewish studies:

> (…) while some work to free Jewish Studies from the hold of the "lachrymose" conception of Jewish history, others believe that it needs to be cut free from the stunting idea that Judaism is first and foremost shaped and governed by the dimension of time. Jewish texts of all periods contain countless descriptions of concrete places where Jews comfortably practiced their rituals, struggled to do so, or were prevented from doing so. The history of Jewish text and ritual is inseparable from the history and nature of such places.[10]

Diaspora, these authors continue, "has often been equated with lack of space, or lack of control over space; the State of Israel has been seen as national territory with a plenitude of sovereignty, as space to be designed as the central Jewish place in the world."[11]

While subjects of diaspora, topography, travel, and architecture do have obvious cross-pollinations in actual space without ever approaching the digital world, this project—one that has considered the interfacings for the internet expressly—makes clear how the best synapses for electrifying the peregrinations and dimensionality of journeying are located in the spaces or slippages between and across them; with each of these terms, the concept of mobility is key. Paradoxes of mobility in virtuality notwithstanding, the web is arguably about mobility, both in the literal sense of our manipulation of the mouse, to the flash screens of software that aim to heighten movement by feverishly shifting our point of reference and sharpening—or as some might lament—weakening the imagination, and where our own creative capacity to connect the dots from one image or screen to the next is, in a sense in our ability to fill in the blanks and close the circle. But more important still, a large part of the intrigue of internet exploration is the promise of dimensionality or the attempt to create (further inventions of) spatial possibilities, continuous ribbons and endless mobility. As Georges Perec has poetically described it,

> This is how space begins, with words only, signs traced on the blank page. To describe space: to name it, to trace it, like those portolano-makers who saturated the coastlines with the names of harbours, the names of capes, the names of inlets, until in the end the land was only separated from the sea by a continuous ribbon of text (…). Space as inventory, space as invention.[12]

Virtual Ending

An open-ended, virtual conclusion—if indeed such a project as this can have one—would ask, what is a specific Jewish experience of place and movement from home to a place elsewhere? What kind of inventory of actual or invented spaces can the internet offer? Can the internet enable a better or different understanding of wandering and Jewish cultural identity and articulate the dimensionality of a diasporic experience? Much like the way looking at maps of cities and places on the internet—particularly Google Earth—has enabled a thoroughly new appreciation of perspective and space, diaspora can be reconfigured through images, text, and sound on the internet, to create another (truly, an-other) kind of spatial experience of place and home.

The value of an internet voyage is that the user can create itineraries through space; or literary musings of movements that track urban patterns where each city is highlighted by its vectors to history (this restaurant, that house, this city). By labeling places or giving names to sites, and identifying places by giving them a visual structure, these sites, instead of non-places or sites of insignificance between other spaces of significance—become "lieux-dit," or said places. Marc Augé designated what he called *non-lieu* or non-place as a site we typically pass through, bypass, and that is void of our engagement with it. "If a place can be defined as relational, historical and concerned with identity," comments Augé, "then a space which cannot be defined as relational, or historical, or concerned with identity will be a non-place.[13] Augé refers to the idea that travel itself captures the sense of a non-place to the extent that there is, what he calls, a "double movement" defined as

> (…) the traveller's movement (…) but also a parallel movement of the landscapes which he catches only in partial glimpses, a series of 'snapshots' piled hurriedly into his memory and, literally, recomposed in the account he gives of them, the sequencing of slides in the commentary he imposes on his entourage when he returns.[14]

Moreover, he points out that much tourism may in fact be the "archetype of non-place."[15] A diasporic journey, here taken as a traveling concept, might suggest then—somewhat peripherally—passing through or lighting upon topographies where there seems to be an absence of engagement or attachment. Yet, upon closer observation, is not a diasporic journey one that brings the sojourner to precisely the obverse: that is, the traveler settles, perhaps even arbitrarily or accidentally, in a place, and takes that place to be the said-place, or the "lieux-dit" by appropriating its culture and structure and by becoming the consumer of and consuming the place itself? For all the indeterminable physicality of virtual space, a spatial topography, a sort of imagined concreteness of place, takes shape in the mind's eye. In this way, with Augé's imperative that travel and tourism is the "archetype of a non-place", it is possible to imagine, with specific reference to diasporic journeys and particularly Jewish diasporic journeys, that travel and tourism can be a place of encounter, creativity and reinvention where identity and meaning take form. Therefore, whereas tourism and travel is considered by Augé to be a non-place, I am suggesting it can be reconceived as a *place*, or a site of relations, history and identity-formation.

Furthermore, in the context of the ethereality of virtual space, this idea of place and place-making is arguably pushed to the extreme. And so it is no surprise that the internet blends seamlessly and interdependently with architecture, which in turn cohabits with literature for this project to label places and create maps. Through its interrelated and layered associations of webpage to webpage, and rollovers to other features that pop up and demand to be noticed, our noses are pushed up against these places in real time where, by clicking on a site of history or memory or fictional fantasy, we gain access to that place today: we have a mapping, a visual display of destinations, pathways and locations sited with colliding chronologies of then and now, that are the substance of the project: sometimes Romania then, sometimes Montreal now. The personal and the private, the public and the shared, are played out on this individual and collective journey, an interactive virtual experience that playfully and provocatively takes up concepts of travel, diaspora, topography and architecture to provide a different and other consideration of the boundaries of time and place.

Notes

1   Robert Rosenstone, *The Man Who Swam Into History: The (Mostly) True Story of My Jewish Family* (Austin: University of Texas Press, 2005), 11.
2   Click on "Pilot Projects" or click on this URL: http://www.yorku.ca/mosaica/man_who/intro/. I am creating this project with a team of undergraduate and graduate students at York University. The primary designer of the site is Kristina Rostorotskaia, a graduate in Design, York University. I want to extend thanks to: Jen Brown, Karen Justl, Allison Ksiazkiewicz, Meaghan Lowe, Marieke Treilhard, and Jenny Western.
3   Rosenstone, *The Man Who Swam Into History*, 15.
4   Henri Lefebvre, *The Production of Space,* trans. Donald Nicholson-Smith (Oxford, UK: Blackwell, 1991).
5   Marcel Proust, *Remembrance of Things Past, Vol. 3* (New York: Random House, 1981), 1098.

6   Philippe Hamon, *Literature and Architecture in Nineteenth-Century France*, trans. Katia Sainson-Frank and Lisa Macguire (Berkeley: University of California Press, 1992), 18–19.

7   These ideas are distilled from two essays by Jonathan Boyarin and Daniel Boyarin, *Powers of Diaspora: Two Essays on the Relevance of Jewish Culture* (Minnesota: University of Minnesota Press, 2002).

8   On virtual Jewish space in the internet-based virtual world of Second Life, see Julian Voloj's article in this volume.

9   Rosenstone, *The Man Who Swam Into History*, 47.

10   Charlotte Elisheva Fonrobert and Vered Shemtov, "Introduction: Jewish Conception and Practices of Space," in "Jewish Conceptions and Practices of Space," *Jewish Social Studies* 11, no. 3 (2005) 1–8; here 4.

11   Ibid., 3.

12   Georges Perec, *Species of Spaces and Other Pieces*, ed. and trans. John Sturrock (1974; repr., London: Penguin Books, 1997), 13.

13   Marc Augé, *Non-Places; Introduction to an Anthropology of Supermodernity*, trans. John Howe (1995; repr., London: Verso, 2006), 77–8.

14   Ibid., 85–86.

15   Ibid., 86.

Part V

Enacted Spaces

15 Foodscapes

The Culinary Landscapes of Russian-Jewish New York

Eve Jochnowitz

I&M International Foods, one of the main food venues for Russian Jewish New Yorkers in Brighton Beach, New York, 2007

The influence of landscape on cuisine has long been a subject of study for ethnographers, food scholars, and ideologues of every stripe. The French famously cite *le goût du terroir*, "the irreplaceable flavor of the soil" or "the taste of place" as being an essential component in the flavor of local foods.[1]

This chapter will address the influence of food on a landscape. I will argue that the foods and foodways of a culture form a landscape of their own. Food makes place. Jews of the former Soviet Union, in bringing their foods and food practices to new homes, have created new places. This culinary landscape, or foodscape, is not just the foods alone; it also includes the traditions of display and performance associated with the food. It includes deliberately and intentionally created aesthetic productions as well as incidental culinary "noise."

For the concept of a foodscape, I am primarily indebted to the musicologist R. Murray Schafer, whose concept of soundscape included deliberately produced "aesthetic sounds" as well as incidental sounds such as overheard conversations, weather, roadwork, and so on.[2] Schaffer's model is particularly useful because of his understanding of the many, possibly infinite, components that combine to create a perceived sensory experience. A foodscape differs from a soundscape in that all senses, rather than just the sense of hearing, are engaged, and in that all these sensory perceptions are related to the ultimate edibility (or inedibility) of the food perceived. I am indebted as well to Jeffrey Shandler, whose definition of a virtual "Yiddishland" provides many useful tools for thinking about how performance creates space and how space itself performs.[3] Work by the geographers David Bell and Gill Valentine on food and space is particularly valuable to anyone seeking to understand the interactions of landscape and eating, and my five-part definition of the foodscape follows their seven-part hierarchy of the geographies of consumption. Bell and Valentine's findings demonstrate that food habits shape one's experience of spatial scale.[4]

Nation and nationalism are central to the French concept of *terroir*, and to the regional foodways of other European countries, but neither Russian nor Jewish nationalism quite fit into the framework of nationalism in the European sense. Russian nationalism is anomalous because Russian nationhood is connected with the multilingual and multi-textured histories of the Soviet Union and before that, the Russian Empire of the Czars. Jewish nationalism, oddly enough, is parallel to Russian Nationalism in this sense, even while the Jewish situation could not be more different from that of an empire. While Jews and Russians may feel deep connection to a place, and powerful impulses toward nationalism, it is not the same phenomenon as the nineteenth century romantic European notion of a consanguineous people united in a contiguous place.

As a diasporic people, Jews are connected culturally, linguistically, and gastronomically to many places, yet are connected to one another by a shared set of dietary laws and a shared liturgical language. Jewish culinary performance, in this sense, can be understood as a medium that provides a virtual presence for people separated in space but proximate in practice. As Barbara Kirshenblatt-Gimblett has argued, the dispersion associated with diaspora, long understood as a pathological condition of displacement from a central or privileged origin, is only half of the diasporic condition. The other half, the production of the local by a rearticulated population in a space of dispersal, is the element of diaspora too frequently neglected in the study of immigrant communities.[5] The commonly voiced concerns about the inauthenticity of the cultural and religious practices of Jews from the former Soviet Union illustrate Kirshenblatt-Gimblett's point. The foods Russophonic Jews are creating in New York are worthy of study. It may be valid to question whether particular innovations, such as "Russian sushi" or whole-wheat versions of traditional breads, are sound or unsound within the context in which one makes them, or to interrogate the reasons for which some home cooks and retailers embrace some innovations while shunning others, but it is irrelevant to argue whether or not they are "authentically Russian" or "authentically Jewish." These phenomena are better understood as an expression of the "independence and stubborn resistance" of what John Bodner calls the "children of capitalism."[6]

The Foodscape

I define a foodscape as consisting of these five separate and partly nested personal sites: the mouth, the body, the kitchen, the table, and the street. My paper will examine each of these culinary landscapes in relation to the ways in which Jews understand and create space, the geographies of Russophonic Jewish New York, and in particular the culinary and place-making practices of the Jews of the former Soviet Union. An examination of Jewish place-making practices, such as the traditions of eruvin and pilgrimage, provide illustrations for understanding the connections of food and space in a larger Jewish context.

This chapter will begin to explain what happens when a phenomenon as inextricably linked to a place as a cuisine continues its existence in a new place entirely. Stemming from Eastern Europe, Georgia, the Caucasus, and Central Asia, the Jewish communities I studied use food to articulate an intense and intimate connection with place, both the place left behind and the place created through sensory and social practices. The cuisines of uprooted and rearticulated Jewish communities selectively retain traditions and innovate at each point in the food system.

The Mouth

All of culture passes through the mouth.[7] As a liminal zone, neither outside nor inside, the mouth is a particularly contested cultural landscape. The mouth is the space within the Jewish tradition that accepts the physical nourishment of food and the spiritual nourishment of Torah. Joel Hecker has found that the mouth has mystical significance in rabbinic culture as the point of entry for both nourishment and Torah.[8] In the Jewish communities of North Africa, a Torah pointer is a teething toy for a child, especially a male child, so that the object that has touched every letter in the Torah can transmit the sacred text to the child when he puts it in his mouth, giving literal meaning to the expression "*Torah she be-al pe*" or "oral law."[9]

The palate is part of the mouth but is also understood as a socially constructed site; that is geographically specific, and possibly even gender- and class-specific. The landscape of one's own mouth, physiological and cultural, determines how all other foodscapes will be perceived. I agree in this case with Amy Trubek's argument that the taste of a place is very much an acclimatization of the palate, and as the agrarian landscape has changed, the landscape of the palate has changed.[10]

While I am not suggesting that there is any difference between a Russian mouth and a mouth belonging to a person from anywhere else, I do feel that I can safely make a case that different groups use and view their mouths differently, or that the mouth occupies a variety of social spaces. Tasting and smelling, chewing and swallowing, speaking, smoking, and spitting are all performances that distinguish the mouthscapes of a culture.[11] Within the community of recent Russian immigrants, smoking is widely considered acceptable among males. Almost no restaurants enforce the city's smoking prohibitions and it is common for men to smoke even while food is being served.

The Body

That food creates the landscape of the body is obvious. All the physical material of one's body and the energy to keep the body functional is made out of food that body has consumed; but one's cells and tendons are not the only components of the body's foodscape. The body, or more precisely, the way in which people see their bodies as needing nourishment (in the form of good food) and protection from harm (through avoidance of bad food) is a culinary landscape that might seem to be unchanged by relocation. In fact, one's body and one's understanding of one's body are socially determined phenomena. Historian Mark Swislocki has noted that a new eating system involves a new concept of what the body is and how it works.[12] In China, for instance, people who adopted the western biomedical understanding of the body and nutrition found themselves inhabiting new bodies. Adopting a new system, however, does not necessarily mean one discards the old. A diner familiar with Chinese food theory can approach a meal and see it both as containing calories, fiber, vitamins, and other elements of food as understood in western medicine, and also as occupying a position on the heating-cooling (Yang-Yin) spectrum, and having various proportions of the five elements as understood by Chinese philosophy. Swislocki describes the people who participate in two eating systems as being "bi-corporal."

"In Russian there is no cholesterol," one restaurateur told me. He was partly joking, of course, but in fact it is no joke. There is a very different Russian body, which requires what it requires according to an internally consistent system. Cholesterol is not one of the components a Russian body encounters when evaluating a food. A popular Russian proverb reminds us that "you can't spoil kasha with butter."[13] While one's Russian body needs fat to be healthy, one's western body needs to avoid it. "Who ever heard there about cholesterol, saturated fats, free radicals, or anything in that vein?" asked one recent immigrant. Similarly, the dieting and health practices that many women commonly followed in the former Soviet Union, such as fasting once a week, or "separated eating" seem alarmingly ineffective and even dangerous to someone familiar with the western biomedical model of the body.[14]

Many instances of bi-corporality exist within the Russian-Jewish foodscapes of New York. The most salient example is that of the social space occupied by fat. This space is very different in Russia, both with regard to fat as an ingredient, something wholesome, delicate, and expensive in the Russian context, unhealthy, cheap, and greasy in the West, and with regard to fat on the human body. Socialist realist art of the Soviet era shows attractive women as being powerful, sturdy, or even plump, while by no means overweight. Within the context of the revolutionary ideal, the valorization of the fuller figure may be related to the primacy given to a woman's work or to the difficulty of maintaining such a figure in times of scarcity, or both. One immigrant ruefully reported: "In Russia we did not have enough to eat because of shortages; here we do not have enough to eat because of calories. What was the point of emigrating?"[15]

Annie Hauck-Lawson has found that women who immigrated to the Americas from Poland have experienced what she calls going "from under-nutrition to over-nutrition." Since emigrating and facing the bewildering choices of a new foodscape, these women find themselves reaching sizes with which they are uncomfortable.[16] Women from Russian-speaking lands face the same difficulty. Their discomfort is due not only to their weight gain, but to the fact that having adopted the bi-corporal model, they are judging themselves against much harsher standards of thinness. Vera Kishinevsky encountered a young woman who can only be understood as bicorporal—

her American body was too fat (the young woman herself reported), while her Russian body was too thin. Her grandmother objected to her weight loss: "You have only your nose left!"[17]

Modern Western Europeans, faced with the prospect of Jewish emancipation and citizenship, imagined a pathologized Jewish body created, at least in part, by Jewish food.[18] This "Jew's Body" was broken, incomplete, incapable of military service, and lacking the productive capacity for full citizenship. To become emancipated, the Jews of eighteenth- and nineteenth century Europe became bicorporal. Judah Leib Gordon's famous prescription is that one ought to be "a Jew in one's home and a man outside," but in some sense a Jew had to understand himself as occupying two bodies both at home and in the street in order to function as an emancipated citizen. Russian popular culture before the revolution also associated Jewish eating, and particularly Jewish consumption of garlic, with the pathology of the Jewish body. Emancipation, however, was not an issue within the Russian empire because no one in Russia, whether Jewish or not, had such an option.

The Kitchen

Culture itself is a kitchen, transforming the "raw," the untamed phenomena of the natural world into the "cooked," the domesticated, the civilized, the cultured. The landscape of the kitchen sees the creation of order out of chaos, and the very creation of this order can reduce the kitchen landscape itself to chaos. For my understanding of the landscape of the kitchen, I follow geographer Maria Elisa Christie, who defines "kitchenspace" as the complex of spaces, both indoors and out, involved in the preparation of food. Activities and relationships, according to Christie, delineate kitchenspace as much as do physical structures.[19] It is in the performances associated with provisioning, food preparation, and cleaning up that the foodscape of a kitchen is created. Deidre Sklar suggests that women (and men as well) perform according to a specific "choreography of the kitchen" when they cook and that this choreography infuses the foods prepared with meaning. Her performance ethnography of the fiesta of Tortugas examines the back-region of the kitchen as well as the front-regions of sacred space.[20]

While perhaps not as intimate as the landscape of one's own body, for many cooks a kitchen is a crucially personal place. For Russian Jewish immigrant women, the kitchen is the seat of one's power and a safe venue to play out one's enthusiasm for innovation while also choosing at times to resist change. One young woman observed, "We try new things we find here and then we make it in our Russian way."

For those old enough to remember the social upheaval that occurred during the middle years of the twentieth century, when the Soviet Union attempted to revolutionize domestic space and work, autonomy in one's own kitchen is not to be taken for granted. The Soviet model of equality for women in the workforce and home required the abolition of the private kitchen entirely.[21]

The crisis that prompted the Soviet Union to attempt to abolish domestic cooking and housekeeping had the opposite effect in the United States, where the work of American housewives in their kitchens was idealized as patriotic, especially in wartime.[22] No one in the former Soviet Union had the choice of working full-time as a homemaker, and women's movements took very different forms from those of the West, where feminism's second wave, heralded by *The Feminine Mystique*, valorized work outside the home as a means to liberation. The modernization of the Soviet kitchen did not lead to less division of labor between the sexes in domestic housekeeping. If anything, males from the Russophonic communities are even more bewildered by the world of cooking than their counterparts in the west.

The kitchen in an urban apartment in Russian-Jewish New York is primarily functional. While many Russophonic immigrants have delightedly adopted ultramodern décor for the rest of the house, they have not followed the Italian-American or the Italian-Canadian practice, noted recently by food scholars and journalists, of outfitting the kitchen so lavishly that any actual cooking and eating has to occur somewhere else, usually a makeshift kitchen and dining room set up in a basement or garage.[23] The Russian kitchen remains a back-region, while the Western kitchen has been moving toward the front of the dwelling since the beginning of the twentieth century, and in the twenty-first century has become the most displayed room in many houses.

The Table

The table is a particularly resonant Jewish landscape. The phrase "the set table" (*shulhan arukh*), is the synecdoche within Jewish practice for the entire complex of Jewish observance. "*Shulhan Arukh*" is best known as the title of the sixteenth-century work that codified Jewish practice, but the term is much older.[24] "A set table" has become a common Jewish vernacular expression describing everyday practice, both sacred and profane.

How the table is to be laid is culturally relevant for Jews of the Former Soviet Union. Non-Russophones frequently remark on the abundance with which Russians set tables at home and in public venues. One Russian-born woman commented: "My husband could not understand why they put more food on the tables than anyone could possibly eat. My father had to explain that you need to feed the eyes—the eyes need to eat as well."[25]

For this woman's Western-born husband, a table laid *à la Russe* is itself an unfamiliar landscape—uncharted territory that is difficult for an explorer to navigate. Her father's comment that "the eyes need to eat as well" illustrates the different geography of the Russian table.

Anthropologist Barbara Myerhoff found performance of precepts rooted in normative Judaism to shape the behavior of the Jews with whom she worked in their struggles and celebrations. A lady Myerhoff called Basha explained that she begins each meal she eats alone in her tiny apartment by laying the table with linen and saying the blessings before eating. "This [what] my mother taught me to do. No matter how poor we would eat off white linen and say the prayers before touching anything to the mouth." Myerhoff argues that Basha is part of a hidden aristocracy. She dines more splendidly on chicken feet than most do on far finer fare. Basha gives an example of acting Jewish that involves both specifically ordained religious actions—in this case, saying the blessing—and the gestural household traditions she learned from her mother. For Basha, the two performances are inseparable and indispensable for a Jewish table.[26]

Within the Ashkenazic context in particular a table thrums with ritual and cultural significance. The *tish*, or "table" is the central event of Hasidic gathering. A *rebe* (Hasidic Rabbi) is said to "preside at a table" (*firn a tish*), when he interacts with his *khsidim*.[27] A *tish* involves food, both real and symbolic, as well as prayer, paraliturgical song, and conversation. On the Jewish holiday of Purim, the table itself becomes the stage on which the Hasidim perform a *pirimshpiyl* (Purim play).[28] "*Es tut zikh oyf tish un oyf benk*" is a common Yiddish expression meaning "Things are really happening." Within Jewish tradition the table is a stage for eating and meal related ritual.

In Russian, the word *stolichny*, or "of the table" can indicate both that a given food or drink is ordinary and that it is indispensable. The preparation of potatoes, meats and vegetables in a matrix of mayonnaise known as *Salat stolichnii*, or *salat Olivier* unites all communities with any connection to the former Soviet Union. The salad itself functions as a matrix, almost a kind of cultural mayonnaise, in the way it binds together peoples separated by language, cuisine and physical distance. *Vodka Stolichnaya*, Russia's most popular vodka, was the most familiar product from the Soviet Union available in the United States during the Cold War years.

The Street

Street food, fast food, cafés, restaurants, and grocery stores comprise a culinary streetscape. When the architect Louis Kahn famously noted, "A street is a room by agreement," he had in mind the synergy created by buildings, businesses, pedestrians, and other travelers and their means of communication. In the neighborhoods of New York, the culinary streetscape engages all senses to bring a walker into a food-world.

In any urban context the street is a significant theater, and in the Jewish context particularly, the street is where observers can most clearly discern the patterns and communications of a neighborhood. The phrase "*di yidishe gas*" (The Jewish street) means not just the street itself, but what is going on inside the homes and even inside the minds of a Jewish community. The expression "*Shikt dayne oyern in di gasn*" or "Send your ears into the streets" means be aware, or find out what is really going on. The many greengrocers' produce spills out onto the streets. *Gastronom* shops, that provide specialty imported foods as well as locally-made pickles and smoked fish, tempt passersby and lure them in with colorful window displays and briny fragrances. Pushcarts sell fried pastries and the many coffeehouses welcome busy pedestrians.

Russian Jewish grocery store on Brighton Beach Avenue/Brighton 5th Street, New York, 2007

Delis and grocers provide immigrant communities with familiar foods and with safe venues to connect with people from the old country as well. Medical sociologist Larissa Remennick has found that it is food businesses in particular that function as social and cultural centers for shoppers seeking more than physical sustenance.[29]

Russian food stores, carrying a wide assortment of familiar groceries, also serve as places of social encounter and information exchange. Many women in my sample said they found work, housing or caretakers for their elderly (if they could afford it) via Russian grocery or bookstores.[30]

In some cases, the imported items are identical to local versions, at least to my untrained eye, but shoppers will choose to select familiar packages.

Jewish Food, Jewish Space

The most essential and indispensable concept in understanding the Jewish approach to space is the unique place making practice of the *eruv*. The word 'eruv', in its most common usage, refers to a boundary between public and private space.[31] Within traditional Jewish practice, the eruv is an important concept in the understanding of space and the creation of place. The Talmud defines three kinds of *Eruvin*, all of them intimately connected to food: the *Eruv hatseroth*, or eruv of courtyards; the *Eruv tavshilin*, or eruv of cooking; and the *Eruv thummim*, or eruv of distance.[32] All three types of eruv require the assembly of various foods. In common usage, the word 'eruv' refers almost exclusively to the *Eruv hatseroth*, which symbolically separates an inner or "private" space from an outer or "public" space. While it has become a convention to use the word 'eruv' to refer to a physical barrier, or a symbolic physical barrier which serves to unite everything that lies within and distinguish this space from the outside, in fact, the word 'eruv' literally means "mixture," and refers to the foods that the residents of an area assemble to unite themselves symbolically as one courtyard. In cases where it might become necessary for a group of householders to define their separate residences and the yards and all the areas between them as one symbolic domain for the purposes of carrying on the Sabbath, neighbors will agree to establish an eruv. It is food that is the essential element for this most Jewish idea of creating space, even while the food collected for the eruv stays in a synagogue or other central location, and is not consumed. As Charlotte E. Fonrobert, a scholar of Talmudic and early modern Jewish culture, has noted:

> The *Eruv* community is first and foremost established by a collection of food. It is food, though not a meal per se that forms the center of the entire ritual system. The food collected is not consumed, but deposited in a suitable location within the confines of the neighborhood. Thus, the point is not actual commensality but a symbolic representation thereof. Indeed, I would venture to suggest that the food operates as a symbolic representation of the community itself. As such, the food serves the purpose of unification and integration of the neighborhood.[33]

The foods that constitute the eruv are defined in the Talmud as being anything other than water and salt, but later commentators make it clear that bread, and only a whole loaf, is the ideal. It is vital to note that the assembly of symbolic foods unites neighbors ritually just as the assembly of actual foods unites them in fact. Whether or not a given Jewish community participates in the establishment of an eruv, the traditional Jewish connection of food to the relationship with one's home and one's neighbors continues to resonate. Neighbors who engage with the foodscape of a place, by buying produce, by adapting their cooking, and by dining out and inviting others in, are creating what I would argue is a secular eruv, an invisible barrier between the public and private, between the sacred and the profane. The foodscape is the territory of the secular eruv.

Foodscapes: The Pilgrimage

Within the Jewish context, food and pilgrimage are first connected in the passages in the Hebrew bible, which prescribe three pilgrimage festivals in the course of a year (Exodus 23:17 and Deuteronomy 16:16). The adult male Israelites who participated in the pilgrimage brought grain, fruit, and meat on the hoof to Jerusalem, where they would subject each foodstuff to the required ritual. They would then eat the food they had brought, minus the portion given to the priests working in the Temple.[34] In the ancient Near East, these pilgrimages served to unite tribes that were geographically scattered and possibly ethnically diverse by creating what Victor Turner understands as "communitas," or a deeply felt bond that transcends the official or formal connections people might have with one another.[35] As in the case of eruvin, pilgrimage unites people by uniting their food.

For the pilgrim, the journey can be as important as the sacred site that is its goal. The novelty of the terrain, including the experience of unfamiliar foods, pleasurable or unpleasant, is part of the process that transports and transforms the traveler: As the pilgrim moves away from his structural involvements at home his route becomes increasingly sacralized at one level and increasingly secularized at another. He meets with more shrines and sacred objects as he advances, but he also encounters more real dangers (…). But all these things are more contractual, more associational, more volitional, more replete with the novel and the unexpected, fuller of possibilities of communitas as secular fellowship and comradeship and sacred communion, than anything he has known at home.[36]

In the modern world, Hasidic tradition ties the concept of biblical pilgrimages to the modern Hasidic custom of making a pilgrimage to a *rebe*. "*Kayn kotsk fort men nisht; kayn kotsk geyt men; vayl kotsk iz dokh bimkoym hamikdesh, un kayn kotsk darf men oyle regel zayn.*" The words of a well-known Hasidic hymn explain "you do not travel to Kock, you walk to Kock, because Kock fulfils the place of the Temple, and pilgrimage to Kock must be made on foot." Journeys on foot, as it happens, are an important element of culinary pilgrimages to Russian-Jewish New York: the culinary walking tours.

Culinary Pilgrimage: The Walking Tour Phenomenon

Walking tours of New York City became popular in the wake of the city's fiscal crisis of 1975–76, when a shared sense of emergency and grievance stirred New Yorkers with local patriotism.[37] One consequence of the civic pride provoked by the crisis was a renewed interest in and curiosity about the city's history and culture among all sorts of laymen for the following thirty years. Culinary walking tours began to multiply shortly afterward. Myra Alperson, who has been leading culinary walking tours and bicycle tours since 1983, reports that people of all ages and backgrounds attend the tours and many repeat their favorite tours several times.[38]

Seth Kamil, who began leading walking tours in 1990, and founded New York's "Big Onion Walking Tours" in 1991, emphasizes the role of the tour in providing broader historical context for a visitor's understanding of a neighborhood while on the other hand providing what he calls a "micro-history" in contrast to the macro-history of text.[39]

Several years ago, in answer to a call from the New School University, I agreed to lead culinary walking tours of New York's Jewish neighborhoods, including "Queensistan," the Central Asian Jewish enclave, located in the Rego Park section of Queens.

The Rego Park tour begins in front of the oldest synagogue in the area with a discussion of the history of the development of the neighborhood and the influence of the New York World's Fair of 1939–1940. Our first stop is a bakery where students sample such treats as walnut *halwa*, fried noodles in honey, *lipeshka* (pronounced lipyoshka), a bialy-shaped *naan* (frequently spelled "non") and *roti-toki*, or shelf-bread, so-named because of its keeping qualities, a matzo-like wafer baked in the shape of a bowl. I have found it is best to schedule eating early and often throughout the tour so that students do not become fatigued and so that they have the feeling of being repeatedly surprised and delighted. After restoring ourselves at the bakery, we take a long walk to 108th Street, Rego Park's central Jewish shopping street. I use the walking time to discuss the history of the Bukharan Jews, how they came to settle Central Asia after leaving what is now Iraq and Iran, and subsequent Muslim and then Soviet domination of the area. On 108th street we visit a Hungarian bakery (a layer of the culinary landscape from a previous wave of Jewish immigration), a *gastronom* (deli/supermarket) and a tableware store and then proceed to lunch.

I was intrigued to learn that my tour is one of scores of well-attended culinary walking tours of New York City neighborhoods. The appeal of a guided tour that puts the culinary in context is clearly widespread among many elements of the population. Pilgrims of all ages and backgrounds enjoy the tours for their safe introduction to a new food-world. They want to sample the flavors of another group, but they do not want to venture alone into an unknown environment. They also do not want to "miss" anything. Ideally, trusted friends introduce one another to new culinary and cultural territory. One student told me, "In New York, I want someone else to tell me what's good." A tour guide in these cases plays the role of a friend from another community. In fact, almost all of the guides leading culinary tours, myself included, are outsiders in all the Jewish communities in which we conduct tours. Seth Kamil suggests that it may even be preferable to have an outsider as a tour guide, because outsiders can provide passion for a neighborhood without the emotional attachment that comes from feeling personally threatened by a neighborhood's changes.[40]

These visitors are seeking more than just a good meal. Folklorist Lucy Long has found that culinary pilgrims seek a deeper and more personal understanding of foodways that are in some way more authentic than those of their everyday lives. Within the Jewish context, the cuisines of the former Soviet Union evoke both places and times of heightened significance. In many cases, the pilgrims' quest is as much for more authentic Judaisms as it is for more authentic foods. Culinary pilgrims to Rego Park are likely to be unfamiliar with the flavors of Bukharan cuisine, and to have no familial connections to central Asia, but for them as well, the tour is a kind of homecoming because they identify with the journey that this new group of New Yorkers has made, recapitulating the phylogeny of the Jewish immigrant experience to the Americas.

Cuisine as Cultural Performance

Cultural performances are the occasions in which as a culture or society we reflect upon and define ourselves.[41] Food is at the center of some of this community's most important events, where collective history and memory are dramatized. Food, in its many manifestations, can be viewed as the prism that refracts all-important cultural concerns into their elemental components. In an unfamiliar setting, as Bell and Valentine have noted, a familiar meal "helps fight off the panics of disorientation."[42]

Even *terroir* itself performs. Gustavo Esteva and Madhu Suri Prakash give us an opportunity to think of virtual *terroir* with the concept of "cultural soil," the virtual soil in which people's experiences are rooted.[43] The imagined rootedness of populations, along with its metaphors, is useful in understanding how people experience food, region, and place. In the case of a population in transition, or to use the titles of two landmark histories of immigration, *The Uprooted* and *The Transplanted*, the food practices of a community throw into relief all of their cultural practices.[44] Taken together, the studies in this article make a clear case that food makes place at least to the same extent that place makes food.

Notes

1   The term *terroir* originally referred to the typical taste of a specific wine region such as Burgundy, Alsace, Bordeaux etc., the French, however, understand *terroir* also in a broader sense encompassing not only the geographical characteristics of a specific wine region and the specific agricultural techniques applied in this area, but also the regional foodways and cuisine. The concept of *terroir* thus integrates a cultural dimension and is tied to regional identity and the European understanding of regionalism.

2   R. Murray Schafer, *The Tuning of the World*, 1st ed. (New York: Knopf, 1977). I am grateful to Lucy Long for bringing this work to my attention.

3   Jeffrey Shandler, *Adventures in Yiddishland: Postvernacular Language and Culture* (Berkeley: University of California Press, 2005).

4   David Bell and Gill Valentine, *Consuming Geographies: We Are Where We Eat* (London, New York: Routledge, 1997).

5   Barbara Kirshenblatt-Gimblett, "Spaces of Dispersal," *Cultural Anthropology* 9, no. 3 (1994): 339–44; here 342.

6   John E. Bodnar, *The Transplanted: A History of Immigrants in Urban America, Interdisciplinary Studies in History* (Bloomington: Indiana University Press, 1985), 216.

7   Susanne Skubal, *Word of Mouth: Food and Fiction after Freud* (New York: Routledge, 2002), 43. Skubal, a literary theorist, finds that the mouth unites "nature and culture, biology and mythology, art and science."

8   Joel Hecker, *Mystical Bodies, Mystical Meals: Eating and Embodiment in Medieval Kabbalah*, Raphael Patai Series in Jewish Folklore and Anthropology (Detroit: Wayne State University Press, 2005).

9   Harvey E. Goldberg, "Torah and Children: Symbolic Aspects of the Reproduction of Jews and Judaism," in *Judaism Viewed from within and from Without*, ed. Harvey E. Goldberg, Suny Series in Anthropology and Judiac Studies (Albany: State University of New York Press, 1987), 107–30; here 114–15.

10   Amy B. Trubek, "Tasting Wisconsin," in *The Restaurants Book: Ethnographies of Where We Eat* (Oxford: Berg Publishers, 2005), 20.

11   Most relevant smelling takes place *inside* the mouth. See Linda Bartoshuk, "The Biological Basis of Food Perception and Acceptance," *Food Quality and Preference* 4 (1993): 21–32.

12   Mark Steven Swislocki, "Feast and Famine in Republican Shanghai: Urban Food Culture, Nutrition, and the State (China)" (PhD diss., Stanford University, 2002).

13   Vladimir Ivanovich Dal', *Poslovitsy Russkogo Naroda: V Dvukh Tomakh* [Proverbs of the Russian people], 2 vols. (Moskva: Khudozh. lit-ra, 1984), quoted in Robert A. Rothstein and Halina Rothstein, "Food in Yiddish and Slavic Folk Culture: A Comparative/Contrastive View," in *Yiddish Language and Culture: Then and Now*, ed. Leonard Jay Greenspoon, Studies in Jewish Civilization (Omaha: Creighton University Press, 1998), 305–28.

14   Vera Kishinevsky, "Survival in the Land of Glamor: The Experience of Three Generations of Women Who Emigrated from the Former Soviet Union (Acculturation in the United States

and Its Influence on Their Perceptions and Lifestyles)" (PhD thesis, New York University, School of Education, 2001), 324.

15   Ibid.

16   Annie Hauck-Lawson, "Foodways of Three Polish-American Families in New York" (PhD diss., New York University School of Education, Health, Nursing, and Arts Professions, 1991).

17   Kishinevsky, "Survival in the Land of Glamor," 292.

18   Sander L. Gilman, *The Jew's Body* (New York: Routledge, 1991).

19   Maria Elisa Christie, "Kitchenspace: Gendered Spaces for Cultural Reproduction, or, Nature in the Everyday Lives of Ordinary Women in Central Mexico" (PhD diss., University of Texas, 2003).

20   Deidre Sklar, *Dancing with the Virgin: Body and Faith in the Fiesta of Tortugas, New Mexico* (Berkeley, Calif.: University of California Press, 2001), 78.

21   Musya Glants and Joyce Toomre, *Food in Russian History and Culture* (Bloomington: Indiana University Press, 1997).

22   Cindy J. Dorfman, "The Garden of Eating: The Carnal Kitchen in Contemporary American Culture," *Feminist Issues* 12, no. 1 (1992): 21–38; here 28.

23   Lara Pascali, "Two Stoves, Two Refrigerators, Due Cucine: The Interplay of Public and Private Spheres in Italian Canadian Homes" (paper presented at the American Folklore Society, Salt Lake City, Utah, October 14, 2004).

24   The code of Jewish law entitled *Shulhan Arukh* by Joseph Caro was published in 1565. A supplement by Moses Isserles (the Remu) covering Ashkenazic practice entitled *Ha-mapah* [The tablecloth] was first included in a published edition of the *Shulhan Arukh* in 1571.

25   Audience member attending the panel "Unorthodox Expressions: Jewish Self-Assertion in Russian Popular and Material Culture" at the Association for Jewish Studies Boston, MA December 21, 2003.

26   Barbara Myerhoff, *Number Our Days: A Triumph of Continuity and Culture among Old People in an Urban Ghetto* (New York: Simon & Schuster, 1980), 22.

27   The term *firn a tish* literally means to "lead a table."

28   Shifra Epstein, "Drama on a Table: The Bobover Hasidim "Piremshpyil," in *Judaism Viewed from within and from Without*, ed. Harvey E. Goldberg (Albany: State University of New York Press, 1987), 195–217.

29   For a detailed discussion of kosher delis and grocery stores in suburban Toronto, see Etan Diamond's article in this volume.

30   Larissa I. Remennick, "All My Life Is One Big Nursing Home": Russian Immigrant Women in Israel Speak About Double Caregiver Stress," *Women's Studies International Forum* 24, no. 6 (2001): 685–700; here 696.

31   For a detailed discussion of the *eruv*, see Manuel Herz's article in this volume.

32   The *Eruv tavshilin* is a plate of two kinds of cooked food prepared before a holiday. The *Eruv thummim* extends the distance a person may walk on the Sabbath.

33   Charlotte Elisheva Fonrobert, "The Political Symbolism of the Eruv," Jewish Conceptions and Practices of Space, *Jewish Social Studies* 11, no.3 (2005): 9–35; here 12.

34   Eleven chapters in the book of Leviticus detail the requirements involved with the ritual offerings, most frequently translated as "sacrifices." While historians differ on whether or not the Israelites did in fact carry out the ritual sacrifices in the manner prescribed in the Pentateuch, the idea of uniting people and food through pilgrimage was clearly the ideal.

35   Victor Witter Turner, *Dramas, Fields, and Metaphors: Symbolic Action in Human Society* (Ithaca: Cornell University Press, 1974).

36   Ibid., 182–83.

37   The front page headline of the New York *Daily News* for October 30, 1975 screamed "Ford to City: 'Drop Dead!'" Milton Glasser's famous and widely imitated "I ♥ NY" logo became ubiquitous shortly thereafter.

38   Myra Alperson, "Welcome to Nosh Walks" (2004); available from http://noshwalks.com (accessed August 8, 2007).

39   Seth Kamil, "Tripping Down Memory Lane: Walking Tours of the Lower East Side," in *Remembering the Lower East Side*, eds. Hasia Diner, Jeffrey Shandler, and Beth Wenger, *The Modern Jewish Experience* (Bloomington: University of Indiana Press, 2000), 226–40; here 232. Interestingly, while Alperson stresses that a walking tour spares the traveler the "hassle" of overseas travel, Kamil points out that a walking tour is more challenging "without the protective walls of a bus or the comfort of watching a videotape at home." (ibid., 230)

40   Ibid.

41   Milton B. Singer, *Semiotics of Cities, Selves, and Cultures: Explorations in Semiotic Anthropology, Approaches to Semiotics* (Berlin: Mouton de Gruyter, 1991).

42   Bell and Valentine, *Consuming Geographies: We Are Where We Eat*, 19.

43   Gustavo Esteva and Madhu Suri Prakash, *Grassroots Post-Modernism: Remaking the Soil of Cultures* (London: Zed Books, 1998), 192.

44   See, for instance, Oscar Handlin, *The Uprooted*, second and enl. ed. (Boston: Little, Brown, 1973) and Bodnar, *The Transplanted*.

16 The Buena Vista Baghdad Club

Negotiating Local, National, and Global Representations of Jewish Iraqi Musicians in Israel

Galeet Dardashti

Yair Dalal teaching Middle Eastern music to the next generation of students at his studio in Jaffa, Israel, 2003

On a cold Monday morning in Pardes Katz, an impoverished neighborhood outside of Tel Aviv, eighteen enthusiasts of Arab music—all Jewish Israelis who immigrated from Iraq and almost all over the age of seventy—gather in one of the Community Center's classrooms.[1] The ambiance is not particularly inspiring for music-making: there is writing on the chalkboard from the group or class that met there last, and only a window-unit air conditioner adorns the plain white walls. Nevertheless, this devoted group has never lacked inspiration, gathering weekly in this room and others like it for over twenty years in order to enjoy some of the musical sounds of their native Iraq. This particular week in 2004, therefore, was much like the group's numerous Monday morning meetings of past years. What did make this gathering different from most, however, was a German television station's filming of the group's session for a special-interest news story on Israel.

Since the meeting that day was already being filmed by a professional television film crew, I was told by Ofer—the only thirty-something regular attendee of the group and the nephew of Elias Shasha, a veteran musician of the "club"—that this would be the perfect day to bring my video camera. In spite of the cameras, the group's behavior did not significantly differ from that of the other two meetings that I had witnessed. There were more in attendance than usual since the appeal of appearing on TV—even in Germany—had attracted some extra friends of the musicians, and those present certainly smiled more and did their best to perform for the camera. But like most other days, both professional and amateur musicians jammed together on some of their favorite classical Arab songs, some much more skillfully than others. Having a TV camera there didn't throw off the group too much since the crew was certainly not the first to demonstrate interest in the musical meetings. Israel's Channels One, Two, and Eight[2] had all paid the group visits in the previous few years, filming their musical meeting and interviewing some of the members. Their stories, however, began over sixty years ago.

***

Some of the Middle East's most highly acclaimed musicians immigrated to Israel in the 1950s, and although some continued performing within the confines of their own ethnic communities, the larger Israeli public never embraced these former celebrities. This reflected the marginalization of Eastern and in particular Arab culture in Israel that pervaded Israeli society. For the first few decades of Israeli existence, the dominant musical media almost entirely excluded most Middle Eastern musical traditions. Many second-generation Israeli Mizrahim (Jews of Middle Eastern and North African background) felt ashamed of their parents' thick Arabic-accented Hebrew and the music they brought with them from their native countries.

During the past twenty years, however, Israelis have become increasingly interested in Middle Eastern music. This paper traverses the physical and discursive landscapes of Iraqi Jewish musicians who immigrated to Israel from Baghdad in the early 1950s and had no choice but to reinvent their musical careers in a completely different urban space. Today, these Mizrahi septuagenarian and octogenarian musicians are only now being asked to share their unique knowledge with newly receptive Israeli students and audiences.

Recently, some of these elderly musicians have become popular topics in the Israeli media. This paper will contrast the discourse of such media concerning the histories of these Jewish-Iraqi musicians with that of their own historical accounts. Utilizing the theoretical musings of Arjun Appadurai, I will address the many complexities implied by reconfigurations of physical space amidst globalization—in this case, the mass immigration of Iraqi-Jews to Israel. Although the emergence of deterritorialized and transient populations is not a new phenomenon, particularly in the Jewish context, the disjunctures caused by recent mass-mediated discourses and practices that now occur throughout and across nation-states as a result of the obsession with constructing locality is new.[3] Most of the films and articles in Israel that address Mizrahi histories are produced by second- and third-generation Israeli Mizrahim with a heightened awareness of ethnic identity politics and a desire to connect to the Arab, Iranian, Turkish, etc. cultural heritage and histories of which they feel they have been robbed. They offer new possibilities for those looking to counter Israeli hegemonic metanarratives that ignore the accomplishments of Mizrahim and the unjust treatment they received upon their immigration to Israel.

The majority of such films, articles, and other publications regarding Iraqi musicians paint a picture of artists who failed and suffered in Israel because the Ashkenazi-dominated Israeli establishment ignored and discriminated against their music. I will argue that this myopic focus on national popularity and discourses of discrimination ignores the "rock-star" status and vibrant musical careers that Iraqis have held within the ethnic spaces that they crafted and maintained for themselves in Israel. Many elderly Iraqi Mizrahim, though delighted that the mainstream has suddenly taken an interest in them, are resentful of their pathetic depiction in the Israeli and now international media.

Drawing on Appadurai's notion of "mediascapes," I will demonstrate how the new opportunities afforded by these new forms of non-state media offer both productive and unproductive potentials.[4] The images of Mizrahi musicians performing in Pardes Katz transmitted onto Israeli televisions throughout the country, for example, produce imagined and blurred notions of localized spaces for a national audience. These images are then picked up by transnational media outlets (such as the German TV station described above) whose audiences are even farther away from the local subjects. A critical problem posed by these new mediascapes, therefore, are the disjunctures between those residing in today's world and the images created to represent them.

Historical Background

Although some of the Baghdadi Jews were religious, most were largely secular and assimilated. In the early to mid-twentieth century, Jews held prominent governmental and economic positions and for the most part maintained good relations with their Muslim neighbors. Jews also held a highly important role in Iraq's musical culture. Of the 250 Baghdadi leading instrumentalists in the 1940s, only three were Muslims. There was a historical reason for this. Islam holds an ambivalent view toward music. During certain periods, Muslims in some countries were prohibited from engaging in music, and it was therefore up to minority groups (Jews, Zoroastrians, Christians) to maintain the musical traditions. The Ottoman government in Iraq in the nineteenth century observed these prohibitions, forbidding Muslims to engage in music. Jews and other minority populations, therefore, particularly flourished musically in Iraq, becoming some of that country's most accomplished instrumentalists.

In 1941, a series of riots known as the "Farhud" broke out in Baghdad in which approximately 200 Jews were murdered[5] and up to 2,000 injured. Many Jews viewed this as an isolated instance of anti-Semitism, and in 1947 there were 118,000 Jews in Iraq. Once Israel achieved statehood in 1948, however, many Iraqi Jews were fired from their jobs and emigration to Israel was declared illegal. In 1950, however, Iraq passed a law allowing Jews to emigrate on condition of relinquishing their Iraqi citizenship. By 1951, 107,603 Jews had renounced their Iraqi citizenship.[6] Many Iraqi Jews were quite successful and wealthy in Iraq but had no choice but to emigrate to Israel, leaving all of their possessions (confiscated by the Iraqi government) behind.[7] Although Israeli immigrant absorption policies sought to promote integration and equality, in reality, Mizrahim experienced inferior housing placement, unequal educational opportunities, and over-representation at the lowest levels of the army and labor force. A disproportionate number of Mizrahim were steered to the squalid *ma'abarot,* the infamous immigrant development towns on the frontier regions of the country, which many never managed to leave.[8] Unlike the majority of Moroccan and Yemenite immigrants to Israel, however, Iraqi immigrants were largely settled in *ma'abarot* relatively close to the center of Israel, including areas such as Pardes Katz.

By the time Mizrahim began immigrating to Israel in the 1950s, Israeli cultural identity had already been "invented" according to the European aesthetics of the Ashkenazim. Mizrahim were "encouraged" to assimilate into the Israeli "melting pot" by abandoning their previous cultural traditions. The exclusion of Middle Eastern cultures was pervasive due to the state-controlled media that Israel established in order to support the formation of the nation-state. In this way, the state could, for example, more successfully imbue an entire nation with a specific musical genre. The early Zionist music was mostly adapted from the styles of Yiddish, Polish, and Russian folk songs, with Hebrew words inserted. "By controlling access to the airwaves, ethnic variation could be minimized, while new national songs helped shape a common national identity."[9]

Many prominent Iraqi immigrant musicians did, however, attain professional positions in the Israel Broadcasting Authority's (IBA) Arabic Orchestra soon after arriving in Israel. This Orchestra was established in 1948 by the IBA's "Voice of Israel (*Kol Israel*) in Arabic" Radio Station. Its Arabic language programming was primarily intended as public diplomacy toward Israel's Arab neighbors across the borders. Its secondary target was its Palestinian population. The Arabic-speaking Mizrahim within Israel was its last priority. Composed of Iraqi musicians and—after 1957—some Egyptian musicians, the IBA's Arabic Orchestra, therefore, served a valuable political function for the state by attracting its listeners to the propaganda programming that followed its high quality Arabic musical interludes.[10] Only those musicians who could perform the more mainstream urban Arabic music from Egypt and Lebanon that had become popular throughout the Arab world, gained steady employment in the Orchestra.

Those musicians whose musical mastery did not serve an expedient national goal, however, found it quite difficult to pursue their music professionally in Israel. A case in point is the history of my grandfather, Yona Dardashti, a nationally renowned master singer of Persian classical music in Iran. In spite of some of the anti-Semitism he experienced early in his career, in the early 1950s Dardashti began performing at Iran's most coveted concert halls, such as *Talare Farhang* and at the Royal Palace for Mohammad Reza Shah Pahlavi, and he garnered a prime-time weekly radio performance with the Iranian National Radio Orchestra. As Iran had only one radio station at the time, this position brought Dardashti instant national fame. He maintained this weekly radio spot for the next fifteen years, until moving to Israel in 1967.

When Dardashti and his wife immigrated to Rishon le-Ziyon, Israel, he found no forum for his music. He did perform on Israeli television and radio once or twice on programs featuring new immigrants, but very seldom performed concerts in Israel. Instead, he traveled back to Iran a few times a year to give performances up until the Islamic Revolution in 1979, after which he could never again return to Iran. Dardashti, therefore, supported himself mostly through various jobs in Rishon le-Ziyon such as gardening and earned a small income giving short performances at Iranian weddings.[11] Many musicians like Yona Dardashti, who specialized in localized musical forms (e.g., Persian, Iraqi, Moroccan) were marginalized, finding few opportunities for performance in Israel, while many of those who could perform the mainstream Arab classical music found jobs through the IBA.

In addition to their work with the Arabic Orchestra, many Iraqi musicians performed at Café Noah, a Baghdadi-style nightclub that opened in the Iraqi neighborhood near Tel Aviv and drew hundreds of primarily Iraqi patrons each weekend. Through their employment on the radio and performances at Café Noah and other such cafes, many of these musicians developed excellent reputations in Israel and abroad, garnering lucrative private gigs in the Iraqi and other Mizrahi communities.

Middle Eastern Music in Israel Becomes En Vogue

In 1997, the Israeli music label Nada—a label that primarily specializes in "ethnic fusions"—released a musical retrospective of the work of Abraham Salman, a Jew born in Iraq in 1931 and considered one of the greatest masters of the *qanun*, the flat wooden Arab zither. This album is one of the many CDs featuring Arab classical music released in Israel beginning in the 1990s. But unlike most of the other recent Middle Eastern-infused world music albums, which were released by a younger generation of Israelis, this one features a 67-year-old musician whose music had never before been available for purchase to the general Israeli public. Salman immigrated to Israel in 1950, and the CD contains music recorded there between 1963 and 1996. Many of the tracks came from the archives of the Israel Broadcasting Authority's (IBA) *Kol Israel* (Voice of Israel) Arabic radio station, as Salman was one of the prominent stars of the IBA's Arabic Orchestra from the time of his immigration to Israel until 1988. The CD, printed in English and Arabic—but not Hebrew, reflecting the CD's primary intended audiences—also includes a page of praise from perhaps Israel's most highly esteemed contemporary musician of Arab music, the Palestinian-Israeli Taiseer Elias, attesting to the "genius," "unbridled imagination," and "extraordinary sensitivity, spirituality and ethereality" characteristic of Salman's musicianship.

A handful of other albums released mostly on Israel's two most successful world music labels (Nana and Magda) from the late 1990s till the present, feature the musical works of other Middle Eastern and North African Jews over the age of seventy. One CD released in 2001 and distributed both in Israel and abroad (with liner notes in English, French, Hebrew, and Arabic) contains the music of Filfel Gorgy, an Iraqi Jew who had already begun a successful career as a singer and composer of Arab classical music when he immigrated to Israel in 1950 at the young age of twenty. He continued his musical pursuits in Israel, quickly becoming a star among Arabic-speaking Mizrahim and Arabs in Israel, as well as among Arabs in Jordan and Iraq, many of whom listened to *Kol Israel* radio in Arabic. He passed away at his home in Israel in 1983 at age 53. This recording marked the first ever collection of his recordings on compact disc.

As of this writing, Magda's most recent release was the 2006 double CD entitled "Their Star Shall Never Fade," a collection of rare recordings by the legendary masters of Iraqi music, the Jewish brothers Saleh and Daoud Al-Kuwaity. During their seventy years of creativity, they composed and performed with some of the most celebrated Arab singers and musicians, including Salima Mourad, Um Kulthum, and Muhammad Abed-el Wahab. When the situation for Jews in Iraq became untenable, the brothers immigrated to Israel in 1950 and performed with *Kol Israel's* Arabic Orchestra. Their musical legacy slowly faded in Iraq, and in 1972, Sadam Hussein officially ordered the Iraqi Broadcasting Authority to erase the names of the Al-Kuwaity brothers from every official publication and from the curricula of the academy of music. From that point on in Iraq, their compositions were labeled "of folk origin."[12] The 2006 CD, which was released in Israel and abroad, offers a broad perspective of their work in cinema (e.g., they recorded music for an Arabic version of *Romeo and Juliet*), court music (they composed and performed a piece for King Faisel's coronation ceremony), and devotional songs. The recordings were restored and re-mastered by the descendants of the two musicians, who live in Israel.

The recent interest in these elderly or deceased Mizrahi musicians is part of a much larger trend of the increasing Israeli interest in traditional and classical Middle Eastern musical styles. As many scholars have noted, Mizrahi-tinged pop music entered the Israeli mainstream in the early 1990s, challenging Eurocentric notions of Israeliness.[13] In the transnational realm, beginning in the late 1980s and early 1990s, music from marginalized communities emerged from its second-class status to take a position at the forefront of the globalized music market in the form of "ethnic" and "world" music. As one Israeli musician told me, "Once the world music craze took hold of Israel, Israelis suddenly discovered that they had their own Arabic, Turkish, and Persian music here." In addition, it became trendy for young Israelis to backpack in the Far East, South America, the Indian Peninsula, or Oceania for several months or up to a year, after their mandatory army service,[14] and there many became exposed to diverse musical traditions. The early nineties was also an optimistic period for peace between Palestinians and Israelis and this, in turn, encouraged an openness toward Arab music. Many Ashkenazi

(European-descended) Israelis began to view the cultures of their Mizrahi brethren—which they had long disparaged—with interest, and many younger Mizrahim began to reexamine their family histories. Since then, a few formal and informal schools have surfaced in Tel Aviv and Jerusalem, offering courses that cover a range of classical Eastern instruments.

In this environment, some of the Iraqi musicians—many of them formerly renowned musicians from Baghdad—became popular subjects in the Israeli media, and to some extent the global media, even being featured in a couple of films. In an interesting conjunction of global space, several of these newspaper articles referred to them as the "Buena Vista Baghdad Club,"[15] a reference to the elderly Cuban musicians of the internationally successful *Buena Vista Social Club* album (1997) and film (1999), which experienced popularity in Israel. In Cuba, it was the American Ry Cooder who claimed to have "discovered" the talented Cuban musicians who had aged and disappeared into obscurity there. Although Cooder asserts in the 1999 film that without his intervention, the musicians would have remained forgotten in Cuba, between 1989 and 1996 alone at least six albums were released featuring the musicians of the *Buena Vista Social Club*.[16] While Cooder did successfully bring the music of the Cuban musicians to an international audience, his claims of discovering a lost national treasure were undoubtedly exaggerated.

In the Israeli case, it was Yair Dalal, a second-generation Mizrahi musician of Iraqi lineage, and Israel's best known emissary of Middle Eastern music in the world music scene who was instrumental in bringing Israel's elderly Iraqi musicians to the wider public's attention. They had taught him the Arab modes and music that eventually brought him fame. As a result, once Dalal achieved stature in Israel and abroad, he wanted to show them his appreciation. Dalal began appearing in high-profile performances with a number of these Iraqis beginning in 2002. The first series of these concerts received a fair amount of media attention.

"Chalrie Baghdad"

In addition to the high profile concerts that the Iraqi immigrant musicians performed with Yair Dalal, their biggest media blitz came from filmmaker Eyal Khalfon, a second-generation Israeli of mixed Mizrahi/Ashkenazi origins, whose 2003 film *Chalrie Baghdad* (Baghdad Bandstand) documents those recent performances in Israel and the lives of some of these Iraqi immigrant musicians. The film was made for Israeli television and was extremely well-received by the Israeli public. It has aired repeatedly and was made available in Israeli movie rental stores such as *Ha-ozen ha-shlishit* (The Third Ear) that carry foreign and other less mainstream films.

The film highlights the lives of six of these Iraqi musicians, who discuss their lives in Iraq and later in Israel after their immigration. The film, however, takes a very clear vantage point in its description of the lives of the musicians it describes, offering little background on the successful careers many of them maintained in Israel. The contrast between the worlds they inhabited in Iraq vs. Israel is stark. The film opens with the following displayed text:

> In the early 50s, during the big wave of immigration from Iraq, several musicians arrived in Israel who were considered stars in the Arabic world of music. These musicians, who later will constitute the basis of the Israeli Broadcasting Authority's legendary Arabic orchestra, will be forgotten as if they'd never been here (…). In the winter of 2002, they reunited — perhaps for the last time — for a series of performances.

One segment early in the film features Avraham Salman, the *qanun* player described earlier. He states:

> In Baghdad, I'm telling you, musicians lived like Kings. Musicians had a good life over there. We used to, but not anymore. I'm an Iraqi. That's it. In Baghdad, the audience understands me, my language. Not here. They don't understand music here. Over there, they'd cry when I played a *muwash*.[17]

These types of comments glorifying the old days in Iraq and disparaging life in Israel are prevalent in the film.

Many contemporary Mizrahi scholars critique the Zionist-European meta-narrative for denying the existence of sophisticated intellectual and artistic cultures among Mizrahim before their immigration to Israel and for failing to acknowledge the victimization of Mizrahim at the hands of the hegemonic Ashkenazi establishment.[18] In line with such Mizrahi scholarship, Khalfon's film critiques and takes this Israeli meta-narrative to task. With a decidedly sad tone, the film underscores the frustrations and disappointments of these master musicians.

Of all the musicians in Khalfon's film, Yousef Shem-Tov features most prominently. He received the nickname Yousef al-Awad (Yousef the oudist) at the age of twelve in Iraq for his virtuosity. The most disheartening part of the film occurs in the middle when the director stages an awkward visit by Salim Al-Nour and his wife to Yousef's home. Al-Nour became known for his compositions of Arab music in Iraq at a young age and later, after his immigration to Israel, composed many pieces for the Arab Orchestra in Israel with which Shem-Tov performed. It is clear that the two musicians have not seen each other in a while. Al-Nour notices that Shem-Tov is not in very good spirits and asks him, "Are you suffering from something?" Shem-Tov responds: "Just from the cigarettes." Al-Nour responds: "We told you to quit." Yousef replies, "I did quit but I'm alone. I have no children. I have no one (…). But at least I have my cigarettes. They are my big brother (…). What will whiskey do for you? Nothing. But a cigarette is a real companion."

Segments like this, which emphasize the miserable fate of the musicians in Israel, feature prominently in the film. A few of the scenes Khalfon chose also show the musicians bickering or screaming at each other in fits of frustration during rehearsals. According to Ofer, the sound engineer for the film and, as mentioned above, also the nephew of Elias Shasha—one of the musical stars of the film—Khalfon had a clear agenda in making the film, one which Ofer did not appreciate:

> Let's say that he was interviewing someone—many of these Iraqis live in respectable houses. He [the filmmaker] would say, we need more noise in the background, more children and cars—as if they live in a depressed neighborhood. It's a manipulation. He wanted to exaggerate the plight of these musicians. This is the story that people want, I guess.[19]

Given the choice to represent an actual Mizrahi space and an imagined stereotyped space, Khalfon chose the latter.

Shem-Tov gets the last word in the film, however, reflecting on his musical career: "I had a bad life in music—not good. I'm not cut out for Israel. I should never have come. I should never have come. This life wasn't right for me. The life I used to have—I was a king. Here, I'm not. I don't feel …" His voice trails off and he lights a cigarette, inhales, and disconsolately looks toward the camera.

The film then cuts to the final sequence featuring part of the celebrated 2002 *Chalrie Baghdad* performance with the film's featured musicians performing "Foug el-nakhil" (Above the palm trees), a lively Arabic love song from the Iraqi classical repertoire that has become popularized throughout the Arab world. The scene moves back and forth between archival footage of the original Israeli Arab Orchestra (with segments featuring some of these same musicians) from the early 1960s and then back to close-ups of the musicians performing in the "Baghdad Bandstand" performance. In spite of the excitement of the performance, the film ends here on a gloomy note. As the beginning of the film stated, after being forgotten for so long, this last flirtation with fame could be their "last performance" together.

Following the film's lead, a 2004 edition of the liberal daily *Ha'aretz*, contains a long article on Yousef Shem Tov. Like Khalfon's film, the article uses the image of cheerless living space to set the mood for later discussion of uncomfortable cultural space:

> Every morning at 10 a.m., Yousef Shem Tov leaves his dismal ground-floor studio apartment and sits on the balcony facing the entrance gate. The balcony isn't exactly a balcony, but a paved area in the backyard of an old building in Ramat Gan—a kind of improvised patio that, thanks to its location, under clotheslines, gets pleasant breezes in the summer. Shem Tov, who will be 80 this year, lights a cigarette and enjoys the breeze.

Lovers of Arabic music still call him by his professional name, 'Yousef al-Oud' (Yousef the oudist), a nickname he received as a wunderkind in Baghdad. Wearing shorts and a short-sleeved shirt, he sits on the balcony motionless, his gaze fixed on the path leading to the backyard. The light in his eyes went out a few months ago, says his sister, Rachel Kamilian, who comes every day and brings him Iraqi delicacies, in an attempt to make her older brother happy.[20]

The article goes on to contrast Shem Tov's terrible plight in Israel with the glory days he recounts in Baghdad. Just as in the film, Shem Tov explains in the article that he had no Zionist inclinations. He was a highly successful musician who performed in the top Baghdadi nightclubs and even on Iraqi radio. But the situation for Jews in Iraq became insufferable after 1947 when the Iraqi government declared the Zionist movement illegal and began arresting Jews. He and his wife and two children were some of the last to leave Iraq in 1952 for Israel and were immediately settled in one of the impoverished *ma'abarot*.

This is where the piece ends. Like the film, the article says very little of the full-time musical career Shem Tov ultimately continued in Israel, or of the pension he receives from the government for his participation in the legendary Arab Orchestra.

Although the film frames the performance at the end as a reunion of many of the musicians who have not performed together in years, as described at the beginning of this article, since 1985, several of these same musicians have gathered weekly on Monday mornings in Pardes Katz. Organized by composer Salim Al-Nur, the group not only performs for its members' own enjoyment, but has also maintained professional engagements in the Iraqi community and at festivals. Yair Dalal learned of the existence of this group a few years after their meetings commenced and went to hear them perform at a Festival in Acco, a city in Northern Israel. Soon thereafter, he called Salim Al-Nur and asked if he could study Arab music theory and *oud*[21] with him. It was not long before he began showing up every Monday morning to make music with them. Today, the group still meets in Pardes Katz, but in a much more informal setting. Guests stop by just to listen, and young aspiring Israeli Middle Eastern musicians periodically drop in to glean some knowledge from these masters.

Omitted Convergences and Disjunctures

Absent from both the film and the article is any mention of how in the context of the recently ethnic-friendly climate in Israel, these elderly Iraqis have indeed become of interest. Aside from having taught Yair Dalal, many have taken on several private students. Zohar Fresco, a thirty-something Israeli and one of the world's most noted Middle Eastern master percussionists, began studying with the older Iraqis in the late 1980s as did the now established singer of Middle Eastern music, Esti Kenan Ofri, an Israeli of Ashkenazi background. Today, many of the musicians born and educated Iraq continue to draw young Israeli students in their twenties.

In 2001, Dalal appeared in a few high-profile performances with several of the noted first-generation Iraqi musicians and the young musician Asaf Zamir. Attending such a concert with the juxtaposition of first-, second-, and third-generation Israelis on stage as well as in the audience is both highly moving and slightly humorous at the same time. Dalal, a second generation Mizrahi, wears fashionable circular spectacles, and he keeps his long, curly black hair pulled back in a fuchsia ponytail holder. His loose-fitting white blouse extends to his knees, and his white flowy pants taper at the ankles with a colorful lining of flowers; his feet are adorned in brown sandals. Seated nearby is the percussionist, Asaf Zamir, a third-generation Iraqi Israeli in his twenties, whose clothing is similar to Dalal's, but all white, with an intricately embroidered white *kippah*[22] on top of his head. Most of the other musicians on stage, however, are first-generation Iraqi immigrants, seated with stern faces and dressed in dark-colored suits and ties, just at they had always dressed for performances throughout their lives.

Like those on stage, audience members represent Israelis from a range of age groups and world-views, including several liberal young Ashkenazim. The performers and audience at the concert capture the unlikely convergence of Israeli hippies and conservative septuagenarians from Arab countries.[23] As El-Ad Gabbai, a young musician of Iraqi background, explained: "The Iraqi musicians are very suspicious about this ethnic thing (…) when they look at the people who are interested in this music they see people with an *oud* in one hand and a joint in the other. This is not what the Sephardim are accustomed to with their traditional ways."[24] In spite of the initial apprehensions of the first-generation Iraqis, many of their students have flourished under their direction. Though the aging of these first-generation master musicians will at some point bring to an end a certain era of Arab music, they have already made their mark on their students and the Israeli music scene.

It is certainly true that the Iraqi musicians were marginalized in the Israeli national realm. The Arab Orchestra did, however, air on Israel's official Arabic radio station, which broadcast regularly both in Israel and throughout the Arab world, where the orchestra was highly regarded. Israel's Arab Orchestra was considered one of the best in the Middle East and many of its musicians will attest that they held rock-star status within many of the Jewish Arab communities in Israel, performing regularly for Jewish and non-Jewish Arabs there and occasionally touring abroad.

The fact that some musicians of classical Arab music continued performing, composing, and in some ways thriving in Israel in an anti-Arab environment should be highlighted rather than obscured. Performing their Arabic music in the private sector in venues such as Café Noah not only played an integral role in mitigating the stress immigrants experienced in Israel,[25] but also defied the Israeli musical hegemony that encouraged Mizrahim to abandon their culture.[26] Mizrahim like Yousef Shem Tov need forums in which to express their feelings of frustration and anger at not being accepted by the Israeli mainstream during those many years when no one listened. But others take offense at being casted as victims by the Israeli media.

Elias Shasha, one of the stars of the "Baghdad Bandstand," who played violin, *oud,* and sang on Iraqi radio before coming to Israel, speaks very fondly of his twelve years in Israel performing with the Inbal Dance Troupe and over twenty years performing with the Israeli Arab Orchestra. As he related to me in Ramat Gan in 2004:

> On one hand, the movie preserved the tradition. People didn't know about me and the others—only through the movie did they learn about us. But in my opinion, there are other things he [the filmmaker] could have focused on. The film depicts us as if our lives were terrible. That's why it emphasizes Yousef Al Awad and his crying. I worked for the Israeli Broadcasting Authority until I retired and now I receive a pension. I earned a living as a musician for forty years. Not an amazing living, but I traveled all over the world and I loved it. I love music. I didn't just do it for the money.

Elias and his friends do talk about how difficult their immigration to Israel was and express some feelings of resentment toward the government for how they and other Mizrahim were treated. But those are not his only memories. He also recounts the bohemian lifestyle he continued living in Israel, where even today women still adore him for his musicianship and charm.

Concluding Thoughts

The myopic focus on national popularity and discourses of discrimination ignores the history of the vibrant musical careers that some Iraqis held within their own communities in Israel, for the history of the Israeli national sphere does not account for all histories of Israel. The fact that Ashkenazim never appreciated the talent of these master musicians was devastating for some, but the lack of approval by the Ashkenazi-dominated society thankfully did not end the careers or feelings of self-worth for many of these musicians. As Ella Shohat has stated, "An essential feature of colonialism is the distortion and even the denial of the history of the colonized."[27] However, what she was critiquing at the time was the way in which academics and the media had depicted Mizrahim as having come from "primitive" societies, when in fact many of them (such as the Iraqis) came from highly urban environments. Such discourse typically contrasted images of their misery in North Africa and the Middle East with the enlightened and blissful faces of the Mizrahim once they arrived in Israel.[28]

What we are seeing today is in some ways a reversal or corrective of this metanarrative. Khalfon's film is a part of a recent trend in Mizrahi cinema in which filmmakers seek to commemorate their characters' Arab culture and exilic past.[29] These films epitomize this second and third generation's own anxious perception that, with the aging of the Mizrahi immigrant generation, a whole chapter of history and culture integral to their own identities will be lost. As in Khalfon's film, many of these documentary films reflect the political vantage point of the filmmaker much more than they do of the characters themselves. As Shemer notes, "at times the film becomes a stage where the subjects and the filmmaker are pulling in two different directions along the connective/separating axis of the Arab-Jewish hyphenated identity."[30] This "imagined nostalgia,"[31] or nostalgia for things that never were, can be quite powerful. In much of the Israeli media today, Iraq of yore is now romantically depicted as a kind of paradise, in contrast to the miserable and neglected lives of Mizrahi Jews of Israel. Neither scenario brings us closer to understanding the reality.

Appadurai's analysis of global cultural flows, particularly his notion of "mediascapes," is of particular relevance here. He refers to the imagined land*scapes* of people and groups spread throughout today's unstable world. "Mediascapes" describe the production and distribution of electronic information (reflecting all different interests) to local, national, and transnational audiences. What is most important about mediascapes is the narrative they convey to their audiences:

> he lines between the realistic and the fictional landscapes they see are blurred, so that the farther away these audiences are from the direct experiences of metropolitan life, the more likely they are to construct imagined worlds that are chimerical, aesthetic, even fantastic objects, particularly if assessed by the criteria of some other perspective, some other imagined world.[32]

An integral problem, therefore, posed by the new landscapes of our globalized existence, are the disjunctures between the residents of today's diverse world, and the mediascapes created to represent them. The mediascapes created from the imaginations of second and third-generation Iraqi intellectuals and artists—many of whom are Mizrahi activists and desperately wish to connect with the Arab heritage from which most feel alienated—do not always correspond to the lives that their parents and grandparents lived in Iraq or Israel.

Why, with over a year's worth of footage of these Mizrahi musicians, did Khalfon choose to take the particular perspective he did in his film? In *Imaginary Homelands*, Salman Rushdie argues that describing history through any artistic medium is itself a political act. He notes that the African American writer Richard Wright was moved to write fiction when he found black and white descriptions of American society incompatible in the 1940s. For writers and artists who often write against the establishment, then, "redescribing a world is the necessary first step towards changing it."[33] This "redescribing" certainly adopts a specific vantage point, for representations of the past are largely shaped by present day concerns. Although one can sympathize with Rushdie's advocation of utilizing creative license for political means, one must acknowledge that these artistic histories sometimes include the lives of real subjects who may find this "redescription" of their lives denigrating. Rather than celebrating the virtuosity of these Iraqi musicians, one that continues to this day, the current "imagined nostalgia" of younger Mizrahim unfortunately depicts them as recently excavated but no longer relevant cultural relics.

Telling the stories of the way in which some of these Mizrahim were able to continue thriving as musicians within a society that denigrated or ignored their music is as important as telling the sad stories of musicians like my Iranian grandfather, Yona Dardashti, who found no opportunities for performance in Israel. As Yehudah Shenhav has noted, "The history of the Arab Jews[34] and their immigration to Israel is complex and cannot be subsumed within a facile explanation."[35] A large number of recent works today that strive to write the histories of Mizrahi immigrants to Israel (whether in the press, presented in fiction, or as the subject of academic inquiry) focus on the stories of their failure in Israel, thereby critiquing the nationalist project as a totalizing force. It is a myth to imagine that the Israeli nationalist project of shaking the diasporic remnants from the immigrants and immediately turning them into homogenized "Israelis" upon their arrival in Israel actually occurred.

As Amnon Raz-Krakotzkin notes, "To write history from the oriental (as opposed to orientalist) perspective means to write against the present still de-Arabized consciousness, while at the same time overcoming romantic-idealistic perceptions of the Orient that inevitably reproduce the dichotomy."[36] Rather than falling back into essentialist conceptions, I argue for a Mizrahi historical paradigm that neither romanticizes the lives of Mizrahim in the diaspora before immigrating to Israel, nor depicts all Mizrahi immigrants within Israel as pitiful victims. It is in this space that some Mizrahi immigrants—in spite of the Israeli Eurocentric hegemonic order—are able to flourish in Israel. It is important to remember, as Shohat, Shenhav, and others have rightly emphasized, that the history of Mizrahi Jews did not begin in Israel even though this representation typified many early Israeli narratives. But a concept of equal importance is that Mizrahi history in Israel did not begin when Ashkenazim began to notice the Mizrahim. Similarly, although the American Cooder helped bring the Cuban musicians of *Buena Vista Social Club* to an international market, his patronizing claims to have unearthed these musicians from obscurity discounts their professional music careers in Cuba. Although the narrators (filmmakers, scholars, journalists) of some of these recent Mizrahi historical mediascapes clearly intend to expose the unjust history of Israeli Eurocentrism, the sometimes narrow scope of their historical lens only further reifies Eurocentric visions of Israeli national history.

***

After the German television station finished documenting the Iraqis' music-making session at Pardes Katz on that Monday morning, they began conducting interviews with some of the individual musicians. The singer Suzanne Sharaban, known as Iman, was interviewed and featured as one of the stars of the film *Chalrie Baghdad*. Although Iman never had the opportunity to sing professionally when she lived in Iraq, she began a vocal career singing in Arabic in Israel. As she told her German interviewer in Hebrew: "When I started singing in Israel, the audiences treated me very well, and the audience is very important for a singer." The interviewer interrupted her, stating, "But you perform here without an audience." Iman then motioned to the other Iraqis sitting in the room, "These are audience members too."

Notes

1   This article draws from the ethnographic fieldwork I conducted on Middle Eastern music in Israel both for my master's thesis and doctoral dissertation through the University of Texas at Austin from 1999–2004. I am grateful for the fellowships I received to support my research and writing from Fulbright-Hays, The National Foundation for Jewish Culture, The Memorial Foundation for Jewish Culture and from the University of Texas at Austin. Many thanks to the numerous Israeli musicians who took the time to tell me their stories during my fieldwork. In addition to the editors of this book, Michal Hamo, Tom Guthrie, Rachel Zaslow, Mason Weisz, and members of my dissertation writing group in New York City all provided invaluable comments, for which I am truly indebted.

2   These three television channels are quite diverse. Channel one is the state-funded public television channel; channel two is a commercial channel; and channel eight is a specialized cable only documentary channel.

3   Arjun Appadurai, *Modernity at Large: Cultural Dimensions of Globalization* (Minnesota: Univesity of Minnesota Press, 1996), 199.

4   Ibid., 194.

5   Sources vary widely on this number, and some list the number as up to 600 murdered.

6   Haim Saadoun, *Iraq* [Hebrew] (Jerusalem: Ministry of Education and Culture and Ben-Zvi Institute, 2002), 38.

7   It is worth mentioning that some Mizrahi activists and scholars assert that the Israeli government played a major role in hastening the surge of emigration of Jewish Iraqis to Israel by setting off bombs between 1950–1951 in Jewish areas to convince Jews that it was no longer safe for them to stay in Iraq. There is still no clear evidence indicating who planted these bombs though the Israeli government has denied any connection to these incidents. See Yehuda Shenhav, *The Arab Jews: A Postcolonial Reading of Nationalism, Religion and Ethnicity* (Stanford: Stanford University Press, 2006).

8   Eliezer Ben-Rafael and Stephen Sharot, *Ethnicity, Religion and Class in Israeli Society* (Cambridge: Cambridge University Press, 1991).

9   Amy Horowitz, "Musika Yam Tikhonit Yisraelit = Israeli Mediteranean Music: Cultural Boundaries and Disputed Territories" (PhD diss., University of Pennsylvania, 1994), 87.

10   Inbal Perlson, "Ha-mosdot ha-musikalim shel ha-mehagrim me-arzot ha-islam ba-shanim ha-rishonot shel medinat israel" [The musical institutions of the emigrants from the Islamic countries during the first years of the state of Israel] (PhD diss., Tel Aviv University, 2000; preface in English); Esther Warkov, "Revitalization of Iraqi-Jewish Instrumental Traditions in Israel: The Persistent Centrality of an Outsider Tradition," *Asian Music* 17 (1986): 9–31.

11   Galeet Dardashti, "Yonah Dardashti," in *Iran*, ed. Haim Saadoun (Jerusalem: Ministry of Education and Culture and Ben-Zvi Institute, 2006), 177; Bruno Nettl and Amnon Shiloah, "Emigration: The Radif in Israel," in *The Radif of Persian Music: Studies of Structure and Cultural Context*, ed. Bruno Nettl (Chicago: Elephant and Cat, 1987), 145–149.

12   Erez Schweitzer, "From the King's Palace to a 'Ghetto' of Oriental Music," *Ha'aretz* (online edition), June 5, 2006.

13   See, for example, Amy Horowitz, "Performance in Disputed Territory: Israeli Mediteranean Music," *Musical Performance* 1, no.3 (1997): 43–53; Motti Regev and Edwin Seroussi, *Popular Music and National Culture in Israel* (Berkeley: University of California Press, 2004; Motti Regev, "The Musical Soundscape as a Contest Area: 'Oriental Music' and Israeli Popular Music," *Media, Culture, and Society* 8, no. 3 (1986): 343–355.

14   On Israeli backpackers in the Far East and elsewhere, see Erik Cohen's article in this volume.

15   See, for example, Aviva Luri, "Buena Vista Baghdad Club," *Ha'aretz* (online edition), June 23, 2000, 82–87.

16   Dan Sharp, "Imperialist Nostalgia and Cultural Nationalism in *Buena Vista Social Club*" (paper presented at the American Anthropological Association Centenial Meeting, New Orleans, LA, November 2002), 15.

17   A *muwash* is a long instrumental improvisatory piece in Arab music.

18   See, for example, Shenhav, *The Arab Jews*; Sami Shalom Chetrit, "Mizrahi Politics in Israel: Between Integration and Alternative," *Journal of Palestine Studies* (Autumn 2000): 51–65; Ella Shohat, "Sephardim in Israel: Zionism from the Standpoint of Its Jewish Victims," *Social Text* 19/20 (Autumn, 1988): 1–35.

19   Ofer, Interview with author, Tel Aviv, 2004.

20   Dalia Karpel, "Art, etc./The Player," *Ha'aretz* (online edition), Oct. 1, 2004.

21   An Arabic lute.

22   'Kippah' is the Hebrew word for the Jewish head-covering worn by religious men (and some contemporary liberal women).

23   Galeet Dardashti, "Discourses of the Middle Eastern Musical Aesthetic in Israel" (MA thesis, University of Texas at Austin, 2001).

24   El-Ad Gabbai, interview with author, Jerusalem, July 2000.

25   Warkov, "Revitalization of Iraqi-Jewish Instrumental Traditions in Israel," 19.

26   Perlson, "Ha-mosdot ha-musikalim shel ha-mehagrim me-arzot ha-islam," 17.

27 Ella Shohat, "Sephardim in Israel: Zionism from the Standpoint of Its Jewish Victims," *Social Text* 19, no. 20 (Autumn 1988): 1–35; 7.

28 Ibid.

29 See for example the documentaries *Cinema mizrayim* [Cinema Egypt], Rami Kimchi, 2003, *Mama Faiza*, Sigalit Banai, 2002; *Fantazya sheli* [My fantasy], Duki Dror, 2001; and the mockumentary *Bayit*, David Ofek, 1994, and features such as *Ha-Mangalistim* [The Barbecue People], David Ofek and Yossi Madmony, 2003 and *Kikar ha-halomot* [Desperado Square], Beni Torati, 2001. For a more lengthy discussion of these films and of Mizrahi Salvage Cinema, see Yaron Shemer's dissertation "Identity, Place, and Subversion in Contemporary Mizrahi Cinema in Israel" (PhD diss., University of Texas at Austin, 2005).

30 Ibid., 149.

31 Appadurai, *Modernity at Large*, 77.

32 Ibid., 35.

33 Salman Rushdie, *Imaginary Homelands: Essays and Criticism 1981–1991* (London: Granta Books in association with Viking Penguin, 1991), 14.

34 For political reasons, Shenhav utilizes the term 'Arab Jew' to replace the term 'Mizrahi'. Although Iranians are not considered 'Arabs', in Shenhav's model, this detail is not crucial.

35 Yehuda Shenhav, *The Arab Jews: A Postcolonial Reading of Nationalism, Religion and Ethnicity* (Stanford: Stanford University Press, 2006), 200.

36 Amnon Raz-Krakotzkin, "The Zionist Return to the West: Zionist Revisionism as a Mediterranean Ideology," in *Orientalism and the Jews*, eds. Ivan Davidson and Derek J. Penslar (Waltham, Massachussetts: Brandeis University Press, 2005), 162–181; here 180.

17 Mini Israel

The Israeli Place between the Global and the Miniature

Michael Feige

Shoreline of Tel Aviv at Mini Israel, 2007

Introduction: Some Mini-Ethnographies

A soldier stands in front of a miniature model of the Israeli Officer School in Mini Israel, where little toy cadets are marching in an endless loop. He jumps to attention and salutes them. His action can be interpreted in two opposite ways. He may be displaying his deep, heartfelt respect and admiration, replacing the toy soldiers for "real" ones; he may also be jesting, thereby subtly criticizing militarism, the practice of saluting, or maybe the model's presumptuousness to truthfully represent social reality.

Religious students visiting Yad Vashem at Mini Israel, 2006

In one of my visits to Mini Israel, the manager showed me a new mini-building, a synagogue about to be "opened to the public." "Too bad you cannot be here tomorrow," he smiled. "There is going to be a great ceremony. A rabbi is coming with his yeshiva students to place a mezuzah on the door." Up until that point I had not been aware that mini-buildings needed (mini?) mezuzot. The manager did not mention a ceremony of placing a miniature Torah book in the synagogue, but is it too far-fetched an idea to imagine? An organized group of religious yeshiva students gathers round the miniature model of Yad Vashem, the Holocaust memorial. One of them places his foot on the concrete platform surrounding the model. "Take your foot down!"

shouts the guide. "Don't you know that the place where you stand is holy?" In Israel, Yad Vashem, albeit a secular institution, enjoys sacred status. What about its miniature model? And the miniature Western Wall? And miniature mosques and monasteries? Should the miniature Knesset and High Court be respected and revered? Is the aura of the large place transformed to small places that look similar? Is there a mini-sacredness and mini-holiness that should be respected by full-sized visitors?

Going around mini Temple Mount, the blurred lines between the macro and the micro, the full-size and the miniature, cease to amuse. There is a transparent plastic cover protecting the miniature Muslim prayers from the vengeance of full-sized, very real Jewish extremists. "They throw rocks at us, I'm going to throw rocks at them," was the excuse given, heightening the absurdity of mixing the real and its miniscule representation.[1] The site's manager told me in frustration that he is tired of replacing the ruined figures, which is a costly affair as well. The non-fanatic visitors are requested to look at the Temple Mount and see the harmony of Muslims praying on the mountain above, Jews praying at the Western Wall below, their recorded chants intermingling, while somehow ignoring the rude intrusion of the "outside" world, symbolized by the protection the site offers to its endangered miniature people.

Mini Israel's slogan, "see it all...small," invites a critical semiotic examination, especially of the meaning of "all" (its Hebrew slogan reads "the entire country in one little city," which is even more enticing). The site is a meta-place, including within its boundaries representations of "all"—space and money permitting—worthy places in Israel. It can therefore inform us of the principles behind place-making in Israel today, the symbolic hierarchies among them, and especially of the formation of national identity in the global age.

The site, located near Latroun, halfway between Tel Aviv and Jerusalem, was opened with great ceremony and after much delay in 2002, and quickly became one of the most popular tourist attractions in Israel.[2] It was conceived and constructed as a private enterprise, linked to a network of miniature cities around the world, of which the most famous is Madourodam in the Netherlands. Mini Israel prides itself on its 300 models of well-known, mostly monumental structures from around Israel, built to a scale of 25:1, and thousands of human figures, some of them animated. The entire compound is shaped like a big Star of David, with miniature airport and kibbutz in the center and various Israeli regions (Tel Aviv, the Coastal Plain, Galilee and the Golan Heights, Jerusalem, the Negev, and Eilat), located in the six tips. Israel's flora is represented through miniature Bonsai trees.

The popularity of the place can partly be attributed to its location, promising easy access from Israeli metropolitan centers. Its playful nature, with toy cars, trains, planes and people, a small football field and basketball court, and cut-to-size buildings, makes it a magnet for families and children.[3] Foreign tourists and pilgrims, to whom the site addresses itself explicitly, use the place to commence or conclude their tour, or, in any case, to get a sense of the state as a whole. Strolling through the grounds, however, one can see how the place affects adult Israeli visitors, and I shall concentrate on this category of visitors.

Map of Mini-Israel (official map of the site), 2007

The site can be understood as part of the post-hiking phenomenon: Israelis, like their counterparts around the world, tend to prefer the invented and artificial, the simulcra in Baudrillard's terms, over actual visits to places considered "authentic" and "real."[4] As hiking was part of nation-building, Zionist rituals,[5] the success of Mini Israel reflects the transformation of Israeli society into a postideological era. In accordance with current theories of authentic reproductions,[6] I want to suggest a complementary interpretation: a visit to Mini Israel constitutes an inner journey into Israel and Israeliness, to a sense of pristine existence for which many Israelis long and feel that they have lost. The former general-manager of the Ministry of Education, Ronit Tirosh, captured this feeling when she wrote in the visitor's book: "Well done on the amazingly beautiful initiative, Zionist in spirit and connecting in soul. This is Israel, the most beautiful of nations. You returned to us the feeling of excitement and pride."[7] This article aims at explaining how this theme park can be understood as a national pilgrimage center in modern Israeli society.[8]

Israel (Mis-)represented

Mini Israel is a monument of and to the present, representing and "commemorating" a state in existence. While visitors come to historical museums to learn about lost civilizations, they come to Mini Israel to learn about a present one, and examine miniature versions of places, which with some effort, they could see "for real." Places no longer in existence are not represented in the mini city, and in one case, a site not yet built, the Egged Bus Museum, is represented. The site museumizes, and therefore glorifies, a present state and society.

Allowing for place and cost limitations, the site promises a true representation of what is important and meaningful in Israel. The general principles used to choose places were formulated and presented by Eran Gazit, the entrepreneur, in the guide book:

> Mini Israel is a vision come true. We wanted to create a unique platform that will express everything beautiful, good, important, and special that we have in this wonderful country of ours. We aspired to create a kind of living 'picture album' that will reflect the reality of life in this small strip of land.[9]

Here is a first selection principle, which is problematic because it presents itself as banal and self-explanatory: "everything beautiful, good, important, and special."

Almost nowhere on the site, nor in the accompanying texts, is there a reflexive discussion of the principles guiding the actual selection of sites, or in other words, how the beautiful, good, important, and special were determined. Through its ignorance of selection procedures and criteria, the site depoliticizes national representation. Assuming that the act of selection is detached from questions of sectarian interests and power struggles, its only criteria are technical, namely, how similar the miniature is to the original. In guide books and films made specifically to be shown at the site, visitors can learn how teams of experts constructed the models, but nowhere can they find a discussion of what makes a particular place worthwhile.

The attempt to depoliticize the site is evident in its general structure: the main routes create a huge Star of David, encompassing the entire grounds.

Similar sites around the world are constructed in the shape of the map of the respective countries, and visitors simulate a downsized walking of the land. The chosen sites in these mini cities are located roughly according to their respective location on the actual map.

Mini Israel, a declaratively copycat site, could not choose this obvious solution for two reasons. One has to do with the concept of a worthwhile site. As 60 percent of Israel's national territory consists of the "site-less" Negev desert, a different solution, allowing more space for the abundance of accepted sites in the Coastal Plain and Jerusalem, had to be conceived. The triangles of the Star of David make it possible to artificially enlarge the parts given to selected areas and minimize regions declared as peripheral. Without explicit declaration, Mini Israel reproduces, and even strengthens, images of center and periphery that have been nurtured and accepted throughout Israeli history. Mini Israel does not represent solely monumental structures and points of interest, but also social conceptions and biases regarding the relative importance of places and regions.

A second reason that "site in the shape of the map" was unworkable is that creating the site in the shape of a map of Israel would require the entrepreneurs to specify their position regarding the most divisive issue in Israeli politics, namely, national borders and the status of occupied territories. Choosing the Star of David, a unifying national and religious symbol, was therefore considered an ingenious, nonpolitical decision, defusing an explosive situation. From a deconstructive perspective, this solution is anything but nonpolitical; based on hegemonic conceptions of the nature of the state, it naturalizes and hides power relations, thereby assisting in making them invisible, taken-for-granted, and thereby, legitimate.

The visitors to the site symbolically enter a Jewish, Zionist, and Israeli ethnocracy, where the land is predefined by an ethno-religious symbol.[10] Muslim and Christian sites are included inside the triangles of the Star of David. While the eventual map of the state is not precisely declared, the principles defining the nation, and therefore constructing the map, are clearly stated.

However, the creators of the site could not avoid making some highly controversial decisions regarding borders and territories. In accordance with Israeli public opinion regarding national borders (yet in total defiance of international law), the Golan Heights are part of the mini city, as are Eastern Jerusalem, including the Old City. The rest of the West Bank and the Gaza Strip are hardly represented, apart from, significantly, the Cave of Machpela (Hebron) and Rachel's Tomb (near Bethlehem). Those holy sites are extracted from their geographical context and float over their harsh geopolitical reality. Today, the original sites are fortified and heavily guarded, thereby reflecting the ethno-religious conflict between Israel and the Palestinians; this change in their appearance is not reflected in the miniature version, which conserves the "peacetime" way that they looked years ago.

Not represented are other holy or monumental sites from the West Bank, such as the Church of Nativity in Bethlehem. Neither are Palestinian cities and towns, refugee camps, or even Jewish settlements. While actual Israel is still struggling with historical decisions regarding its future, its miniature version has already, so it seems, disengaged from the territories. Right-wing journalist Meir Uziel jokingly suggested that the Jewish settlers imitate their actual political behavior by sneaking into Mini Israel during the night and establishing illegal miniature outposts on one of the mini hills.

Larry Abramson, in a pioneering article about the site, observed that the seemingly apolitical decisions reached by the site's entrepreneurs is in actual fact a clear declaration of the possibilities open in Israel's future. To his mind, a viable "complete" Israel can only exist within the boundaries of the Green Line, and the site, unwittingly, exemplifies this principle. The absurdity of the occupation is beyond representation. He claims that Mini Israel sacrifices Greater Israel in order to hold on to a logical notion of a state:

> Against the paradigm of borderless "Maxi Israel", we can see "Mini Israel" as a suggestion for an alternative theoretical model, where the border returns to serve as a constructive principle: a model of an Israeli utopia, whose existence is dependent upon a clear boundary (…). On the ground, as much as in cognition, 'Mini-Israel' offers the utopia of a Jewish democratic Israel as a possibility that can exist in actual fact only in terms of a return to the restrictions of defined and fenced borders.[11]

Over and above exploring the basic structure of the site, each and every miniscule model can raise questions of selection and representation. The most meaningful of these is probably the placement of Holocaust memorial Yad Vashem, as it embodies the centrality of the Holocaust catastrophe in Israeli public life. Two main issues make the transformation from the full size to the miniature problematic. The first is the soothing experience of standing in front of a miniature, deemed inappropriate when faced with the symbol of unimaginable horror. As Susan Stewart explained in her exposition of the concept of longing,[12] grand buildings represent religious and national grandeur and evoke awe, while miniatures create a relaxing, cozy, homey feeling. The "real" Yad Vashem was created with the intent of instilling a sense of religious-like reverence toward the memory of the Holocaust, and its size was an indispensable part of its design. A miniature of the same buildings loses this significance. No wonder that the yeshiva student who placed his foot on the frame around the model was not aware that he had engaged in an act of profanation.

A second problem is that by placing Yad Vashem, and by implication, the memory of the Holocaust, in the small triangle dedicated to Jerusalem,

Mini Israeli marginalizes and localizes the nation-constituting event. The geographical center of Mini Israel is not Jerusalem, the state's capital, but rather, maybe significantly and maybe not, the airport. To somehow counter the problem, Yad Vashem was "pushed" toward the center of the mini city, elevated above its environment, and surrounded by a concrete wall. While miniature Yad Vashem is still within the space allotted to Jerusalem, it is also "extra-territorial," overlooking the rest of the small land. Responding to some small extent to the problem of representing great significance using miniature structures, the site forces visitors to make a mild climb in order to reach Yad Vashem, where they are encapsulated in a somewhat closed space, enabling reflection and solitude (not that many visitors choose to use the opportunity).

As Mini Israel became, in a sense, a pantheon of symbolically important Israeli places, having a model placed in the site became a proof of worth, and the point was not lost on heads of institutions and city mayors, who requested or demanded a fitting representation of their place, even being willing to subsidize the costly construction of their favorite miniature.[13] The site's manager, working with another set of criteria regarding entry standards to his site, turns down most requests, thereby reproducing accepted notions of symbolic value in Israel. Major Israeli cities, such as Bat Yam or Holon, are ipso facto defined to be faceless and devoid of national interest.

The selection of sites is guided by the complex interconnection of politics and aesthetics, and much of Israel's social diversity is relegated to the backstage.[14] No deprived development towns can be seen, no Arab towns, (neither within the Green Line boundaries nor in the occupied territories), and for that matter there is no militaristic presence other than the Mitzpe Ramon officer's school, practically demilitarized, where cadets march in their endless loop, never actually graduating. Deviance and social control are nonexistent in the mini city. In choosing to highlight certain aspects regarded as positive, while hiding the negative and problematic, Mini Israel follows the logic of mass tourism and other mechanisms of national representation.

Still, being represented does not mean being represented "correctly," and the city of Beer Sheva can serve as one good example. As is the case with other Israeli cities, major landmarks picked to represent the city are extracted from the cityscape, while the city itself, with its main political institutions and sociological characteristics, practically "disappears" from view. Its Bedouin past was selected to represent Beer Sheva. Bedouins live mostly in poor towns and unrecognized villages outside the city limits, while the city itself has a predominantly Jewish population, which is served by some monumental buildings, such as those of the local university. Mini Israel vouched for the exotic, "forcing" on the city an identity alien to most of its inhabitants. Ironically, the Bedouin sites do not even represent contemporary

Bedouin society, most of whose members live elsewhere and whose modes of existence have changed considerably under Israel's rule.

According to its manager, Mini Israel does not give into the pressures to provide representation according to local patriotic interests. It may allow itself this partial autonomy due to generous financial support from private firms, and those are indeed represented handsomely. The only hotel portrayed along the Tel Aviv shore belongs to the Dan chain, and on its wall there is a list of all the other Dan hotels; the only newspaper and book publisher in the mini city is Yediot Aharonot; industrial plants exemplifying Israeli industry belong to sponsors Coca Cola and Tambur; and the only model that has no referent outside Mini Israel is the Mini Israel bus company's Egged Museum, whose buses are also the only ones seen on the mini city's roads. For their support, Israeli and international firms receive publicity that is hard to quantify: belonging to the pantheon of canonical Israeli sites, on a par with the Western Wall, the Knesset and Massada. Larry Abramson comments that the representation of giant corporations in Mini Israel is "a symbol of Israel's economic might and its desire to merge in the global capitalist market, but no less a symbol of the power of the giant corporations themselves, which by endorsing Mini Israel imply their endorsement of the state as a whole."[15]

Abramson went even further in 2006 when he arranged and curated an alternative exhibition, also called "mini Israel," for the Israel Museum in Jerusalem. As against the centralized organization of the tourist site, Abramson's exhibition brought together 70 models of forty-five different artists. Instead of showing "everything beautiful, good, important, and special," the exhibition selected places and events according to the interests of the artists and curator, some openly critical of contemporary Israeli society. For example, Mayan Strauss's work, "Settlement evacuation in playmobile," shows miniature bulldozers and cranes destroying a house, with settlers and policemen in the background. Ravit Cohen-Gat and Moshe Gerstel created a wall 10 feet high, which is one-third the height of the barrier wall erected roughly along the Green Line and through Jerusalem, and called their model "Next year in built Jerusalem." This subversive exhibition, presented in the Israel Museum of all places, exemplifies an alternative route, one not taken by the creators of Mini Israel.

Heterotopia and Homotopia: Representing a Purer Israel

Historian Pierre Nora wrote of *lieux de mémoire*, sites of memory, not living yet not quite dead, existing in the margins of the contemporary world, representing the past in the present.[16] The Mini Israel Park is also cast betwixt and between: it is a city defined by its creators and operators as "living," yet its life is based on technological and mechanical principles. Its liminal position provides the imagination with a wide space within which to operate.

The tension between life and death is most apparent in the decontextualization of place. Taking only the monumental, and representing it in the minutest details, the national miniature city surrenders claims of reproducing a semblance of mundane social life.[17] The Knesset and the High Court, for example, are in "true life" working institutions, not only monumental buildings; the labor done there and the meaning of these places to the Israeli public can hardly be represented properly out of scale and out of context. The Particle Accelerator and President Weitzman's tomb in the city of Rehovot are presented, but the city itself is missing. Public buildings exist without their public. Therefore, the implied hierarchy between the various sites is in terms of architectural values, such as which is more elaborate, or modern, or simply higher and larger. In its reproduction, Israeli society is reduced to its bare physicality.

The decontextualization is most evident in the heterotopic spaces,[18] which in theory should connect various locations and cut across conflicting narratives. According to Michel Foucault, heterotopia is capable of juxtaposing in a single real place several spaces, several sites that are in themselves incompatible. Among his examples were the brothel, colonies, and the ship, and to attest to their importance, he claimed that in a civilization without boats dreams dry up and the police takes the place of the romantic pirates. More relevant to our case are the airport and the highway. In the Mini Israel airport, planes wander aimlessly, never leaving the grounds, not connecting the city with other cities. On the highway, cars and buses travel in endless loops. Everything repeats itself, doomed to exist in a circular world, even though the essence of the site, especially the heterotopic spaces it includes, are transient in nature.

Playing freely with Michel Foucault's insightful term, heterotopia is replaced by homotopia. If a heterotopia is a place that includes and juxtaposes various places, thus enabling, if not enforcing, dialogue, hybridity, and reflexivity, homotopia is the place closed within itself, not allowing other places to penetrate its hermetic wholeness. Mini Israel is at its best in constructing stable, unchanging, uniform locations, populated with the appropriate people. The Arab is dressed in his traditional attire and prays at the Mosque, and the Haredi Jew is dressed accordingly and lives in his designated neighborhood. Nearing the Kibbutz, the visitor will hear cows and folk songs and see dancing in the barn, as if the Kibbutz had never undergone industrialization and privatization. The Bedouins are in their traditional tents, where in actual fact very few still live. Everyone in this city plays according to his or her stereotypic self. Mini Israel allows very

*Boy visiting old
Tel Aviv at Mini
Israel, 2006*

little space for the strange and hybrid, social experimentation and marginal figures, let alone the deviant. It is a celebration of the representative.

Mini Israel, through its overly respectful attitude toward typical fixed identities, portrayed its symbolized object as, first and foremost, a liberal, multicultural society. As the various religious and ethnic groups—Jewish and others—are represented through their folkloristic attributes rather than by their socioeconomic dynamics and political demands, the cultural diversity is seen to hold no threatening potential, especially as the buildings and people are so small, and each locked within its designated venue. Even hegemonic Jewish Israel, religious as secular, is folklorized, and limited to being a mere homotopia. The large secular city of Tel Aviv, the Kibbutz, and the Haredi neighborhood are each characterized by buildings, typical behaviors, and distinctive attire. Rather than publicly displaying conflicting cultural and opposing political positions, they are presented as identity capsules where natives happily reside. Mini Israel depoliticizes actual Israel, dismantles conflicting claims on land, history, and identity, and, through the lack of working heterotopias, precludes the possibility of inner change.

Guides assure visitors that each can find his or herself in one of the thousands of human figures. Few of the visitors, however, possess such a shallow, one-dimensional identity. Unlike the pretentiousness of the site, typical visitors would hold multiple, at times contrasting identities. Ironically, the place itself is a classical Foucauldian heterotopia in more than one sense. More than any other place in Israel, it combines conflicting places in close proximity, and the different sounds emerging from them intermingle in the ears of the visitors. It is also a heterotopia in the sense that it brings different visitors—Israelis and tourists—to the same location, where they can meet and interact. The site, therefore, creates a third space of meeting and encounter, while assuming that other places do not have similar characteristics.

The Nation and its Image: "The Picture of Dorian Gray" in Reverse

In *The Picture of Dorian Gray*, Oscar Wilde tells the story of a man and his representation, a portrait made by a painter.[19] As Dorian Gray turns to crime, deviance, and corruption, the picture permutates, adding gruesome features representative of his moral decline. The actual person, having the picture to absorb his misdeeds, remains as young and handsome as ever. The allure of Mini Israel is the reverse idea of Oscar Wilde's disturbing tale: while the state and nation is considered to be in constant crisis, mired in corruption and conflict, morally suspect and with an uncertain future, the picture painted by Mini Israel remains as perfect as the people of Israel would want it to be.

Credibly representing Israel (or any other state for that matter), with all its inner stresses, ethnic tensions, and political strivings is never easy, especially not today. In the past, Israel was portrayed through the progressive Zionist narrative, using stories in which the desert bloomed, immigrants were absorbed, and kibbutzim thrived. With the weakening of the Zionist ethos, these symbols could no longer be used with unequivocal confidence, as they fail to represent Israel's multicultural social reality. Zionist ideology has been criticized on various grounds, and the basis for representation used in the past is seen today as naïve, if not insensitive to historical wrongs done to various social groups. Israel's position vis-à-vis the Palestinians is an inherent hitch in a simplistic nationalistic representation. In the eyes of many around the world, including a certain portion of Israeli citizens, Israel is defined through antagonist terms such as military occupation, colonialism, and aggressive settlement. Another factor problematizing representation is developed identity politics, whereby various social groups attempt to redefine the Israeli symbolic center according to their worldview. Consequently, decisions regarding communal representation are always suspect of either neglecting or misrepresenting some part of society, and a common accepted narrative, never an easy task, is currently more difficult to come by.

Mini Israel caters to the longing most Israelis share of returning to a simpler, more pristine reality, when the unequivocal and unproblematic positive representation of the beauty of the land and nation were still deemed possible. Obviously, it is a nostalgically imagined past; even when Israel could, with some conviction, present a pioneering ethos, a future-oriented spirit, values of social equality and moral righteousness vis-à-vis its neighbors, these images were strongly contested within and outside of Israel, and part of the criticism was the coercion and hypocrisy needed to hold on to them. Mini Israel's great achievement is in being able to connect to the desires of Israelis while still holding a defensible claim to viable representation. This achievement comes at a price.

Mini Israel expresses respect towards Israel's diversity, but without referring to its potential for eruption, and without connecting social groups to their expressed goals of transforming society. When the promising or threatening possibility of change is inherent in the structure itself, such as in a West Bank settlement, the model is simply absent. In other cases within the mini city, homotopia insures that the identities represented in the site, such as the Muslim-Arab or the Haredi, stay within the confines of their designated spaces, and remain unthreatening and benign.

At the Temple Mount complex, Jews and Muslims pray in close proximity. In "real life," this explosive situation brought, on some occasions, major eruptions of violence; in Mini Israel, the encounter is stripped of its aggressive potential and portrayed as a utopian co-existence. The miniature site demands that its visitors suspend their time perspective, not to remember occasions when harmony was brutally shattered, and not to fantasize about periods when the others' presence shall cease to exist. The possibility of a diachronic progressive narrative, of Judaization or Arabization of the land, is not hinted at on the site. While it is no surprise that the Palestinian narrative of return receives no mention, the Zionist progressive story is also no more than hinted at for those attentive enough to notice.[20] A perpetual present is the dominating order.

Mini Israel's Israel looks more "perfect" than the original, cleaner, more peaceful, and better at representing a utopian liberal vision. The conflicts of mere actual Israel are left outside the gates of the model. A far cry from modern Israel, as critically described by its citizens, the place is run efficiently by a young, energetic entrepreneur. "We have no car accidents, no strikes, no drought, no long lines at the Ministry of the Interior," Gazit, the manager and entrepreneur, half joked.[21] Little wonder that visitors feel that they are visiting a "real" Israel, one that suits their hopes and dreams better than the actual state of which they are citizens.

As Mini Israel is assumed to be perfect, and holds no conflict among competing identities, some visitors, holding a different vision of the state, "volunteer" to complete the missing parts. Desecrating the holy places of Muslims and Christians—as if the miniatures were actual holy shrines—introduces Israeli identity politics at its harshest into the site and precludes its attempts to create an extraterritorial space, symbolically outside and overlooking the actual state of Israel. While the plastic shield protecting the Muslim prayers is a security device that visitors are supposed to disregard, it represents Israel to no lesser extent (possible even more) than the rest of the park. It gives rise to the question of whether in Israel, or the Middle East for that matter, there can be an escapist space exempt from the violent logic of national ideology and conflict.[22]

To conclude, let us return to the basic fascination the site holds for Israelis. What makes the site so popular is the tension between entering and exiting the land. On the one hand, the ticket buyers receive a technological fantasy à la Disney and a tourist attraction like the ones so enjoyed abroad. On the other hand, they embark on a patriotic visit—or maybe revisit—to their beloved homeland, and while the site's staged authenticity cannot be concealed, it communicates an essence that is more real than reality itself. In Mini Israel, the Israeli visitor finds the beautiful, eternal, unthreatening face of the Promised Land, where religions and ethnic groups live together in harmony, holding their clear and fixed place and identity, all under the canopy of the Star of David. Israel is presented as a peaceful member of a friendly world, without surrendering its uniqueness. As Israelis enter Mini Israel they concomitantly leave behind the real Israel, with all of its unsolved political, economic, and security troubles, and make pilgrimage to a true and pristine Israel, the one in which they would choose to live. If soldiers salute, Rabbis place mezuzot, and right-wing extremists feel the urge to desecrate Muslim holy shrines, Israeli visitors take their visit there very seriously indeed.

Notes

1 Jaimal Yogis, "Big trouble in Mini Israel," Columbia News Service, April 19, 2005, http://jscms.jrn.columbia.edu/cns/2005-04-19/yogis-miniisrael/ (accessed August 8, 2007).

2 According to Ministry of Tourism statistics, in the year of its opening, attendance was second only to Hamat Gader in the Golan Heights; however, Hamat Gader serves medical tourism as well. Meanwhile, attendance has dropped dramatically. See "Dun's 100: Israel' Largest Enterprises 2006," http://www.dundb.co.il/duns100heb/tour.asp (accessed August 8, 2007).

3 This should be stated with some reservation: with today's hyper-real technologies and images, children and young adults tend to be easily bored with the place, and the managers struggle with ideas to keep them occupied. The safe environment makes the site appealing to caretakers of young children, especially schoolteachers on a one-day trip.

4 Jean Baudrillard, *Selected Writings*, ed. Mark Poster (Stanford: Stanford University Press, 1988), 166–184.

5 On hiking in Israel, see: Orit Ben-David, "Tiyul (hike) as an act of consecration of space," in *Grasping Land: Space and Place in Contemporary Israeli Discourse and Experience*, eds. Eyal Ben-Ari and Yoram Bilu (New York: State University of New York Press, 1997), 129–146.

6 The term 'authentic reproduction' was suggested by Edward Bruner for living history museums, where actors reenact historical events and everyday behaviors in a reconstructed setting. As these tourist sites claim to be selling authenticity, Bruner debates the meaning of the term. See Edward M. Bruner, "Abraham Lincoln as authentic reproduction: A critique of postmodernism," *American Anthropologist* 96, no. 2 (1994): 397–415. For further examples, see: Erik Cohen, "Authenticity and Commoditization in Tourism," *Annals of Tourism Research* 15 (1988): 371–386. Richard Handler and Eric Gable, *The New History in an Old Museum: Creating the Past at Colonial Williamsburg* (NC: Duke University Press, 1997); Ning Wang, *Tourism and Modernity* (London: Pergamon, 2000).

7 Written on December 8, 2005. All translations in this article are mine.

8 The research is based on several ethnographic visits to Mini Israel and collection of material from various sources, including publicity brochures, newspapers, visitor book, and the internet. As part of a course on Israeli places, students visited the site and handed in a report. I am especially thankful to my MA student Nirit Zilberstein, whose thesis was on Israeli hiking, for a long ethnographical account. Since many Israelis have visited the sites, I received numerous anecdotes and insights from family and friends, and all deserve thanks.

9 This English translation appears also on the site's internet site. See: http://www.minisrael.co.il/home_en.html.

10 The term 'ethnocracy' was developed in the Israeli context by Oren Yiftachel. See: *Ethnocracy: Land And Identity Politics in Israel/Palestine* (Philadelphia: University of Pennsylvania Press, 2006).

11 Larry Abramson, "Mini Israel: Ha-ir ha-idealit ha-israelit" [The Israeli ideal city], in *Ha-ir ha-israelit: Ha-ir ha-ivrit ha-aharona?* [The Israeli city: The last Hebrew city?], eds. Oded Hilbruner and Michael Levin (Tel Aviv: Resling, 2006): 212.

12 Susan Stewart, *On Longing: Narratives of the Miniature, the Gigantic, the Souvenir, the Collection* (Baltimore: The John Hopkins University Press, 1984).

13 Told to me in conversation with manager Haim Rugatga, February 6, 2006.

14 The front and backstage in this context are terms taken from Dean MacCannell, *The Tourist: A New Theory of the Leisure Class* (Berkeley: University of California Press, 1973).

15 Abramson, "Mini Israel."

16   Pierre Nora, "Between Memory and History," in id. and Lawrence D. Kritzman, *Realms of Memory: Rethinking the French Past* (New York: Columbia University Press, 1996).

17   Other miniature city versions, such as Legoland or the virtual Simcity, tend to pay attention and even concentrate on the mundane social functions, such as gas stations, local shops, and regular homes.

18   Michel Foucault, "Of Other Spaces," *Diacritics* (1986): 22–27.

19   Oscar Wilde, *The Picture of Dorian Gray* (London: Oxford University Press, 1974).

20   There is a building site and one can make the claim that the very beauty of the site is a proof for the progress brought by Zionism. For a similar argument in a different context, see: Dani Rabinowitz, "The Visualization of a National Narrative of Space," *Visual Anthropology* 6 (1994): 381–393.

21   "Atar hadash: Mini Israel," *Ynet* (Tayarut), http://www.ynet.co.il/articles/1,7340,L-2115375,00.html (accessed August 8, 2007).

22   I have addressed this question in a different article, in reference to shopping malls and the "exterritorial" status Tel Aviv wishes to hold vis-à-vis the Israeli-Palestinian conflict. See: Michael Feige, "The Names of the Place: New Historiography in Tamar Berger's Dionysus at the Center," *Israel Studies Review* 19, no.2 (2004): 54–74.

Epilogue

18 Virtual Jewish Topography

The Genesis of Jewish (Second) Life

Julian Voloj

Opening of the exhibition, "Forgotten Heritage," in the Tachles Gallery at the 2Life Building (screenshot), 2007

The word "topography" describes the configuration of a surface and the relations among its human-made and natural features. Derived from the Greek words for "place" (topos) and "writing" (graphia), topography is a precise, detailed study of a region, down to and including local history and culture. I therefore understand a Jewish topography as the study of local Jewish history and culture; a minority culture that occupies a space between assimilation and tradition.

Jewish topography is often the topography of a metropolis: New York's Lower East Side, Thessalonica, the "Jerusalem of the Balkans," or Barrio Once in Buenos Aires. While these places are widely known, there are also less well-known places with exotic and obscure names such as Cuscus, Nessus or Choi, virtual spaces that are part of a futuristic and at the same time real Jewish topography.

These places are graphic animations in the world of Second Life, and they are a challenge for topographical research. How does one describe a place that does not "really" exist and that can be changed by a simple mouse-click? And how does one describe a culture in transition?

As an artist, Jewish topography and the exploration of the ways Judaism reinvents itself are recurrent themes in my work.[1] In September 2006, when a friend introduced me to Second Life just before he became its Reuters correspondent,[2] the idea of a virtual, parallel world sounded very strange to me, and to a certain extent it still does, but out of curiosity I decided to explore this Second Life.

Even if I did not expect to find anything, I was still curious to see if there was something like virtual Judaism in Second Life, and I soon discovered Temple Beth Israel, Second Life's only synagogue at the time. Needless to say, I was fascinated by the existence of a virtual Jewish community and it was clear to me that I had to document this phenomenon. The timing could not have been better since my introduction to Second Life coincided with the introduction of Judaism to the virtual world. I was therefore able to witness the development of Second Life Judaism from its genesis onwards, and became an integral part of the community.

This chapter is a first retrospective—analyzing the phenomenon of virtual Judaism—and a description of its status quo, creating what in German is call a *Momentaufnahme*, close to the first anniversary of Second Life's Jewish community.

What is Second Life?

In order to better understand the phenomenon of virtual Jewish identity, one has first to understand the phenomenon of virtual worlds in general and of Second Life in particular.

A virtual world is a three-dimensional Web simulation, a kind of virtual community in which people interact with each other through self-created virtual alter egos; so called "avatars,"[3] who communicate with each other via "instant messaging." Second Life is not the only virtual world; other examples are Entropia, the Sims, or vLES [virtual Lower East Side]. Their 3-D animations give virtual worlds a fascinating, realistic feeling, and the online-encyclopaedia Wikipedia compares the phenomenon with the early days of the internet.[4] According to Misha Kobrin, a 32-year-old computer programmer from Russia who lives in Cologne, Germany, the success of virtual worlds has to do with their concept.

> It is a step ahead of traditional chats. Because of the anonymous character of chats, people are more open and it is easier to communicate with strangers. But in [virtual worlds], you are not anonymous; you create a different identity. It does not matter how you look or where you come from. You reinvent yourself.[5]

Linden Labs, based in San Francisco, created the virtual world of Second Life in 2003 and has continued to develop it. The user software allows its "residents" to modify their world independently, and in principle Second Life works just like real life. Second Life has its own currency, the Linden Dollar, which can be converted into "real" currencies, and one can buy items (i.e., clothing, accessories, and/or real estate), travel, start businesses, meet people, etc. Each resident is a co-developer of this world.[6]

"The only thing I say to people (…) is that you're not going to have an opportunity to hide from this phenomenon,"[7] said Linden Lab founder and CEO Philip Rosedale in an interview with *Rolling Stone Magazine*, and the numbers seem to underline Rosedale's analysis. Second Life reached its tipping point on October 18, 2006, when the population had reached one million. By the end of 2006 the population had doubled. In June 2007, close to eight million users were registered, and in terms of square miles, Second Life was about the same size as New York City.

Second Life managed to turn a 3-D animated chat-room into an alternative reality in which Mia Farrow held a virtual press conference to stop the genocide in Darfur,[8] the French right wing party, Front National, infamously became the first European party to open an office in Second Life,[9] and in which a German school teacher started to buy and develop virtual real estate and became the first Second Life entrepreneur to achieve a net worth exceeding one million US dollars.[10]

A Torah Got Carried Away: The Genesis of Second Life Judaism

Judaism in Second Life can be traced back to a single creation: Temple Beth Israel, the Second Life Synagogue created in August 2006 by Beth Odets, the avatar of then 33-year-old Beth Wolanow-Brown, a mother-of-two living in Houston, Texas. Brown/Odets describes the events that led to the creation of her synagogue in *2Life Magazine*:

> [S]omeone found a Torah in Second Life. When I found out about it, I was curious to see it. It was a very weak replica. I don't know where it came from. It did not have textures or anything indicating its creator. I only knew that I could do better, and started building. Before I knew it, I was on the Web looking for a Hebrew texture and half an hour later the Torah was done. Now, having a Torah, I needed a place to put it, and I started to build the Ark. Long story short, what started as "correcting" a Torah became the whole Temple Beth Israel, the Second Life Synagogue (SLS). I have often said that SLS is a "Torah that got carried away," and that's what it is.[11]

With the creation of the synagogue, Brown/Odets brought Judaism into Second Life. The origin of the virtual Torah was never discovered, but this virtual replica inspired the creation of the first Jewish space in Second Life, and gave the Jewish community of Second Life its own foundation myth. With the creation of Temple Beth Israel, Beth Odets became the Matriarch of the Jewish community in Second Life.

Virtual Geography

The establishment of a structure alone could, of course, not lead to the foundation of a Jewish community in Second Life. In order to understand the importance of Temple Beth Israel, one has to understand the geography of the virtual world of Second Life.

The world of Second Life is laid out on a grid of little parcels, regions the residents in general refer to as Sims. The name stands for "simulator" since originally one server or simulator held one entire region. The content of the Sims can be listed in a directory where avatar names, interest groups, and events can also be found. Earlier versions of Second Life had so-called "hubs" from which users were able to teleport themselves from one Sim to another. Good "neighborhoods" were therefore determined by "public transportation," meaning proximity to a major hub, or place of interest, which would guarantee business owners more traffic. The location of virtual real estate mattered: the closer to a hub a place was located, the more accessible and easier to find it was, and therefore the more expensive.

By summer 2006, it was possible to teleport yourself directly from one location to another by just finding a place in the Second Life directory and clicking on the "teleport" option. Easier traveling via teleportation changed the perspective on geography in Second Life. With the option to teleport oneself to any place in the directory, real estate value and — more importantly — location were no longer key factors. It became important only that a place be listed in the directory.

Beth Odets in Temple Beth Israel (screenshot), 2007

Odets listed her synagogue in the Second Life directory shortly after the establishment of Temple Beth Israel, and within a day the first avatars "discovered" the synagogue. In order to better communicate with visitors, Odets decided to create a free membership by establishing a Second Life Synagogue (SLS) group, the membership functioning as a mailing list for distributing information. In less than two months, the synagogue had more than one hundred members.

The First Jewish Events

"When Sukkot arrived," Odets recalls, "I decided to build a Sukkah and to host there the first Jewish event, probably the first Jewish event ever in Second Life. Shortly after Sukkot, there were already one million residents in Second Life."[12] In December 2006, Odets put up Hanukkah menorahs. "We lit each menorah at a different time to allow the international crowd accessibility." The Hanukkah lightings were a "huge success," with dozens of avatars participating, and became the forerunner for the "Shabbat Around The World" candle lighting ceremonies that Odets started in January 2007, and which has since become a recurring Friday event.

According to Second Life blogger Drown Pharaoh, in real life a Muslim from West Yorkshire named Yunus Yakoub Islam who is documenting religion in Second Life, the success of the synagogue had "something to do with the increasing self-confidence of Second Life avatars. When you're in Second Life for a while, your avatar becomes more and more your real self, and you start looking in Second Life for a place that really means something to you."[13]

The introduction of Jewish life into the virtual world was therefore part of a general trend to initiate spiritual experiences in Second Life, where Buddhist temples, cathedrals, mosques, and other places of "real life" religions already existed alongside religious parodies such as the Church of Elvis.[14]

Growing into a Community

Inspired by the success of the Second Life Synagogue, GruvenReuven Greenberg—offline Reuven Fischer, a Hasidic Jew in his mid-forties from rural Pennsylvania—created a replica of Jerusalem's Western Wall in January 2007, where visitors could download the weekly *Parasha* and get information about the teachings of Chabad Rabbi Menachem Mendel Schneerson. In March 2007, then 33-year-old Keith Dannenfelser (as his avatar, Carter Giacobini) created the Jewish Historic Museum and Synagogue, a complex that also included Second Life's first Holocaust Memorial.[15] In April 2007, Australian Jeremy Finkelstein, in Second Life known as Cryptomorph Lake, held the first virtual Seder in Second Life.[16] On April 15, 2007, Yom Ha-Shoah, the first virtual Holocaust commemoration was held at the Holocaust Museum.[17] Second Life's Jewish community started to boom, and as a reaction, the Jüdische Medien AG (JMAG) Switzerland offered me the opportunity to launch a publication dealing exclusively with Jewish arts and culture in Second Life.[18] In January 2007, I had published the first article ever written about Second Life Judaism for the Swiss magazine *Tachles*, a JMAG publication, bringing wider attention to the small community.[19]

With every article, the interest in Second Life's virtual Judaism grew, as well as the membership of Temple Beth Israel and other virtual Jewish venues. The creation of a Jewish Second Life magazine was therefore a logical consequence and, as JMAG editor-in-chief Yves Kugelmann explained, an "investment in innovation and the future."[20] "2Life Magazine," the name a play on the words "Second Life" and the Hebrew expression *l'haim* ("to life"), became the central voice of Second Life's Jewry, and through the magazine the now generally accepted term for Jewish Second Lifers, Javatars, was introduced.

Who are the Javatars of Second Life?

Second Life's "Virtual Diaspora,"[21] which also includes a significant number of Israelis, is dominated by Americans.[22] The dominance of Americans can be partially explained by the roots of Second Life: Linden Lab, the company that invented Second Life, is based in California, and Second Life was initially an American phenomenon, with English as the *lingua franca*. On the other hand, the strong presence of Americans has to do with the general advancement of internet use in North America, which consequently explains why there is such a high percentage of Israelis using Second Life, since Israel is known as a high-tech country.[23] The main reason for the high percentage of North Americans, however, lies in the nature of the Jewish Diaspora. With more than five million Jews living in the United States, it makes sense that American Jews have taken a leading role in Second Life's Jewish community.

But since a computer with internet access is the only requirement for joining Second Life, the Jewish community has a uniquely international spirit, with Jewish participants coming from around the globe. On any given Friday, Second Life users from Australia, Brazil, France, Germany, or Switzerland can be found at Temple Beth Israel.[24]

When at the end of the nineteenth/beginning of the twentieth century Jewish immigrants arrived in New York's Lower East Side, they created so-called *Landsmanshaftn*, mutual aid immigrant societies based on their places of origin. In Second Life, new "immigrants" don't see the need to create these *Landsmanshaftn,* but become part of a truly global society with English as the *lingua franca* and Judaism in its diversity the common cultural ground.

California-based Tamara Cogan (avatar name: TamaraEden Zinnemann) is fascinated by the diversity of Second Life's Jewish community:

> The beautiful thing about Second Life's Jewish community is that I've met people of literally every stream of Judaism. We have Reform and Conservative. We have Modern Orthodox and traditionalists. We even have, which surprised me most, people from ultra Chasidic communities who come and explore and interact with people with whom, in their real

lives, they would never have the chance to interact with. Perhaps a place like Second Life will be the start of many communities, from both ends of the spectrum, to reach out and step outside their worlds, embracing and learning about the most beautiful part of Judaism: our diversity.[25]

Second Life's Jewish community is more diverse than any Jewish community could be in the real world, but there are also many religious seekers who use Second Life as a tool to explore their own roots, many of them with little or no Jewish educational background; and there are interestingly enough also many non-Jews. While there are Jewish users of Second Life who choose not to be Jewish in the virtual space, we find at the same time non-Jews for whom Second Life offers the opportunity to explore Judaism in virtual space. Judaism in Second Life is a choice.

Twenty-six-year-old Derek Goldman from Austin, Texas, online known under his alter ego of Avram Leven, chooses to be Jewish in Second Life:

> Jewish life in Second Life is very important to me. I am constantly amazed by the different social aspects of the virtual world, especially how personal faith is expressed and celebrated. I am not someone who lives his life any differently in real life versus Second Life, so being Jewish in SL is a natural thing for me. My "virtual Judaism" is simply an extension of my RL practices and beliefs.[26]

Similar also is the identity of RebMoshe Zapedzki, in real life Moshe Dreyfuss, an Orthodox Jew from Baltimore: "I wouldn't know how *not* to be a Jew *anywhere*." Asked about a difference between his real and virtual identities, he answered:

> Maybe I lost those pounds in SL that my RL doctor advised me to lose, but otherwise no real difference. In fact, I went out of my way to create my real self in Second Life. I am 54, bearded with glasses, balding with grey hair, and almost always wear black pants and a vest over a white shirt. I want people to see me as I am and have no hidden identities within the avatar they see. I suspect that it surprises many people when they realize I really am a grandfather of five that practices a fairly strict religious life.[27]

Building Jewish Quarters

Unfortunately, the more prominent the Jewish community in Second Life became, the more the number of anti-Semitic incidents increased.[28] On April 20th, 2007, the anniversary date of Adolf Hitler, the replica of the Western Wall, the "SL Kotel," was littered with swastikas.[29] As a reaction to this attack, GruvenReuven Greenberg moved the SL Kotel at the end of April 2007 from Choi to Nessus, in proximity to Temple Beth Israel and the 2Life headquarters.

Although the move of the SL Kotel had a different motivation, it fits into a general trend of building Jewish quarters. Traditionally, Jewish quarters in large cities developed out of a religious necessity; they meant proximity to kosher food and walking distance to synagogues. In the virtual world distance is relative since everything is basically just one mouse-click away. While Jewish life began with isolated Jewish sites spread across the Second Life grid, these sites have now moved together to build "urban" Jewish areas. The two Jewish centers in Second Life are the areas around Temple Beth Israel, the first synagogue of Second Life, and the area around the Jewish Historical Museum and Synagogue (JHMS).

In early May 2007, Carter Giacobini, the creator of JHMS, announced the creation of a Jewish island. On May 20, 2007, *Ir Shalom*, "the City of Peace" (a reference to Jerusalem) was officially opened in Second Life. Many of the isolated Jewish sites in the proximity of JHMS moved to the island.[30] Ir Shalom is the first Jewish island in Second Life, and the future will tell if it can maintain itself as a Jewish-themed island.

SL Judaism: Status Quo, June 2007

Now, in June 2007, less than a year after the introduction of Judaism to Second Life, dozens of Jewish sites and Jewish-themed groups exist in the virtual world. The most important ones are situated in the two Jewish quarters, TMA and Ir Shalom.

In TMA (named after the SL group Tragically Misunderstood Artists), the area around Temple Beth Israel, one can find other creations by Beth Odets (among them an Orthodox synagogue, a mikveh, a community center and a Judaica shop), as well as GruvenReuven Greenberg's SL Kotel, and the 2Life headquarters, — a two-story structure inspired by Tel Aviv Bauhaus architecture. The 2Life building contains Second Life's Jewish art gallery (the Tachles Gallery), and a café (the Aufbau Café) in the tradition of the European café house culture as a space for dialogue and discussion, — even if in a slightly different way.[31]

Ir Shalom's main attraction is Temple Beit Shalom, Second Life's largest synagogue, a creation by Australian Jeremy Finkelstein a.k.a. CryptoMorph Lake. Other attractions on Ir Shalom are the Jewish Community Center, the Jewish Historical Museum and Synagogue, and the Holocaust Memorial Museum.

Among the various other Jewish sites on Second Life's grid, the most ambitious and interesting is Beit Binah, Second Life's Intentional Jewish Community, aiming to create a place for "meaningful work and practice that can be carried back into our real lives, overcoming old negative thought patterns and behaviors and replacing them with a pervasive sense of radical amazement," according to its creator Malachi Rothschild, offline twenty-three year-old Benjamin Dauer from Allston, Massachusetts. Beit Binah is to him a place where we can come together to learn and study, to support each other on our journeys by means of structured and informal sharing, to explore how far we really can push this Wild West of a virtual world into a tool for spiritual growth, and as something that can facilitate real change and make a difference.

This virtual Kibbutz is "a democratic, participatory community of people sharing a space together because we want to cleave to God or simply because we want to feel that yearning to seek the holy until it fills our lives like the water fills the sea."[32]

Most sites have their own group mailing lists that allow communication with other members of the same group. Other groups are inspired by real life groups, such as the Jewish Defense League, an organization fighting anti-Semitism in the virtual world, or simply groups with specific interests, such as the Jewish Land Owners Group. Most Javatars join general groups such as "Jews of Second Life," "Second Life Synagogue," or "Ir Shalom" in order to be informed about Jewish activities in this virtual space, and regularly visit Jewish sites like the Jewish Museum to meet other Jewish residents.

What Will the Future Bring?

Despite the current boom of Second Life in general and its Jewish community in particular, the future is uncertain. The phenomenon of "virtual worlds" is booming. Even if Second Life now holds the leading role in the virtual world market, the question remains whether Linden Lab can maintain its position. Because of the freedom it offers its users, Second Life was the first virtual world that established a Jewish community, but others are already following suit. In May 2007, a group of Jews in the virtual world of "There" announced the plan to establish a synagogue named "Makom Kavuah." Whether the same phenomenon found in Second Life repeats itself in "There" is uncertain.

The creative freedom offered by Linden Lab also attracted subcultures that made negative headlines. The problems with virtual hate groups and pornography, to give just two examples, remind us of the early days of the internet with its discussions on freedom of speech. But the technical difficulties also remind us of the early days of the World Wide Web. Right now, it is not possible to move from one virtual world to another. Just as with the early online services (CompuServe, AOL etc.), users exist in separate communities. The challenge will be to open up the individual virtual worlds into a connecting network such as the World Wide Web.[33]

Wherever the virtual world is located, Reuven Fischer a.k.a .GruvenReuven Greenberg sees the strength of online Jewish communities in their potential to allow users to discover "the richness of Judaism on their own terms. Some of these folks have not thought about their Jewish heritage in years, but by virtue of being online are being drawn back home. (…) The real power I see is in (…) bringing virtuality into reality one step at a time."[34]

Of course, a virtual Jewish community does not replace real life activities, but it adds a new layer to Jewish identity. Second Life Judaism, therefore, forms the base for a unique inter (or intra) cultural dialogue within various streams of Judaism, within various Diasporas and Israel, and between Jews and non-Jews. Judaism in Second Life is a mélange of various identities, in which age, origin, gender, and even religious affiliation are unimportant. It is an experiment with an uncertain outcome, but with obvious potential for new and creative ways to explore culture, heritage, and identity.

Notes

1   Previous works include documenting formerly Jewish neighborhoods of New York, such as Harlem and the Bronx, exploring the transformation of space and identity; a project on Hasidim in Brooklyn, exploring their paths between tradition and modernity; and a series of portraits of African Americans practicing Judaism, discussing the interrelation of race and identity. For more information about my work, go to http://www.julianvoloj.com.

2   Adam Pasick, known online by his avatar name Adam Reuters, became the first full-time correspondent for a real life news agency in Second Life, Andrew Adam Newman, "The Reporter Is Real, but the World He Covers Isn't," *The New York Times*, October 16, 2006. Reuters has its own website dealing with Second Life economy: http://www.secondlife. reuters.com.

3   The word 'avatar' derives from Sanskrit and means "incarnation." In Hindu texts it refers to the deliberate descents of an immortal or divine being into the mortal realm for a special purpose. As used for a computer representation of a user, the term dates (at least) as far back as 1985, when it was used in the "Ultima" series of computer games. American author Neal Stephenson popularized the word 'avatar' in his cyberpunk novel *Snow Crash*, in which the term was used to describe the virtual simulation of the human form in the Metaverse, a virtual-reality version of the internet. The novel is generally seen as the role model for the creation of Second Life, Paul Carr and Graham Pond, *The Unofficial Tourists' Guide to Second Life* (London: Boxtree, 2007), 19. For more about avatars, see: Robbie Cooper, *Alter Ego: Avatars and their Creators* (London: Chris Boot, 2007).

4   Wikipedia, "Virtual World," http://en.wikipedia.org/wiki/Virtual_world (accessed June 15, 2007).

5   Julian Voloj, "Virtual Sanctuary: Spirituality in an Alternate Reality," *The Forward*, February 16, 2007.

6   Michael Rymaszewski et al., *Second Life: The Official Guide* (Hoboken, N.J.: John Wiley, 2007), 23–46.

7   David Kushner, "Inside Second Life," *Rolling Stone Magazine*, April 27, 2007.

8   Mia Farrow, "Avatarized for Darfur," *New World Notes*, December 7, 2006; "Live from Second Life: Crisis in Darfur with Mia Farrow and Guests," *LCMedia: The New Public Media*, February 11, 2007.

9   Oliver Burkeman, "Exploding Pigs and Volleys of Gunfire as Le Pen Opens HQ in Virtual World," *The Guardian*, January 20, 2007, http://www.guardian.co.uk/technology/2007/jan/20/news.france (accessed August 9, 2007).

10   Robert D. Hof, "My Virtual Life: A Journey into a Place in Cyberspace Where Thousands of People Have Imaginary Lives," *Business Week*, May 1, 2006, //www.businessweek.com/ magazine/content/06_18/b3982001.htm (accessed August 9, 2007).

11   Beth Odets, "The Second Life Synagogue: A Torah Got Carried Away," in: *2Life Magazine* 1 (April 2007): 4; http://www.2lifemagazine.com.

12   Ibid.

13   Voloj, "Virtual Sanctuary."

14   Cathy Lynn Grossman, "Faithful Build a Second Life for Religion Online," in: *USA Today*, April 1, 2007; http://www.usatoday.com/tech/gaming/2007-04-01-second-life-religion_ N.htm (accessed August 9, 2007).

15   Marc Shoffman, "Virtual Shoah museum opens," in: *Totally Jewish*, March 3, 2007; http:// www.totallyjewish.com/news/national/?content_id=5769 (accessed August 9, 2007).

16   Joshua Levi, "Aussie Hosts Seder in Cyberspace," *Australian Jewish News*, April 13, 2007; http://www.ajn.com.au/news/news.asp?pgID=2971; Cryptomorph Lake, "Creating Order from Chaos: Building a Seder in the Virtual World," *2Life Magazine* 2 (May 2007): 3.

17   Brandon Catteneo/Carter Giacobini, "Yom Hashoah: Remembering the Holocaust," *2Life Magazine* 2 (May 2007): 9.

18   Daniel Treiman, "Virtual Judaism," in: *The Forward*, April 13, 2007, 2; Heiko Ostendorf, "Es geht um alles Jüdische/Ein virtuelles Schalom aus dem zweiten Leben," *Münstersche Zeitung*, April 14, 2007.

19   Julian Voloj, "Willkommen im zweiten Leben," *Tachles*, January 26, 2007, 18–20. In February 2007 an article for the New York *Forward* ("Virtual Sanctuary"), and then in March 2007 another article for the German newspaper, *Jüdische Allgemeine* ("Schabat bei den Avataren," *Jüdische Allgemeine*, March 1, 2007, 9) followed.

20   Harald Neuber, "Second Life wird nicht besser als das First Life," in *Telepolis*, April 14, 2007.

21   Rachel Fletcher, "My Life on Second Life," *The London Jewish Chronicle*, April 27, 2007.

22   Take, for example, the founding echelon of Jewish Second Life: Beth Odets, the creator of Temple Beth Israel, is in real life from Texas, GruvenReuven Greenberg from Pennsylvania, Carter Giacobini from Iowa, and Malachi Rothschild from Massachusetts. The only exception is Cryptomorph Lake, who is Australian.

23   "Who Uses Second Life?", *2Life Magazine* 3 (June 2007): 2.

24   Julian Voloj, "Schabat bei den Avataren," in: *Judische Allgemeine*, March 1, 2007, 9.

25   "The Last Word: Five Questions for TamaraEden Zinnemann," *2Life Magazine* 2 (May 2007): 11–12.

26   "The Last Word: Five Questions for Second Life Resident Avram Leven," *2Life Magazine* 1 (April 2007), 7.

27   "The Last Word: Five Questions for RebMoshe Zapedzki," *2Life Magazine* 3 (June 2007): 17.

28   Kafka Schnabel, "Virtual Hate—Real Danger?," *2Life Magazine* 2 (May 2007): 7–8.

29   Manta Messmer, "'Nazi' Attack on Jewish Sim," in: *The AvaStar* 22 (May 18, 2007): 6; http:// www.the-avastar.com/slife/microsite/service/epaper/archive.

30   Miralee Munro, "A New Homeland For The Jews," *SL Newspaper*, May 22, 2007.

31   Yves Kugelmann, "A Place to Meet and Discuss," *2Life Magazine* 1 (April 2007), 2; *Tachles*, April 13, 2007, 5.

32   Malachi Rothschild, "Beit Binah: Second Life's Jewish Intentional Community," in: *2Life Magazine* 3 (June 2007): 11.

33   "Online Gaming's Netscape Moment?," *The Economist*, June 7, 2007.

34   GruvenReuven Greenberg, "Bringing Virtuality into Reality: One Step at a Time," *2Life Magazine* 3 (June 2007): 3–4.

List of Illustrations

List of Contributors

Miriam Lipis, an architect, artist, and cultural studies scholar, currently teaches at Humboldt University, Berlin. Her research interests include the intersection of gender and architecture, the combination of creative practice and theoretical analysis, and the interdependence of the material and symbolic aspects of space. Her doctoral dissertation investigated symbolic houses and symbiotic places, such as the *sukkah*, focusing on the role of spatial practices and symbols in the construction and maintenance of Jewish religion, culture, and identity.

A practicing architect with offices in Basel and Cologne, **Manuel Herz's** current projects include the Jewish Community Center in Mainz, Germany. He is the head of Research and Teaching (Institute of the Contemporary City) at the ETH Studio Basel, as well as visiting lecturer at the Graduate School of Design at Harvard University. His major fields of interest and research include the relationship between Judaism and space, specifically the *eruv* as a model of urbanism, and urban transformations and the city of the twenty-first century. Among his publications are the articles "Institutionalized Experiment: The Politics of 'Jewish Architecture' in Germany" in *Jewish Social Studies* (2005) and "Court Jester: Politics of 'Jewish Architecture' in Germany" in *Jewish Conceptions and Practices of Space* (2004). He lives in Basel.

Haim Yacobi is a lecturer in the Department of Politics and Government at Ben-Gurion University of the Negev, Israel. His academic work focuses on the geopolitics of cities and postcolonial urbanity, specifically the production of urban space, social justice, migration, globalization, and urban planning. He is the editor of *Constructing A Sense of Place: Architecture and the Zionist Discourse* (2004) and the author of *The Jewish-Arab City: Spatio-Politics in a Mixed Community* (forthcoming, 2008). Haim Yacobi lives in Beer Sheva, Israel.

Kenneth Helphand is Professor of Landscape Architecture at the University of Oregon. His major fields of interest and research are garden history and theory, contemporary American landscape, Israeli landscape, landscape and film, and design criticism. His most recent book is *Defiant Gardens: Making Gardens in Wartime* (2006), which recently won several prestigious awards (from the Foundation for Landscape Studies and the American Society of Landscape Architects, among others), and his numerous articles include "'My garden, my sister, my bride': The garden of the Song of Songs" in *Gender and Landscape* (2005). Kenneth Helphand lives in Eugene, Oregon with his wife Margot and is a frequent visiting professor at the Technion in Haifa, Israel.

Independent scholar **Etan Diamond** currently works as a researcher for the YMCA. His areas of research interest are geography, religion, urban and suburban history, and the history of religion. His 1996 doctoral dissertation, a study of Orthodox Jews in suburbia, which received the Urban History Association's Best Dissertation Award, was published as *And I Will Dwell in Their Midst: Orthodox Jews in Suburbia* (2000). He is coeditor of *Sacred Circles, Public Squares: The Multicentering of American Religion* (2005) and author of *Souls of the City: Religion and the Search for Community in Postwar America* (2003). Etan Diamond lives with his wife and three children in suburban Toronto.

Eve Jochnowitz is a lecturer in Jewish Culinary History at the New York City University, The New School. She worked for several years as a cook and baker in New York and is currently writing a doctoral dissertation in the Department of Performance Studies at New York University on the subject of Jewish culinary ethnography. She has lectured internationally on food in Jewish tradition, religion, and ritual as well as food in Yiddish performance and popular culture. Among her articles are "Flavors of memory: Jewish food as culinary tourism in Poland" in *Culinary Tourism* (2004) and "Holy Rolling: Making Sense of Baking Matzo" in *Jews of Brooklyn* (2002). Eve Jochnowitz is the Chocolate Lady.

Susan Gilson Miller is Director of the Moroccan Studies Program at Harvard University and Senior Lecturer in the Department of Near Eastern Languages and Civilizations, where she teaches courses on Maghribi history and urbanism. Her book, *Disorienting Encounters: Travels of a Moroccan Scholar in France in 1845–46* (1992), was recently awarded the Ibn Batuta Prize by the Abu Dhabi Foundation for Culture. She coedited *In the Shadow of the Sultan: Culture, Power, and Politics in Morocco* (1999) and has contributed articles on Maghribi social and cultural history to, among others, the *Journal of North African Studies*, *Muqarnas*, and the *Encyclopedia of Women in Islamic Civilization*. She divides her time between Cambridge, Massachusetts and Inverness, California, where she and her husband cultivate a spacious organic garden that feeds three generations of her family.

Joachim Schlör is Professor for Jewish/non-Jewish Relations at The Parkes Institute, University of Southampton. His main interests are German-Jewish history, urban history, migration, and Jewish cultural studies. Among his books are *Das Ich der Stadt: Debatten über Judentum und Urbanität* (*The Self of the City: Debates about Judaism and Urban Culture*, 2005) and *Endlich im Gelobten Land? Deutsche Juden unterwegs in eine neue Heimat* (*At Long Last in the Promised Land? German Jews on their Way to a New Homeland*, 2003). He lives in Southampton and Berlin and originated the idea of the research group Makom at the University of Potsdam.

Yael Zerubavel is a professor of Jewish Studies and History and the founding director of the Allen and Joan Bildner Center for the Study of Jewish Life at Rutgers University. Her major fields of interest are memory and identity, Israeli culture, Hebrew literature, war and trauma, and the Jewish immigrant experience. She is the author of *Recovered Roots: Collective Memory and the Making of Israeli National Tradition* (1995), which won the 1996 Salo Baron Prize from the American Academy for Jewish Research. Her recent articles include "Transhistorical Encounters in the Land of Israel: National Memory, Symbolic Bridges, and the Literary Imagination" in *Jewish Social Studies* (2005) and "Antiquity and the Renewal Paradigm: Strategies of Representation and Mnemonic Practices in Israeli Culture" in *On Memory: An Interdisciplinary Approach* (2007). She is currently working on a book, *Desert in the Promised Land: Nationalism, Politics, and Symbolic Landscapes*, her interest in the desert having developed during her military service in the Sinai. She currently lives in East Brunswick, New Jersey.

Educator and architectural historian **Gilbert Herbert**, Professor Emeritus, Technion-Israel Institute of Technology, has authored eight books, contributed to numerous volumes, and published countless papers in scholarly and professional journals, mainly on the history and theory of architecture. His latest research focuses on two topics: the architectural design of passenger liners (*Symbols of a New Land: Architects and the Design of the Passenger Ships of Zim*, 2006), and the troubled friendship of two pioneer European architects during the first half of the twentieth century (*Erich Mendelsohn, Hendricus Theodorus Wijdeveld, and the Jewish Connection*, forthcoming, together with Liliane Richter, 2007). Together with his family he lives in Haifa, Israel.

Haya Bar-Itzhak is the Head of Folklore Studies in the Department of Hebrew and Comparative Literature and the Head of the Israeli Folktale Archives at the University of Haifa. Her research area is the Jewish folk narrative, combining the ethnographic and poetic aspects of the subject. She has published many articles and five books, including *Israeli Folk*

Narratives: Settlement, Immigration, Ethnicity (2005), and is the editor of the *Encyclopedia of Jewish Folklore* (forthcoming, 2008). Her interest in space, time, and local legends had already developed by the time she wrote her thesis, *Space and Time in Shivhei Ha'Besht*. Bar-Itzhak currently lives in Harrisburg, Pennsylvania.

Samuel Kassow is Charles Northam Professor of History at Trinity College in Hartford Connecticut. His interests include the social history of East European Jewry, Polish urban history, Russian intellectual history, and comparative historiography. His publications include *Who will Write Our History: Emanuel Ringelblum, the Warsaw Ghetto and the Oyneg Shabes Archive* (2007) and *Students, Professors and the State in Tsarist Russia* (1989). He is coeditor of *Between Tsar and People: The Search for a Civil Society in Tsarist Russia* (1993).

Erik Cohen is George S. Wise Professor Emeritus of Sociology in the Department of Sociology and Anthropology, The Hebrew University of Jerusalem. His current fields of interest are the sociology and anthropology of tourism, folk arts, folk religion, and Thai society. His recent major publications include *Contemporary Tourism: Diversity and Change* (2004), *The Chinese Vegetarian Festival in Phuket* (2001), and *The Commercialized Crafts of Thailand* (2000). Together with Chaim Noy he coedited *Israeli Backpackers and Their Society* (2005). He presently lives and conducts research in Thailand.

Shelley Hornstein is the Walter L. Gordon Fellow and Associate Professor of Architectural History & Visual Culture at York University, Toronto, Canada. She has published widely on the examination of place and spatial politics in architectural and urban sites. Some of the themes she is currently exploring are "starchitecture" and Jewish architectural tourism. Her edited books include *Impossible Images: Contemporary Art after the Holocaust* (2003) and *Image and Remembrance: Representation and The Holocaust* (2002). As Executive Director and cofounder of www.mosaica.ca, she codeveloped the first online contemporary space devoted to Jewish culture, virtual space, and diaspora. She lives in Toronto.

Galeet Dardashti is a PhD candidate in Anthropology / Cultural Studies at the University of Texas at Austin and is completing her doctoral dissertation on the cultural politics of performing Middle Eastern and Arab music in Israel. Her primary research interests are Israeli cultural politics, music and globalization, and religious and ethnic identity in national contexts. Her most recent publication is "The *Piyyut* Craze: Popularization of Mizrahi Religious Songs in the Israeli Public Sphere" in *Journal of Synagogue Music* (2007). She performs Middle Eastern and Arab Jewish music internationally with her group *Divahn,* and has performed as a soloist both in the US and Israel. She currently resides in New York City where she was recently awarded a Six Points Fellowship to pursue her project *Voices of Our Mothers: A Middle Eastern Musical Midrash for Today.*

Eszter Brigitta Gantner is a PhD student at Humboldt University in Berlin, and a lecturer at the ELTE in Budapest and at Touro College in Berlin. Her major fields of interest and research include Jewry in the Habsburg Empire of the 19th- and 20th-centuries, and Jewish identity in formerly socialist countries. Her publications include the articles "Berlin szerepe a modern magyar kultúra megteremtésében" (The role of Berlin in Establishing Modern Hungarian Culture) in *Múlt és Jövö* (2007), and "Zsidniland: A zsidó kulturális tér Közép-Európa városaiban. Egy új értelmezési lehetőség" (Zsidniland: Jewish Cultural Space in Central Europe), with coauthor Mátyás Kovács in *Antropolis* (2006). She lives in Berlin and Budapest.

Mátyás Kovács is spokesperson for the Deputy Lord Mayor for Urban Development, Management, and Social Affairs, City of Budapest, and a member of the Research Group on Ethnoregionalism and Cultural Anthropology at the Hungarian Academy of Sciences. His fields of interest include public and media relations as well as contemporary Jewish studies. His publications with co-author Eszter B. Gantner include "A kitalált zsidó" (The Constructed Jew) in *Conference Yearbook of the Center for the Studies of the Culture and History of East European Jews* (2006). He lives in Budapest.

A sociologist at the Ben-Gurion Research Institute, Ben-Gurion University at the Negev, Israel, **Michael Feige's** primary academic interests are collective memory, social movements, and social identities in Israeli society. His book, *One Space, Two Maps,* which examines the discourse concerning the territories occupied by Israel in the 1967 War, was published in 2002 in Hebrew. His book on ideological settlers, *Settling in the Hearts,* is forthcoming. He is currently working on the construction of the image of David Ben-Gurion, Israel's first Prime Minister, in Israeli national memory. He built his home in the desert, at Midreshet Ben-Gurion, where he lives with his wife and children.

Photographer and writer **Julian Voloj** is founder and creator of JWalks, an organization promoting Jewish heritage awareness through various cultural outlets such as walking tours and photography. His major fields of interest and research are Jewish heritage preservation and Jewish identities. Voloj was the first to document the existence and growth of Judaism in the virtual world of Second Life. Voloj has been the editor of *2Life,* Second Life's Jewish magazine, since April 2007. He lives in New York City.

Political scientist **Julia Brauch** currently teaches at the Free University in Berlin. Her research focuses on the history of Israel and on political language, and after joining the graduate seminar Makom as a postdoc, she began examining the state of Israel as a point of reference for Jewish identity and collective self-understanding. In 2004 she published the book *Nationale Integration nach dem Holocaust: Israel und Deutschland im Vergleich* (National Integration after the Holocaust: A Comparison Between Israel and Germany), a study of ideas about parliamentary unity in the proceedings of the Knesset and the Bundestag. Her most recent publication is "Medinat Israel" in *Makom: Orte und Räume im Judentum* (Makom: Place and Space in Judaism, 2007). She lives in Berlin.

Anna Lipphardt is a postdoc fellow in cultural studies at Centre Marc Bloch, Berlin, and the codirector of the Centre's research group on Nazism. A member of Makom, her dissertation on Vilne, 'the most Yiddish city in the world,' and its diaspora in New York, Israel, and Vilnius after the Holocaust, won the Klaus Mehnert Prize from the German Association for East European Studies (DGO) in 2007. She coedited *Der Ort des Judentums in der Gegenwart, 1989–2002* (The place of Judaism in present times, 2004), and has published on the dialectics of migration and home, as well as on place-related cultural practices and on spatial concepts central to Jewish thought, such as 'diaspora' and 'ghetto.' She currently lives in Berlin and is working on a cultural and social history of the circus in Germany.

Alexandra Nocke, Berlin, is an independent scholar, curator, and researcher in cultural studies. Her major fields of interest include the development of identity in contemporary Israeli society, and art and culture in Israel, with a special focus on photography. While a scholarship recipient in the graduate seminar "Makom: Place and Places in Judaism" at the University of Potsdam, in association with the Center for Mediterranean Civilizations Project at Tel Aviv University, she completed her dissertation on *Yam Tikhoniut: The Place of the Mediterranean in Modern Israeli Identity* (forthcoming, 2008). She has worked as a curator of diverse exhibitions and is the author of *Boris Carmi: Photographs from Israel* (2004) and *Israel heute: Ein Selbstbild im Wandel* (Israel today: A changing self-image, 1998). Her most recent publication is "Yam Tikhoniut" in *Makom: Orte und Räume im Judentum* (Makom: Place and Space in Judaism, 2007).

Index